A SHARK GOING INLAND IS MY CHIEF

Also by Patrick Vinton Kirch

The Evolution of the Polynesian Chiefdoms
Feathered Gods and Fishhooks
Anahulu: The Anthropology of History in the Kingdom of Hawai'i
The Wet and the Dry
The Lapita Peoples
On the Road of the Winds
How Chiefs Became Kings

Patrick Vinton Kirch · # A SHARK GOING INLAND IS MY CHIEF

The Island Civilization of Ancient Hawai'i

University of California Press

Berkeley   Los Angeles   London

University of California Press, one of the most distinguished university presses in the United States, enriches lives around the world by advancing scholarship in the humanities, social sciences, and natural sciences. Its activities are supported by the UC Press Foundation and by philanthropic contributions from individuals and institutions. For more information, visit www.ucpress.edu.

University of California Press
Berkeley and Los Angeles, California

University of California Press, Ltd.
London, England

© 2012 by Patrick Vinton Kirch

First paperback printing 2019

Library of Congress Cataloging-in-Publication Data

Kirch, Patrick Vinton.
　A shark going inland is my chief : the island civilization of ancient Hawai'i / Patrick Vinton Kirch.
　　　p. cm.
　　Includes bibliographical references and index.
　ISBN 978-0-520-27330-6 (cloth : alk. paper); 978-0-520-30341-6 (pbk. : alk. paper)
　　1. Hawaii—Civilization. 2. Hawaii—Antiquities. 3. Hawaii—History—To 1893. 4. Hawaii—Environmental conditions. I. Title.
　　DU624.5.K57　2012
　　996.9'02—dc23　　　　　　　　2012010491

Manufactured in the United States of America

21　20　19
10　9　8　7　6　5　4　3　2　1

For Thérèse

CONTENTS

List of Illustrations   ix

Preface   xi

Acknowledgments   xv

PROLOGUE   Islands out of Time   1

PART ONE   VOYAGES

1   A Trail of Tattooed Pots   21

2   East from Hawaiki   38

3   Follow the Golden Plover   50

4   Voyages into the Past   68

5   The Sands of Waimānalo   82

PART TWO   IN PELE'S ISLANDS

6   Flightless Ducks and Palm Forests   99

7   Voyaging Chiefs from Kahiki   112

8   Māʻilikūkahi, Oʻahu's Sacred King   131

9   The Waters of Kāne   143

10   "Like Shoals of Fish"   156

PART THREE   THE REIGN OF THE FEATHERED GODS

11   ʻUmi the Unifier   173

12   ʻUmi's Dryland Gardens   187

13   The House of Piʻilani   202

14   "Like a Shark That Travels on the Land"   217

15   The Altar of Kū   231

16   The Return of Lono   247

17   Prophecy and Sacrifice   265

EPILOGUE   Hawaiʻi in World History   288

Alphabetical List of Hawaiian Historical Persons   303

Glossary of Hawaiian Words   311

Sources and Further Reading   317

Index   335

ILLUSTRATIONS

PLATES

Plates follow p. 142

1. The twin-hulled voyaging canoe *Hōkūleʻa* at dawn.
2. The windward coastline of Molokaʻi Island.
3. A newly planted irrigated pondfield *(loʻi)* on the floor of Waipiʻo Valley, Hawaiʻi Island.
4. A Hawaiian chief in feathered cloak and helmet.
5. The Kohala field system as seen from the summit of Puʻu Kehena, Hawaiʻi Island.
6. A feathered *mahiole* (helmet) in the collection of the Bishop Museum, Honolulu.
7. Three war canoes of King Kalaniʻōpuʻu arriving at Kealakekua Bay, Hawaiʻi Island.
8. Feathered image of the war god Kūkāʻilimoku.

FIGURES

1. The interior of a *heiau* (temple) at Waimea, Kauaʻi Island.   *11*
2. The author at the completed excavation of the Lapita stilt house remains at Talepakemalai, Mussau Islands.   *29*

3. An elaborately decorated ceramic cylinder stand from Talepakemalai, Mussau Islands. *32*
4. The University of Hawai'i excavations at the Bellows Dune site, Waimānalo, O'ahu Island. *88*
5. The *pānānā* (sighting wall) at Hanamauloa, Kahikinui, Maui Island. *114*
6. View of Waipi'o Valley, Hawai'i Island. *119*
7. The ruins of the ancient temple of Hikinaakalā, Kaua'i Island. *125*
8. The sacred birthing stones at Kūkaniloko, O'ahu Island. *133*
9. The terraced front face of Pi'ilanihale temple in Hāna district, Maui Island. *205*
10. The fortress hill of Ka'uiki in Hāna, Maui Island. *211*
11. The main façade of Hikiau *heiau* at Kealakekua Bay, Hawai'i Island. *256*
12. Kealakekua Bay as seen from the lava flats at Ka'awaloa, Hawai'i Island. *264*
13. Impression of a warrior's foot in volcanic ash at Kīlauea, Hawai'i Island. *266*
14. The sepulchral temple Hale o Keawe at Hōnaunau, Hawai'i Island. *271*
15. Haleki'i *heiau,* the old royal center of the Maui Island kings. *279*
16. The stone foundations of the great war temple at Pu'ukoholā, Hawai'i Island. *282*
17. 'Ahu'ena *heiau* at Kamakahonu, Kailua, Hawai'i Island. *285*

MAPS

1. The main Hawaiian Islands. *4*
2. O'ahu Island with important places mentioned in the text. *132*
3. Hawai'i Island with important places mentioned in the text. *176*
4. Maui Island with important places mentioned in the text. *237*

PREFACE

Hawai'i is the most isolated archipelago on Earth. It is astonishing that Polynesian explorers in double-hulled canoes—lashed together with coconut fiber and propelled by sails of woven mats—discovered and settled these islands roughly a thousand years ago. They came upon a verdant island chain with a subtropical climate, rich soils, and abundant natural resources. Nurtured by this salubrious environment, their descendants multiplied, founding an island civilization that remained unknown to the rest of the world. Independently of what was happening in China or Japan, in Mesoamerica, or in Europe, the Hawaiian people constructed their own unique society.

This island civilization in many respects mirrored early states that arose in other favorable zones in both the Old World and the New. From a small founding population, over the course of several centuries a hierarchical society emerged, supported by a robust agricultural economy. A distinct class of chiefs depended on and managed a vast population of farming and fishing commoners. An elaborate system of rules and obligations—the *kapu* system—governed the relationships between the chiefs and the people. At the pinnacle of society were the island rulers, *ali'i akua* (literally, "god-kings"), whose prerogatives included royal incest and human sacrifice. In these practices, the Hawaiian kings resembled the pharaohs of Egypt and the Inca of Peru. Yet Hawaiian culture arose entirely independently in this most remote, most isolated of all places on Earth. How and why did this happen?

I have spent two-thirds of my life seeking the answers to these questions. Born on Oʻahu, I grew up in the lush Mānoa Valley, favored residence of the ancient chiefs. Perhaps it was the stone walls and terraces that I encountered while hiking barefoot as a youth in the shady ravines of upper Mānoa that first piqued my curiosity about old Hawaiʻi. I remember those ruins as if it were yesterday—I could probably still lead you to them. I became an archaeologist, studied at Penn and Yale, received my academic credentials. Returning to Hawaiʻi, I became a researcher at the storied Bishop Museum, expanding my studies beyond Hawaiʻi to other islands and archipelagoes of the Pacific. There I traced the deep roots of Hawaiian and Polynesian culture in the sands of Talepakemalai, Lolokoka, and Toʻaga. Since 1989 I have taught and carried out my research at the University of California at Berkeley. But I continue to spend time, every year, in the islands that I love, continuing my research into their ancient past.

My quest has led me across the breadth of the Pacific, from the Mussau Islands in the west to Rapa Nui (Easter Island) in the east. Always, however, I returned to work in Hawaiʻi. For while the story of Hawaiʻi's unique civilization is linked to that of Polynesia and the greater Pacific, most of the answers to our questions must be sought in Hawaiʻi itself. The rise of this island civilization, its unwritten history, has had to be painstakingly reconstructed from the subtle traces sedimented on the islands' landscape. Fragments of pearl-shell fishhooks in the sand dunes of Waimānalo, scraps of charcoal teased out of the clay soils of ancient irrigation terraces in Mākaha, lichen-encrusted rock walls of temple foundations on the windy slopes of Kahikinui, the crisscrossed patterns of dryland field embankments in the uplands of Makiloa: these are the building blocks upon which my history of ancient Hawaiʻi has been constructed.

But Hawaiian history is not just about ruined temples and scraps of bone and shell from ancient middens. In spite of horrific population declines and the debilitating impacts of Western colonialism, Hawaiian culture with great resilience has survived, recently undergoing a major renaissance in language and arts. During the nineteenth century, several Native Hawaiian scholars wrote down the great sagas of their past, which had originally been passed down orally from generation to generation. These texts, collectively known as the *moʻolelo o Hawaiʻi*, open a window into the Hawaiian past that complements the archaeological record. This indigenous history speaks directly to the human emotions, passions, and ambitions that helped to drive the rise of Hawaiian civilization. At times dismissed by anthropologists as mere "myth" or "legend," the *moʻolelo* in fact encode a real history, one that we ignore to our loss. One aim of this book is to integrate the knowledge gained from archaeology with the indigenous record of history preserved in the *moʻolelo o Hawaiʻi*.

As I write this preface, I am sitting on the veranda of a little cottage high on the slopes of Haleakalā, at ʻUlupalakua. The sun set an hour ago, its reddish-orange globe filtered through the fingered leaves of a breadfruit tree. It has been a good day of fieldwork out in the backlands, the *kuaʻāina* country of Kahikinui. Today I mapped two previously unknown temple sites, *heiau*, fascinating to me in the details of their construction, and in their particular orientations to distant cinder cones. It is through such incremental bits of data that we increase our understanding of how life was ritually ordered on this arid lava landscape. For four decades, I have slowly accumulated such fragments of knowledge, inscribed in my field notebooks, recorded on film and now digitally, captured in my maps and plans of stone ruins. My field research has encompassed every major island in Hawaiʻi with the exception of Niʻihau.

Over the years, I have published the results of my field explorations and laboratory studies in more than three hundred scholarly articles, and in two dozen books and monographs. Written for other scientists and scholars, these are couched in the coded language of academia, for this is how we are trained to write, and how we train our students. The academic world is a self-contained and self-perpetuating guild. It is not easy, I have come to realize, for a layperson to pick up one of these articles or monographs and comprehend what is being related. Too many technical terms get in the way, not to mention an assumed background level of knowledge. As with any guild, this is the way scholars shield themselves from would-be intruders on our turf.

This book is my effort to break out of the straightjacket of academic prose, to relate what I and other archaeologists have learned about the deep history of Hawaiʻi. Without watering down the thorny debates, or glossing over the controversies of historical interpretation, I seek in these pages to trace the big themes of Hawaiian history—from the initial migrations of the Polynesians and their discovery of the Hawaiian archipelago, through the development of uniquely Hawaiian cultural patterns, to the rise of a society that rivaled other early civilizations anywhere in the world. It is a saga that deserves to be told in a manner accessible to all who are interested.

Mine is, as it must be, a personal recounting of this history. Others would tell this history differently, though one hopes without too much contradiction. I have tried to tell the story of Hawaiʻi's deep past from two perspectives. The first is from my own, personal encounter with the material—the archaeological—record of this unique island civilization. I therefore privilege particular sites and places where I have explored, discovered, excavated, and interpreted what happened in the past. Others would doubtless choose different venues and examples. History is always an individual construction.

The second perspective is the one offered by the Native Hawaiian traditions, the *moʻolelo*. Invoking these ancient sagas brings the Hawaiian actors themselves onto the stage of history, their voices ringing out once again, whether in a chant of joy at their arrival in a new land, in the muttered grumblings of dissatisfied chiefs plotting against their king, or in the insults and challenges tossed by warriors at each other across a battlefield. By weaving the Hawaiian traditions into my history, the past becomes populated by real, named people.

Most of all, I have tried to be faithful to the evidence of the past—whether archaeological or traditional. History is never entirely objective; the facts do not speak for themselves. The past is a foreign country, it has been said, and any voyager into history brings along his or her own intellectual and cultural baggage, no matter how lightly we try to pack. But I have tried to tell this remarkable story of an island civilization as directly as I can, leaving it for others to pronounce on the philosophical implications.

NOTE ON ORTHOGRAPHY

Throughout this book, Hawaiian words are spelled according to Mary Kawena Pukui and Samuel H. Elbert, *Hawaiian Dictionary*, rev. ed. (Honolulu: University of Hawaiʻi Press, 1986). The spelling of place-names follows Mary Kawena Pukui, Samuel H. Elbert, and Esther T. Mookini, *Place Names of Hawaiʻi*, rev. ed. (Honolulu: University of Hawaiʻi Press, 1974). Older sources of Hawaiian traditions frequently do not indicate glottal stops (marked with the *ʻokina*) or vowel length (marked with the *kahakō*) when spelling Hawaiian personal names, sometimes resulting in two or more variants. I have endeavored to check the spelling of personal names against the most reliable modern sources.

ACKNOWLEDGMENTS

This book encapsulates the fruits of a lifetime of research. Over nearly five decades, my life as a student and scholar of Hawai'i and Polynesia has benefited from the help and generosity of mentors and teachers, colleagues and collaborators, students, friends, and acquaintances. To list them all would be impossible; I can only single out a few whose role cannot go unmentioned. Yoshio Kondo had the temerity to take a thirteen-year-old Punahou student into his summer program at the Bishop Museum, thus launching my career in science. To Kondo I will forever owe the *ong* about which he spoke so often with reference to his own mentor and patron, malacologist C. Montague Cooke Jr. Kenneth P. Emory, Yoshiko Sinoto, Lloyd Soehren, Marion Kelly, William (Pila) Kikuchi, Peter Chapman, and others of the Bishop Museum's archaeology program gave me early opportunities for field research in Hawai'i during the 1960s, as did Richard Pearson of the University of Hawai'i. Douglas E. Yen and Roger C. Green, then also of the Bishop Museum, encouraged my early efforts and helped to enable my transition from student to professional researcher. Yen and Green later became close colleagues and collaborators; many of my ideas were formed out of discussions, arguments, and debates with them.

Numerous individuals have contributed to my several major Hawaiian research projects. In our early work in Hālawa, Moloka'i, Tom Riley and Gil Hendren shared the excitement and challenges posed by that remarkable valley. Pilipo Solotario opened his family home in Hālawa to us and shared his deep insights into the valley's traditions. Charles Pili Keau worked with me at several sites on Maui, imparting his *mana'o* regarding Hawaiian culture and protocol along the way. Marshall Sahlins graciously invited me to collaborate on research in the Anahulu Valley, O'ahu, sharing with me his vast knowledge of Hawaiian and Polynesian ethnohistory. Dorothy Barrère, a collaborator in the Anahulu project, shared her boundless comprehension of the Hawaiian Kingdom's archives. Marshall Weisler participated in the Anahulu excavations, and worked with me on the archaeological survey of Kawela and Makakupai'a, Moloka'i. Rose Schilt and Jeff Clark similarly managed extensive highway-corridor salvage archaeology projects under my direction at Waimea-Kawaihae and Kona, Hawai'i Island, adding greatly to our knowledge of that island's archaeology. In my long-term research at Kahikinui and Kaupō, Maui, I have been aided by my Berkeley doctoral students Sidsel Millerstrom, James Coil, Lisa Holm, Cindy van Gilder, Kathy Kawelu, and Alex Baer. The members of Ka 'Ohana o Kahikinui, and especially Mo Moler, along with ranchers Pardee Erdman and Sumner Erdman, Bernie and Andy Graham, and Jimmy Haynes, and the Department of Hawaiian Home Lands are especially thanked for opening up those vast *kua'āina* lands to my research team. John Holson has been a stalwart field team member not only of the Kahikinui and Kaupō projects, but during our challenging 2000 expedition to Kalaupapa, Moloka'i. Since 2001, I have learned a great deal from my collaborators in the Hawai'i Biocomplexity Project, especially Peter Vitousek, Oliver Chadwick, Shripad Tuljapurkar, Thegn Ladefoged, and Sara Hotchkiss. Julie Field ably directed our 2007–2009 fieldwork in Kohala, Hawai'i Island. Parker Ranch, Pono Holo Ranch, Kahua Ranch, and the State of Hawai'i Department of Land and Natural Resources opened their Kohala lands to our research team.

Others (living and deceased) to whom I extend my *mahalo nui loa* for assistance, shared fieldwork, exchange of data or ideas, or other *kōkua* include Jane Allen, Melinda Allen, Emmett Aluli, Pia Anderson, Steve Athens, Pat Bacon, William Barrera, Alan Carpenter, Francis Ching, Carl Christensen, Steve Clark, Bonnie Claus, June and Paul Cleghorn, Sara Collins, Ross Cordy, Neal Crozier, Dorothe and Dave Curtis, Bert Davis, Mahealani and Glenn Davis, Michael Dega, Greg Dening, Boyd Dixon, Tom Dye, Tim Earle, Ben Finney, James Flexner, Sam Gon III, Ward Goodenough, Michael Graves, P. Bion Griffin, Jo Lynn Gunness, Rose Guthrie, Hal Hammat, Toni Han, Edward S. C. Handy, Sara Hotchkiss, Robert Hommon, Terry Hunt, Helen James, Sharyn Jones, Elaine (Muffett) Jourdane, Adrienne Kaeppler, Jennifer Kahn, Solomon Ka'ilihiwa, Betty Kam, Ray and Shiela Kawaiaea, Kathy Kawelu, Alison Kay, Michael Kolb, Eric Komori, Ed

Ladd, Charlotte Lee, Sam Low, Steve Lundblad, Mark McCoy, Patrick McCoy, Holly McEldowney, Art Medeiros, Peter Mills, Gail Murakami, Buddy Neller, T. Stell Newman, Cordelia Nickelsen, Aimoku and Lehua Pali, Cedric Puleston, Debra (Connelly) Prentice, Storrs Olson, Jack Randall, Roland Reeves, Paul Rosendahl, Clive Ruggles, Laura Schuster, Will Shapiro, Warren Sharp, David Sherrod, John Sinton, Aki Sinoto, Matthew Spriggs, Donna and Leon Sterling, Elspeth Sterling, Harold St. John, Catherine (Cappy) Summers, Jillian Swift, Margaret (Makaleka) Titcomb, Myra Tomonari-Tuggle, Dave Tuggle, Randy and Victoria Wichman, and Alan Ziegler. If others have been omitted here, it is only because the passage of so many years has dimmed my memory, and I beg their forgiveness.

At the University of California Press, I thank Blake Edgar for his continuing editorial stewardship. Peter Mills went beyond the normal duties of an external reviewer, and his extensive suggestions greatly improved the final manuscript. David Cohen prepared the individual island maps.

Finally, my wife, Thérèse Babineau, has been an endless source of inspiration throughout the long process of writing—encouraging me, reading drafts and rewrites, offering suggestions on both content and style. It is a small token of my gratitude to dedicate this book to her.

PROLOGUE · **Islands out of Time**

It was mid January 1778. General Washington's beleaguered army faced frostbite and starvation at Valley Forge, in the third year of the American colonists' war of independence against the tyranny of Great Britain. On the far side of the world, two diminutive ships of His Majesty's Royal Navy tacked their way northward around latitude 21° north, in the yet-uncharted vastness of the north-central Pacific Ocean. The *Resolution* and *Discovery*, stout Whitby barks built to haul coal, rather than fighting ships of the line, had departed England in the summer of 1776, just as the uppity Americans were proclaiming their Declaration of Independence in Philadelphia. Each less than one hundred feet from stem to stern at the waterline, the two ships held a combined complement of 182 officers, midshipmen, and able-bodied seamen. They were cramped together with only five feet, seven inches between decks. Now at sea for a year and half, the British sailors knew nothing of the war raging between England and her colonies. They had been dispatched to the farthest corners of the globe on a mission so secret that its orders had been kept sealed until the *Resolution*'s commander, Captain James Cook, received them from the Admiralty's Sea Lords on the eve of departure from Plymouth Sound.

Captain Cook, just short of his fiftieth birthday, was a celebrated veteran of two voyages to the Great South Sea. In 1768 he had sailed at the behest of the Royal Society of London to observe the transit of Venus from the newly discovered island of Tahiti. Cook's superb management of that voyage convinced the Admiralty to send him out again, in 1772, to ascertain the supposed existence of a great southern continent, the fabled Terra Australis Incognita. Though of humble origins, Cook possessed skills in navigation and command that had won him acclaim in aristocratic circles; he had been introduced to King George at Buckingham Palace. Now he had agreed to yet a third voyage around the world. The Sea Lords' secret instructions laid out their challenge, daunting even to seasoned Cook. He was commanded to sail once more into the vast Pacific, by way of the Cape of Good Hope and thence to Tahiti, where he would replenish food, wood, and water. From Tahiti, Cook was to "proceed in as direct a Course as you can to the Coast of New Albion, endeavouring to fall in with it in the Latitude of 45° 0' North."

New Albion was what Europeans then called the western coast of North America. Mindful that Cook might be distracted by new islands in his path, the Sea Lords cautioned him not to "lose any time in search of new Lands," for the expedition's main objective was paramount. That goal was nothing less than to find an entrance to the fabled Northwest Passage—should it exist—a waterway long imagined by the geographical savants of Paris and London. These armchair philosophers, ensconced in their comfortable salons, theorized the existence of a waterway connecting the Atlantic and Pacific oceans over the north of the Americas. In an era when the Panama and Suez Canals were not even dreamed of, the nation that could discover and control such a passage—assuming it was ice-free and navigable—would have an enormous geopolitical advantage in the rapidly expanding global trade networks of the late eighteenth century.

Following the Sea Lords' instructions to the letter, Cook arrived at Tahiti in August 1777. Drawing on his prior good relations with the local chiefs, Cook quickly filled the ships' larders. He then allowed himself the luxury of exploring the smaller islands of the Society Islands archipelago. In December, as Washington's troops settled into their bleak quarters at Valley Forge, the *Resolution* and *Discovery* bade farewell to Ra'iatea's coconut palm–fringed lagoons, setting course for New Albion. Long days at sea passed with the rhythm of the watches and working of the sails. On the morning of December 25, Christmas Day, a low coral atoll was sighted, populated only by millions of seabirds, which, along with turtles captured in the shallow lagoon, provided fresh fodder for the ships' galleys. Cook named the atoll Christmas Island.

Northward they sailed, these fragile, creaking constructions of wood, rope, and canvas. Cook was eager to make landfall on the coast of North America while the ice floes were in summer retreat. At daybreak on January 18, a Sunday, the lookout reported an island bearing northeast by east; soon after a second land was glimpsed. Unlike low-lying Christmas atoll, these were high islands resembling

Tahiti. The twin volcanic peaks rose higher over the horizon, but the ships' progress was thwarted by "light airs and calms." As their limp sails glowed umber in the sunset, the British sailors could only gaze expectantly toward the distant islands.

Dawn broke clear and bright on the nineteenth, illuminating a third island. With a fine breeze from the northeast, Cook steered his ships toward the two westerly islands. During the night his crew had laid bets on whether these lands were inhabited; speculation was soon put to rest when the lookout sighted canoes coming offshore. Cook brought the *Resolution* to, allowing the canoes to approach.

The first shouted exchanges between the poop deck and the spray-splashed canoes confirmed that the denizens of these new lands were related to the Tahitians, even though thousands of miles separated the two archipelagoes. Lieutenant King scribbled in his logbook, "What more than all surprised us, was, our catching the Sound of Otaheite [Tahiti] words in their speech, & on asking them for hogs, breadfruit, yams, in that Dialect, we found we were understood, & that these were in plenty on shore." Cook noted that the islanders were timid, refusing to venture aboard his ship, even though they traded fish for some iron nails.

Cook worked his ships close to the larger, more verdant island. The bo'sun rhythmically heaved his lead-weighted sounding line as they approached the coast, wary as always of running aground. More canoes surrounded the *Resolution* and *Discovery*, and some barkcloth-clad men got up the courage to clamber aboard. The seasoned captain had experienced such moments of first contact many times before, yet he was struck by "the wildness" of his visitors' countenances, and how astonished they were by their strange surroundings. "Their eyes were continually flying from object to object," he later wrote. One brave fellow tried to make off with the sounding lead and line.

Lieutenant Williamson, out in the ship's small boat reconnoitering for a landing place, returned to report that an anchorage might be had on the island's lee side. The natives called the place Waimea (Red Water), for here a river meets the sea, its outflow stained reddish-brown by silt and sediment carved from the steep canyon slopes below Mount Waiʻaleʻale. The sailing master shouted orders to furl sail and drop anchor. Cook, eager to set foot on this new landfall, led three boats ashore with a party of armed marines. An excited crowd waited on the black sand spit separating the beach from the estuary.

First landings were always tense, fraught with danger. Earlier, Williamson's boat had been so thickly surrounded when he attempted to land that he felt obliged to fire musket shot to disperse the crowd. A native had been killed. Would that

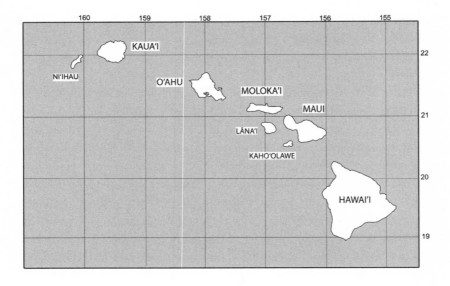

MAP 1.
The main Hawaiian Islands.

incident now provoke more violence? The boat's keel crunched against the gravelly black sand. Cook hesitated momentarily, and then leaped onto the shore. In that instant, to his complete astonishment, the crowd "all fell flat on their faces, and remained in that humble posture till I made signs to them to rise." Then, in a manner equally strange, the people began to present Cook with many small pigs, and with stalks of banana trees, making it clear that these were gifts, not items of trade. Cook could not grasp the full cultural significance of it all, although he realized that the islanders were acting "in a ceremonious way."

Cook and his crew had unwittingly stumbled upon one of the last "pristine states" to have arisen in the course of world history. In total isolation from the outside world, over the course of centuries the Hawaiians had developed a unique civilization. Displaying many similarities with earlier archaic states that had developed in the Near East, Egypt, China, Mesoamerica, and the Andes, Hawaiian civilization at the time of Cook's arrival was based on principles of divine kingship. Hawaiian society was divided into two great classes—a vast majority of commoners who worked the land and provided the economic underpinnings of society, and a smaller cadre of elites who included warriors, priests, and chiefs. At the pinnacle of society was the divine king, the *aliʻi nui*.

On that January morning, as he stepped ashore onto the black sands of Waimea, Cook was accorded the deference owed to an *aliʻi akua,* a divine king. The Waimea people were performing the *kapu moe,* the prostrating taboo, lying facedown in the sand. This gesture was obligatory in the presence of the highest-ranking chiefs and king, on pain of death. For the moment, however, the island's rulers were curiously absent, perhaps uncertain of Cook's intentions. Indeed, at this moment in its history, the Hawaiian archipelago was engaged in a series of interisland wars of territorial conquest. The larger islands of Maui and Hawaiʻi—of which Cook was as yet unaware—had for a number of years been locked in intense fighting with neither one gaining decisive control. Here, at the western end of the main island chain, having the advantage of distance, the Kauaʻi kings nonetheless were rightly wary of any intruders into their island realm.

· · ·

Eight centuries earlier, Polynesian voyagers had been the first humans to make this landfall, following much the same course as Cook. Arriving in double-hulled canoes propelled by sails of woven mats, they had followed the flightways of the golden plover in search of unknown lands. The canoes held a precious cargo necessary to reestablish life on new landfalls: bundles of living taro shoots, breadfruit seedlings, and slips of sugarcane, along with other plants to establish new gardens. A floating menagerie of pigs, dogs, and chickens wallowed about in the bilge water. Little rats the Polynesians called *kiore,* the size of a house mouse, either hid themselves in the cargo or may have been taken along on purpose as a food source. A Polynesian Noah's Ark, bound for the unknown.

These "Vikings of the sunrise," as Maori anthropologist Te Rangi Hiroa once dubbed them, suffered cold and wet, day after day, night after night, for a month or more, led on only by sheer faith that new lands would greet them at the next daybreak. And they were not disappointed. The majestic peak of Haleakalā, House of the Sun, may have been first to rise above the horizon, if some legends are to be believed. The weary voyagers named this part of Maui Kahiki-Nui (Great Tahiti) because of its resemblance to Tahiti Island to the south. They bestowed the ancestral place-name, Hawaiki, on the huge island across the channel from Kahiki-Nui, with its snow-capped volcanoes. Hawaiki was the remembered homeland whence their ancestors had set out generations before to conquer the vast eastern Pacific. Over time, in the way that languages subtly change, the ancient *k* sound would be dropped in place of a simple closure of the windpipe, as in a Cockney accent. Thus Hawaiki became Hawaiʻi.

The Hawaiian Island chain stretches over more than fifteen hundred miles across the central North Pacific. Cook had landed on Kaua'i, and sighted O'ahu in the distance. These are two of the eight volcanic islands making up the main, eastern part of the chain. Beyond them to the southeast lay the smaller islands of Moloka'i, Lāna'i, and Kaho'olawe, along with majestic twin-peaked Maui. At the far end loomed the giant of them all, Hawai'i, abode of the fire goddess Pele, with majestic snow-capped Mauna Kea and Mauna Loa towering more than thirteen thousand feet. With this vast archipelago to themselves, the voyagers claimed the most fertile places to set up their homesites. One early settlement was established along the windward shore of O'ahu. Here the reefs and lagoons teemed with fish, while the gently sloping inland valleys offered fertile volcanic soil in which to plant their gardens.

These new islands were good to the Polynesian settlers. There were no debilitating diseases or poisonous animals, indeed few natural hazards at all. The population multiplied rapidly, initially unimpeded by the "Malthusian brakes" of hunger and famine. By the time Cook's ships hove to off Kaua'i, the archipelago's Polynesian population numbered at least four hundred thousand, quite likely more. From their first tentative foothold, the Hawaiians had gradually expanded over the land, occupying the lowlands and valleys of all the main islands. Eventually, as the better lands became scarce, the population pushed into the hot and dry leeward regions where a living was not to be so easily wrested from the rocky lava slopes.

The British explorers in 1778 pierced the veil of isolation surrounding Hawai'i to come face-to-face with the largest and greatest of all Polynesian civilizations. Not only were the Hawaiians more numerous than their cousins in any other archipelago in the great South Sea, but their society had also achieved a high level of economic, political, and religious sophistication and complexity. The detailed accounts by Captain Cook, Lieutenant King, and other members of his expedition make it clear that Hawaiian civilization at this time of first contact comprised four competing "archaic states." Archaic states are what anthropologists and archaeologists call the political entities that arose in such areas as the Fertile Crescent of Mesopotamia, along the Nile River valley of Egypt and China's Yellow River, and in the New World in Mesoamerica (the Olmec and Maya civilizations) and the Andes (the Wangka and the Inca). These are sometimes called pristine states because they arose independently, rather than through interaction with other already existing states, as occurred later in world history.

Joyce Marcus and Gary Feinman, two archaeologists who have studied the rise of archaic states, point to several criteria that distinguish archaic states from other kinds of complex societies such as chiefdoms. Whereas social organization in chiefdom societies is far from egalitarian, an ideology of *kinship* nonetheless pervades the society. People may be of higher or lower rank based on their birth order, genealogy, and other factors, but they all see themselves as related to one another through bonds of kinship. Even the highest chief is still a kinsman to his people. In archaic states, for the first time in human history, the ideology of kinship was replaced by the notion of *kingship*, specifically divine kingship. Archaic states were headed by kings who regarded themselves as descended from the gods, and whose intercession with the gods was deemed essential to the continued well-being of society. Commoners had separate nondivine origins. Archaic state societies were strictly divided into commoner and ruling classes.

Along with the distinction between commoners and elites came other kinds of social changes, such as a three- or four-level decision-making hierarchy (with the king at the apex). This hierarchy was reflected in the society's settlement pattern. The king and his court resided in a royal center or city, surrounded by two to three levels of subsidiary towns or settlements, all bound to the center by lines of communication and tribute. Social control in archaic states depended heavily on religious ideology. Thus royal centers or palaces were closely linked with the main state temples, presided over by full-time priests. Taxation, usually in the form of food and produce but also to some degree of prestige wealth items, was collected as tribute in the name of the state gods and stored in warehouses adjacent to the palace and temple complex. And in early states, the king and elites who surrounded him maintained their control over society through a monopoly of force. Divine kings held the right to decide life or death; their own royal burials were not infrequently accompanied by numbers of sacrificial victims.

Hawaiian society in the late eighteenth century met all of the criteria that anthropologists recognize as essential to the world's earliest states. Like the pharaohs of ancient Egypt, the Inca rulers of Peru, or the Tennō (Heavenly Sovereign) of Japan, the Hawaiian rulers traced their descent directly from the gods. The Hawaiian kings married their own sisters and half-sisters to keep their bloodlines pure, a pattern also seen in early Egypt and among the Inca. The commoners accorded these rulers the respect owed to divine kings, prostrating themselves in obeisance. Cook had just experienced such abject deference firsthand. In contrast, the Hawaiian commoners, or *makaʻāinana* as they were called, did not even keep genealogies that extended beyond their immediate parents and grandparents. A commoner who dared to claim a genealogy faced likely death, his body offered up as a sacrifice on the altar of the king's war temple.

Fond of metaphor, the Hawaiians compared their chiefs to sharks. *He manō holo 'āina ke ali'i* goes an old Hawaiian proverb: "The chief is a shark that travels on land." The feared, man-eating tiger shark was a natural manifestation of the king. Just as the king was at the pinnacle of society, so the tiger shark was at the top of its food chain, able to devour anything it wanted to. In the words of a chant honoring the king,

> A shark going inland is my chief,
> A very strong shark able to devour all on land.

At the time of Cook's arrival Hawaiian kings—ruling over four independent kingdoms that fought with one another when not forging alliances through marriage of their sons and daughters—were the masters of a self-contained universe. Resplendent in flowing cloaks and crested helmets gilded with thousands of red and golden bird feathers, surrounded by courtiers and warriors, these god-kings controlled a highly developed economic system based on intensive farming and aquaculture. The society revolved around an annual cycle alternating between a season for war and a season for peace and harvest (and taxation, it might be noted). From early November through spring, Lono, god of agriculture, held supreme and the king sent forth priests to collect tribute in his name. As the harvest ended, the war temples were rededicated, their wooden altars set upon giant platforms of lava rock. Kū, god of war, then demanded human sacrifices, usually recent transgressors against the strict code of *kapu* (from which our English word *taboo* is derived).

Over a period of about eight centuries—in complete isolation from the rest of the world—the Polynesians who discovered and settled Hawai'i developed this unique civilization. It is remarkable that this oceanic civilization had avoided contact with the expanding world of Western colonialism. After all, the Spanish first crossed the Pacific with Magellan in 1519–1521. Soon after, Spain's Manila galleons began making annual expeditions from Acapulco to the Philippines and back, exchanging the gold and silver of the New World for Asian silks and spices. The Spanish ships were cumbersome, and their pilots cautious. They followed a slow but predictable northern Pacific route back from Asia that took them past Japan and far north of the Hawaiian Islands, to descend homeward along the coast of New Albion. By the time they had returned to Acapulco—if they had not fallen victim to one of the fierce typhoons of the North Pacific—their crews were often as not half dead from scurvy. Claims have been made, based on obscure and inaccurate maps, that the Spanish discovered Hawaiʻi. Hawaiian traditions do mention a few foreigners being shipwrecked on their shores, but these may have been Japanese or Korean fishermen whose craft were blown away from the Asian coast by those same storms, and drifted on the currents to Hawaiʻi. Certainly, there is no firm evidence that the Spanish ever called in the islands before Cook.

Although their ancestors ranked among the greatest seafarers and navigators the world had ever seen, the Hawaiians of 1778 lived in a cloistered world, untouched and unaware of what lay beyond the circle of the horizon. Hawaiian orators preserved traditions—memorized and passed on from generation to generation—of famous navigator priests who had once plied the seaways between "Kahiki" and this archipelago, bearing the founding ancestors. Hawaiʻiloa, Moʻikeha, Kila, and others were the names of their great navigator ancestors. Pāʻao had brought the cult of Kū, and his thirst for human sacrifice, from the distant lands of Kahiki. But by Cook's day such voyaging was long past, a thing only of memory and chant. Centuries had passed without a sail piercing the endless circle of the horizon. Meanwhile the descendants of those early navigators had prospered in their subtropical, verdant archipelago so far from other islands or continents. Each year, when the star cluster Makaliʻi (Pleiades) first twinkled in the northeastern sky after sunset, the Hawaiian priests offered prayers to Lono, he who had returned to Kahiki. They entreated Lono to revisit their islands, bringing rain, fertility, and bountiful harvests. And always, Lono came as rain clouds, as thunder, and in other signs interpretable to the priests.

The *Resolution* and the *Discovery* were another matter. Lono was a theological concept; these floating structures were real, their decks and rigging crawling with men (the Hawaiians could not make out any women on board the ships), dressed in strange garments. They had curiously pale skins and spoke an unintelligible tongue. The splendid isolation enjoyed by the Hawaiians for roughly four hundred years was about to be shattered. Cook, seeking a theorized northwest passage, had by accident stumbled upon the greatest of Polynesian archipelagoes. Cook could have had little inkling, pacing across the *Resolution*'s great cabin off Kauaʻi, that the landfall ahead would be the greatest discovery of his career. Nor did he harbor any foreboding that he would meet his fatal destiny in these islands, felled by a Hawaiian warrior defending his divine king.

· · ·

Recovering his composure—the seasoned explorer was more used to a hostile display of spears and sling stones than having welcoming natives fall flat on their faces in the sand—Cook eyed his surroundings. The black sand beach rose up gracefully to give way to a pleasant village of thatched houses, shaded by coconut palms. To his right, the river mouth promised a source of fresh water to replenish the ships' stagnant casks. Reassured that the populace was friendly, even to the point of adulation, Cook felt at ease to explore inland. He prudently left a party of marines to guard the boats. Some of the sailors were given instructions to set up a trade for hogs and sweet potatoes, in exchange for nails and pieces of iron.

Something had caught Cook's eye while coasting offshore in search of an anchorage. At every village he had observed "one or more elevated objects, like Pyramids," or perhaps obelisks, shining white in the subtropical sunlight. One of these rose from a point of land east of the river. Cook signaled to his Hawaiian guide that he wished to see it close up. Unable to cross the swift river, the man took them to another, similar structure on the same bank, slightly inland. To reach this "pyramid," Cook's party passed through intricately laid-out gardens, watered by canals and planted densely with taro, the favored staple of Hawaiian cuisine. Cook later wrote admiringly of these irrigated plots, "sunk a little below the common level so as to contain the water necessary to nourish the roots." Intensive irrigation was another hallmark of early civilizations, and in this regard, too, the Hawaiians excelled.

The white "pyramid" proved to be the focal point of a temple, which Cook called a "morai," using the Tahitian word *(marae)* for such a sacred place. The Hawaiians called these temples *heiau*. There were literally thousands of all configurations and sizes of *heiau* distributed throughout the archipelago, dedicated to a pantheon of gods and their specialized cults. On closer inspection, the so-called pyramid proved to be a rectangular tower, built of wooden sticks, about twenty feet high, wrapped around with white barkcloth. Called 'anu'u, this tower was where the officiating priest entered to receive the god's oracle. The English officers explored the temple precincts, with its thatched houses or sheds and various upright carved boards, as well as the image of a female deity. Meanwhile, the expedition's artist, John Webber, sketched the scene. Later engraved and printed as a lithograph to embellish the Admiralty's official journal of the voyage, Webber's view of the Waimea *heiau* is the first depiction of a Hawaiian temple (see figure 1).

His curiosity satisfied, Cook returned to his flagship, intending to explore the island further on the following day. His plans were thwarted by winds and high surf that arose during the night, obliging Cook to move the *Resolution* off-

FIGURE 1.
The interior of a *heiau* (temple) at Waimea, Kaua'i Island, as drawn by expedition artist John Webber on Captain Cook's voyage in 1778. The high tower was one of the "obelisks" observed by Cook from his ship. From James Cook, *A Voyage to the Pacific Ocean Undertaken by the Command of His Majesty for Making Discoveries in the Northern Hemisphere* (London: G. Nicol, 1784, Atlas). Collection of P. V. Kirch.

shore. While the *Discovery* tarried a bit longer at Waimea, the *Resolution* crossed the channel to the smaller, arid island of Ni'ihau, where they traded with the Hawaiians for yams. On Monday, February 2, the resupplied squadron "stood away Northward close hauled with a gentle gale Easterly." New Albion beckoned; the secret instructions of the Sea Lords could not be put off, no matter how alluring the newly discovered islands and their fascinating culture.

With Cook's arrival at Kaua'i, centuries of isolation from the outside world came to an end. Nothing would be the same again. Cook would return the following year after an exhausting and fruitless search for the elusive Northwest Passage, seeking replenishment and rest. For now, alone in the great cabin of the *Resolution*, the seasoned explorer reflected on his latest discovery. The ship's wake hissed through open stern windows; Cook dipped his quill in ink and began his logbook entry. He began by noting that this new archipelago had at least five islands; the Hawaiians had told Cook that yet other islands lay to the southeast. Exercising the discoverer's privilege, Cook named them the Sandwich Islands in honor of his patron, the Earl of Sandwich, First Lord of the Admiralty. Choosing his sentences carefully as the feather quill scratched across the logbook, Cook found himself teased by a nagging question. Once again, he had stumbled onto a branch of the same people, the "same nation" as he called them in the idiom of his day. These people inhabited not only Tahiti and her adjacent islands, but also the Friendly Islands (Tonga) to the west. The natives of New Zealand in the far southern Pacific were also part of this "nation," as were the inhabitants of Easter Island, tucked far away in the southeastern Pacific. Cook had voyaged to all those islands, had charted their coasts, entertained their chiefs in his cabin, traded with their peoples, listened attentively to their speech, and learned some of their words and phrases. That first night off Kaua'i, he immediately knew that he was once again in the presence of this same vastly distributed "nation." Their common language, as much as their physical appearance and the many similarities in their customs and manners, made that clear. Unknown to Cook, Charles de Brosses, a French scholar, had already coined a term for these island peoples: *Polynesian*, combining the Greek words for "many" *(poly)* and "island" *(nesia)*.

Alone with his thoughts, reflecting on long years of exploration, engulfed by the reassuring sound of the *Resolution*'s timbers creaking about him, Cook penned these words: "How shall we account for this Nation spreading it self so far over this Vast ocean?" How, indeed? The European nations had only begun to map the Pacific, its continental margins, and its countless islands. Yet everywhere he sailed in the central and eastern parts of this vast ocean, Cook found the islands already well populated. Ancestors of the people he encountered had obviously discovered these far-flung islands many generations before Magellan had crossed the Pacific in his fetid galleons. The question had been nagging at Cook for some time. This was not the first time he had remarked on it in his journals. He had first posed the question while at New Zealand in April 1770. "We find them from New Zealand to the South, to these islands to the North and from Easter Island to the Hebrides," he wrote as Kaua'i slipped below the horizon. In the precise manner of a master cartographer, Cook calculated the sheer distances involved: "twelve hundred leagues north and south" and an astonishing "sixteen hundred and sixty leagues east and west."

Here was an intellectual puzzle that an Enlightenment mind such as Cook's could grapple with: whence, and by what means, had these "Noble Savages" discovered and settled every habitable island over such a vast ocean? Cook posed the question, struggled with it, but could not find an answer. Humble sailor that he was, Cook was content to set down his practical observations in the logbook, leaving further speculation to the savants at London's Royal Society.

. . .

I rolled over in bed, facing the open lanai door, my nostrils catching the sea breeze. The low sonorous roar of breakers rolling onto the Waimea shore filled the room. My wife, Thérèse, slumbered softly beside me. Moving quietly so as not to disturb her I pulled on my shorts and T-shirt, slipped out the door, and stepped barefoot onto the damp grass. Dawn was breaking over distant Mount Wai'ale'ale. As I walked out to the beach, my only companions at that early hour were some sandpipers hunting for their breakfast of surf crabs. I waded into the foamy arc of a wave's tongue, letting the soft Pacific lap at my ankles. To the south, the bright lights of Gay & Robinson's sugar mill on the far point signaled that the cane crushing season was still in full swing—the last vestige, I mused, of an industry that once dominated the Hawaiian landscape, only to succumb to the vagaries of globalization. Closer, on the northern side of the sandbar that partly blocks the Waimea River mouth, a long wooden pier juts out into the surf. Built almost exactly on the spot where Cook's cutter put ashore—where the crowd of Hawaiians prostrated themselves at the feet of "Lono"—the pier once served to load sacks of sugarcane onto interisland steamers. Today the economic mainstay of Kaua'i is tourism, with jets arriving at Līhu'e airport from Honolulu, San Francisco, and Los Angeles. The old pier has become a thing of history, frequented by a few retired plantation workers who fish with pole and line from its rickety wooden planks. I could see that two of these fishermen were already out at the pier's end, and decided to stroll down to see if they had had any luck.

"*Aloha kakahiaka*" (Good morning), I called out as I approached the stooped figure in his deck chair, grasping a fishing pole with line extending out into the surf beyond the pier. "*Aloha,*" he mumbled back, his weather-beaten face revealing with a twinkle that the old gentleman might be just a little more friendly than his gruff reply suggested.

"Good place for fish?" I queried, pointedly not asking whether he had actually caught any.

"Sometimes," came the response. "Sometimes get *pāpio*, sometimes even *ulua* get."

"Yeah? *Ulua?* Dat's good kine for make '*inamona*,*"* I replied in the lilting pidgin style, referring to the traditional Hawaiian concoction of raw fish, seaweed, salt, and roasted *kukui* nuts. Sensing, I suppose, that this wasn't the usual Haole tourist out for an early walk, he asked where I came from.

"California, da Bay Area," I said, and for a moment he must have doubted his prior conclusion, but then I hastily added, "but, I stay born and raised Honolulu. Local Haole."

"Oh, dat's how come you know about *'inamona*."

"Yeah. I get one old friend, stay Kaupō, Maui. He make da best *'inamona*. Broke da mouth."

We talked for a few minutes about the fishing, and then about life in the "old days" in Waimea. For this man, who I guessed was in his late seventies, the "old days" were the days of the plantations, when the Gay, Robinson, and Faye families controlled the land and sugar was king. We chatted a bit about work on the plantation, and about the cattle ranching over on Ni'ihau Island. On Ni'ihau a small community of Native Hawaiians still speak their language using the *t* rather than the *k* sound, which had made Captain Cook so aware that they were related to the Tahitians. I pointed to the rocky point of land across the river, where the golden rays of the early morning sun now illuminated a series of stacked stone ramparts. "Know anything about dat?" I asked.

"Yeah, dat's the old Russian fort. Was built when the Russians try take Kaua'i from our king, Kaumuali'i."

"That's right," I said, "but did you know that it was built on top one *heiau*?"

"Nah . . . I nevah hear dat. You sure?" The old man gave me a funny look, and I could hear him thinking, *This Haole is* pupule, *crazy*. Everybody knows that pile of rocks is just the old Russian fort. The King Kamehameha visitor's bureau sign on the highway even says that. He turned his attention to his fishing pole. I sensed that our conversation was over, wished him luck, and strolled back down the pier.

Russian fort it had been indeed, I reflected, stopping to gaze across the river at the reddish boulders that had been heaped up in a classic European-style star-shaped fort under the watchful eye of a Bavarian mercenary, Georg Schäffer. For a brief period from 1816 to 1817, Schäffer tried to create an outpost of the Russian Empire on Kaua'i. But the fort had a deeper and more complex history, one that tied it to Captain Cook, though this was unknown to most people, even to long-time residents of Waimea like my fisherman. In fact, this spit of land was where Cook had seen the "pyramid" that so intrigued him, although being on the far side of the river his guide had been obliged to take him to a different, but similar, site upriver. Peter Mills, while a student of mine at Berkeley, had excavated the structure for his doctoral thesis. Peter concluded that the Russian fort of 1816 was either built directly on, or incorporated the stonework from, a *luakini heiau*, a temple of human sacrifice, called Pāʻulaʻula o Hipo. Peter argues convincingly that King Kaumualiʻi regarded the construction of the fort as a new kind of war temple, one that would lend him supernatural assistance in his struggles with the Hawaiʻi Island king Kamehameha for hegemony.

Resting for a moment against the weather-beaten wooden railing, I found myself reflecting on several decades of a career in Polynesian archaeology. All of my life's work, spanning four decades and taking me literally across the face of the vast Pacific Ocean, from Palau to Easter Island, has convinced me of the special place that Hawaiʻi holds, in the history not only of the Pacific, but of the world. Isolated in the central North Pacific, populated by humans only late on the stage of world history, the Hawaiian Islands might be dismissed by some as a peripheral byway, of interest only to Polynesian specialists. In fact, it is this very isolation that makes Hawaiʻi so interesting, that gives it a special quality for understanding how human societies develop and change over time.

Nearly everywhere else in the world where archaic states first emerged—as in Mesopotamia, Egypt, China, Mesoamerica, and the Andes—this happened long before there were detailed historical records. No one ever observed an archaic state in the process of formation. The existence of such political formations has to be inferred from the archaeological record. In Hawaiʻi, however, we have a unique case of several archaic states that emerged so late on the stage of world history that they were historically observed and recorded—in the annals of Captain James Cook and other European voyagers at the close of the eighteenth century.

Decades of archaeological research have given us a fairly coherent outline of how Hawaiian society gradually changed from one organized around the bonds of kinship to a class-stratified society with divine kings at its apex. We are now in a position to tease out not just the dry facts of how this society changed over the course of several centuries, but to ask *why* a society should reorganize itself so completely. The origins of political states have engaged philosophers since at least Plato and Aristotle. Social philosophers of the Enlightenment, including Hobbes and Rousseau, grappled with the question, as did later theorists such as Morgan, Marx, Durkheim, and Weber. But only with the rise of modern anthropological archaeology have scholars had access to empirical evidence about how societies have changed and evolved over long time spans. Archaeological data from Hawaiʻi offer us a unique opportunity to test our theories of social change and state formation. The answers may give us clues not only to what happened in this remote archipelago, but also to how societies change in general.

Precisely because Hawai'i was so thoroughly isolated from the rest of the world, its transformation from chiefdom to archaic state cannot have been influenced by external forces. This is not a case of "secondary" state formation, as in some of the indigenous states of sub-Saharan Africa or Southeast Asia. In these islands, the processes of social change were wholly endogenous, self-contained. Thus Hawai'i offers an especially good opportunity to understand the conditions—environmental, demographic, economic, social, ideological, or some combination of these—that led to the rise of archaic states, along with their most salient feature, divine kingship.

The bright morning sun breaking through a cloud bank stirred me from my intellectual musings. I walked landward, soon reaching the beginning of the pier. Facing me stood a bronze monument of Captain Cook, set on a stone pedestal in the neglected park. A surfer couple eagerly waxed their boards down by the shore, ready to test the day's break. Beyond the explorer's statue, the little town of Waimea was coming to life. Pickup trucks pulled in to the Big Save for ice and supplies before heading down to the fishing dock at Kīkī a Ola. I wondered how many of these people ever paused to ponder the rich history of this place, a history that stretched back eight centuries before Cook. Sensing that Thérèse would be up and ready for breakfast, I put the river to my back and headed up the shore to our cottage.

· · · ·

My thoughts that morning on the old Waimea pier led to the book you now hold in your hands, about the rise of a complex civilization in the world's most isolated islands. It is a story at once unique and with broad relevance to understanding how human societies everywhere develop and change. In these pages I offer my answer not only to Cook's enduring question, "How shall we account for this Nation spreading it self so far over this Vast ocean?" but to other, equally fascinating questions. How did Hawaiian society grow and develop over the hundreds of years that people made their lives in these islands before the Europeans and others invaded their lands? How did people organize themselves over the geologically and ecologically varied island landscape? Just how large was the indigenous population when Cook first arrived? Why and when did the great Polynesian navigators stop sailing between Hawai'i and Tahiti to the south? Did they ever reach the shores of South or North America? How did Polynesian chiefs become divine kings, in a manner that resembled the leaders of other early civilizations? When did human sacrifice first come to be practiced in Hawai'i, and what role did sacrifice play in the emerging state cults? What were the roles of religion and ideology in Hawaiian political evolution? In the chapters that follow I attempt to answer these and other questions, based on my years of archaeological research, and on the collective work of many colleagues. The answers are sometimes incomplete, and tempered by yet other questions, still to be addressed. Such is the nature of scientific history—the job is never done.

The story of ancient Hawai'i as told in these pages is not based on written documents, as with conventional history. Hawaiian culture before Cook was an oral-aural culture, with no written script. This does not mean that Hawaiians lacked a sense of history. To the contrary, like all Polynesians they had a strongly developed sense of historical time and placed great emphasis on preserving the traditions of their ancestors. But these traditions and stories were carried on through memorization and telling, from one generation to the next. Time was indexed by the genealogical succession of ancestors, especially of high-ranking chiefs and chiefesses. In late eighteenth-century Hawai'i, the royal courts on each major island supported specialists whose job it was to memorize these genealogies and the histories associated with them. Through them, the traditions were passed on to selected *punahele* children of the next generation.

When Protestant missionaries arrived in 1820, their first major task was to create an orthography, an alphabet, in order to begin translating and printing the Bible. The Hawaiian people eagerly took up the new literacy. By the second half of the nineteenth century the Hawaiian Kingdom had a higher literacy rate than the United States. One of the first things the Hawaiians began to do with this new tool of writing was to commit their oral traditions and histories to paper. Men like Samuel Kamakau and David Malo, trained in the missionary school at Lahainaluna, wrote extensive compilations that, fortunately, have preserved the old Hawaiian traditions. I will draw upon this rich body of oral tradition as I tell the "unwritten" story of Hawai'i's pre-missionary past.

While Hawaiians had their own unwritten accounts of their past, Western scholars have developed an entirely different approach to uncovering the deep past of those who are sometimes called the "peoples without history." This is archaeology, the science of finding, mapping, excavating, analyzing, and interpreting the material remains left on and sedimented in the landscape by ancient peoples. Archaeology, like Captain Cook, was a child of the Enlightenment, born initially of curiosity about the remains of Classical Greece and Rome. In the twentieth century, archaeology matured into a modern scientific discipline. A plethora of techniques allow us to recover valuable information from seemingly innocuous scraps of evidence. Pollen grains inform us about ancient vegetation and plant use; fish bones from old midden sites reveal the details of fishing methods and diet. Of course, not everything we study is minute, for example, the monumental architecture of old temple platforms and house sites.

My account of what happened in Hawai'i's deep past draws equally on these two lines of evidence—indigenous Hawaiian traditions, and the physical remains and artifacts unearthed by archaeologists—to tell the story of this archipelago from its discovery and settlement by Polynesians, until the fateful encounter with Cook and the West. My account thus joins these "insider" and "outsider" perspectives. The insider perspective, as recounted in Hawaiian traditions, their *mo'olelo*, reflects the indigenous cultural viewpoint. It tells of the voyages of famous chiefs, to Kahiki and back again; of the love affairs and marriages, and jealousies, of chiefs and chiefesses; of wars of conquest and terrible defeats, and of kings offered up for sacrifice on the altars of their rivals' temples. It is a history acted out by real people, whose names are still honored in the *mele* (chants) of the sacred *hula* dance. The outsider perspective, derived from archaeology, is based on the methods of Western science. Archaeology cannot conjure up the names of those who built the ancient houses, fishponds, and temple platforms; it cannot reveal their emotions. But it does have the power to inform us greatly about the daily lives of those who made Hawaiian civilization what it was. The two approaches are complementary; each augments and enhances the other. And when these insider and outsider viewpoints are combined, a richer, more textured and nuanced history emerges.

The story of Hawai'i's ancient past, of the rise of this island civilization, is one of the great sagas of world history. But its ultimate beginnings must be traced back in time more than four thousand years ago to the distant shores of other islands, in the warm waters of the South China Sea. Hawai'i's story is one of voyages, voyages across the wide sweep of the Pacific, and voyages back into time.

PART ONE · VOYAGES

ONE · A Trail of Tattooed Pots

In August 1985, after a wearying journey from Seattle via Honolulu and Sydney, I stepped off a Qantas jet in Port Moresby, the dusty capital of Papua New Guinea. I was en route to Kavieng, a small town in the country's New Ireland province, several hundred miles to the north. There, I would meet up with the ship *Dick Smith Explorer* and members of the Lapita Homeland Project. Our goal was to find the origins of the Lapita culture, believed to be the ancestors of the Polynesians. The *Dick Smith Explorer* would take me to my final destination, the Mussau Islands on the outer arc of the Bismarck Archipelago. Before continuing on my journey to Kavieng and Mussau, however, I wanted to visit Papua New Guinea's National Museum. I needed to have a firsthand look at some Lapita finds that had been recovered a decade earlier in remote Mussau.

The museum's curator, Pam Swadling, greeted me warmly and introduced me to Johnny Saulo, a staff member who came from Mussau. With Johnny, I walked through the public gallery filled with imposing Sepik River cult house carvings and crocodile-headed dugout canoes. Johnny turned a key and we ducked through a doorway leading back into the storage room where the archaeological collections are kept. Traipsing down a musty aisle illuminated by bare light bulbs, we located several wooden trays where Brian Egloff's Lapita specimens were stored. Egloff had made a brief expedition to Mussau in 1973 after local missionaries had found pottery sherds while clearing a small airstrip. He later excavated a few squares;

from one of these he had obtained a radiocarbon date of about 1900 B.C. It was the oldest date known for Lapita pottery.

Johnny and I laid the trays on a workbench. In front of us were a few hundred small pottery sherds, many covered with characteristic fine-toothed stamped designs. The ancient Lapita culture had first been recognized because of these unique designs. In 1952, at a place called Lapita on the northwestern coast of New Caledonia, Professor Edward Gifford of the University of California at Berkeley had found this kind of pottery. Gifford, a senior professor in the waning days of his career, had gone to New Caledonia to seek the origins of the Polynesians. A few years earlier he had dug in Viti Levu in the Fiji archipelago, uncovering a rich succession of pottery types that pointed to a western origin for the people who had migrated into this part of the Pacific. Turning to the large island of New Caledonia—virtually unexplored up to that time by archaeologists—he hoped that he might find traces of even earlier settlements.

At Lapita, toward the end of his six-month-long expedition, Gifford hit pay dirt. In a series of trenches excavated into the sandy earth not far from the shoreline, Gifford and his student Dick Shutler found quantities of distinctively decorated pottery sherds. A comb-like tool with rows of tiny "teeth" had been impressed into the clay surface of the pots, in complex patterns. (Archaeologists call the technique dentate stamping.) Gifford had seen such sherds before, in Tonga, where in 1920 as a young fieldworker he and his colleague W. C. McKern had dug into kitchen middens. He also knew that similar pottery had been found by the Dutch archaeologist Van Stein Callenfels on the island of Sulawesi in Indonesia. Gifford suddenly realized that he had a clue to the long-standing problem of Polynesian origins: a chain of evidence linking Southeast Asia with the Polynesian island of Tongatapu, via New Caledonia. Gifford became even more excited when, after returning to Berkeley, he received the first radiocarbon dates of charcoal he had excavated along with the potsherds. The charcoal gave an age of about 2,800 years before the present, plus or minus 350 years, making these among the oldest artifacts then known from any Pacific island. Gifford had opened up a new window onto the Polynesian past. He was on the trail of the Polynesian ancestors, including those who eventually discovered and settled Hawaiʻi.

After Gifford and Shutler made their remarkable discoveries on New Caledonia, other archaeologists began to pursue the Lapita trail. Jack Golson and his students from the Australian National University returned to New Caledonia to pick up where Gifford had left off. They also dug sites in Tonga, and on the island of Watom in the Bismarck Archipelago. Roger Green, with Honolulu's Bishop

Museum, turned his sights in the early 1970s to Santa Cruz and the Reef Islands at the far eastern end of the Solomon archipelago. He too found rich Lapita deposits, again full of pottery with those distinctive toothed, tattoolike decorations.

By 1984 a significant body of new data was emerging about what archaeologists were now calling the Lapita cultural complex. Green argued that Lapita represented the ancestors of the oldest Polynesian cultures, found in Tonga and Samoa. In his view, Proto-Polynesian culture had evolved out of this Lapita ancestor during the first millennium B.C., in the Tongan and Samoan islands. Later, according to Green, their descendants again voyaged to the east, where they settled the islands of Tahiti, the Marquesas, the Tuamotus, and ultimately Hawai'i. What especially excited the anthropologists was that Lapita spanned the boundary between Polynesia and Melanesia. It thus appeared to be the founding culture throughout the entire southwestern Pacific. Lapita was the key to the puzzle of Polynesian origins.

As Johnny and I turned over in our hands the small, dusty potsherds from Egloff's Mussau excavations in the Papua New Guinea museum's storeroom, I pondered how little was still known about the earliest phases of the Lapita culture. Some scholars even questioned whether Lapita was a part of the greater diaspora of Austronesian-speaking peoples out of Taiwan and Southeast Asia. These contrarians argued that Lapita had an independent origin in the New Guinea region. The Lapita Homeland Project, which I would soon join, was designed to resolve these questions. We needed to know more not only about the origins of Lapita, but also about the nature of this early culture. Were they, as some had suggested, "strandloopers" who subsisted merely by exploiting the bounty of the region's tropical reefs and seas, or did they possess a fully developed horticultural economy? Were they highly mobile, or did they occupy large, permanent villages? What was the nature of their social organization? Was theirs an egalitarian society or did they have some form of inherited rank and status? These and many other questions remained to be answered.

It was clear to me that the fragmentary sherds in these dusty drawers in the National Museum were not going to give me any new answers. New and carefully targeted field research was needed. A few days later, after clearing all of the government formalities in Port Moresby, I boarded my Air New Guinea plane to Kavieng, the capital of New Ireland Province. At the small Kavieng airstrip I was greeted by Jim Allen, a professor at Australia's La Trobe University and organizer of the international Lapita Homeland Project.

"The *Dick Smith Explorer* is a couple of days behind schedule, Pat," Jim informed me. "But this will give us time to go down to Panakiwuk, where you can have a firsthand look at our recent finds." Jim was referring to a limestone rock shelter about four hours' drive south of Kavieng, where he had been excavating with Rhys Jones and Chris Gosden of the Australian National University. Piling my gear into the Land Rover, we headed down the dirt track that serves as the principal highway on New Ireland. The next morning, Jim led me up a slippery trail into the island's central limestone spine. Panakiwuk is a smallish overhang on one face of a great sinkhole, a place where more than twenty thousand years ago people took refuge from the elements while hunting marsupials and foraging for wild tubers or fruit. Jim's excavations had opened a new window into the truly deep past of Near Oceania, showing that humans had occupied these islands thousands of years before the Austronesians arrived with their distinctive pottery. It was all exciting and fascinating, and made me more eager than ever to get up to Mussau and begin my own fieldwork.

The following day Jim drove me back up to Kavieng, where the *Dick Smith Explorer* had now anchored in the harbor. On the wharf I met Pru Gaffey and Sally Brockwell, two Australian archaeology students who would assist me in Mussau. We quickly rounded up additional supplies in the little Chinese trade stores in Kavieng town, loaded them in the hold of the *Dick Smith Explorer*, and prepared to weigh anchor.

· · ·

The Lapita story is part of an even larger saga of human migration that begins on the shores of Fujian and Guangdong provinces along the South China coast, and on the nearby island of Taiwan. In this coastal region rich in resources and blessed with a subtropical climate, an early maritime culture flourished around the fourth to third millennia B.C. Recently, Chinese archaeologists have unearthed numerous traces of these people, who made distinctive earthenware pottery often decorated by cord-marking. Cord-wrapped paddles were used to beat the surface of the pottery vessels and thin them before firing. My former professor and adviser at Yale University, the famous Kwang-Chih Chang, had excavated at the site of Tap'enk'eng, not far from Taiwan's capital of Taipei. K. C., as he always referred to himself, was one of the first to recognize that these archaeological sites containing cord-marked pottery marked a major stage in the evolution of Southeast Asian cultures.

The pioneering work of K. C. and others who followed him showed that Tap'enk'eng and similar cord-marked pottery sites around Taiwan and along the coasts of south China represent the emergence of early Austronesian culture. (The Greek roots of this term translate as "southern islands.") *Austronesian* refers to a group of people speaking related languages who are dispersed from Madagascar off the African coast to remote Easter Island, an astounding distance of more than thirteen thousand miles. There are about twelve hundred modern Austronesian languages, all members of a single language family that gradually diversified from an original proto-language spoken in the Taiwan–South China region about six thousand years ago. Hawaiian is an Austronesian language, as are all of the other languages of Polynesia and most of those in island Melanesia and Micronesia. The indigenous languages of the Philippines, and those of Indonesia, are also Austronesian. Peoples who speak these languages all trace their roots back to the shores of Taiwan and Fujian, where their distant ancestors lived in small villages like Tap'enk'eng, making simple earthenware pottery and experimenting with a new way of life that paired horticulture with sophisticated knowledge of the sea.

These Proto-Austronesians were part of a great cultural transformation that anthropologists call the Neolithic revolution: the transition from a hunting-and-gathering mode of existence to a settled village life based on the domestication of plants and animals, and on the dependable surplus production these could provide. The early Austronesians cultivated rice along with other roots and tubers, such as taro and yams, and tree crops such as bananas. They raised pigs and chickens, and probably also dogs. Living along the subtropical shores of Taiwan and China, these people possessed an intimate knowledge of the rich resources of the region's bays, estuaries, and reefs. They knew the habits of the diverse kinds of fish and shellfish that teemed in these waters; they invented various kinds of fishhooks, spears, and nets, and used natural plant poisons to catch fish. Early Austronesian archaeological sites are full of fishbones and shellfish remains, along with fishhooks made from *Trochus* shell and net weights and line sinkers of stone, all testifying to the abilities of early Austronesians to capture the natural bounty of their coastal waters.

Most important, the early Austronesians invented one of the most remarkable technologies of the premodern world: the outrigger sailing canoe. Chinese archaeologists have dug up preserved parts of simple wooden canoes from swampy deposits at the site of Kuahuqiao, dated to 6000 to 5000 B.C. The existence of regular maritime traffic between Taiwan and the Chinese coast can be traced through the transport of stone adzes (an implement similar to an ax in function, but hafted so that the blade cuts in a stroke toward the user). Historical linguists, who reconstruct ancient languages based on the systematic comparison of words in the many languages descended from a common ancestral tongue, tell us that the early Austronesian vocabulary was rich in words for the outrigger canoe, or *wangka* as the Proto-Austronesians called it.

The ancient Austronesian *wangka* had a single hull, presumably hewn with the stone adzes traded across the Taiwan Straits. By adding an outrigger float to one side of the canoe, they achieved greater buoyancy and stability in open seas. The outrigger was secured to the hull with wooden thwarts. In later Austronesian designs, these supported a small platform to hold people and cargo; even a small thatched house could be lashed to the platform between the hull and outrigger. What made the *wangka* such a marvelous instrument of exploration and expansion was not just the outrigger but the addition of a mast and sail, harnessing wind power for propulsion. The early Austronesians made their sails from the leaves of the pandanus tree, in the same way that they wove mats. Strips of woven matting were sewn together to form sails, lashed to the mast with ropes made from the tough fiber of coconut husks. With a large steering paddle at the stern, these canoes could transport entire families and their cargo along coasts and between islands. The *wangka* propelled the early Austronesians out of their original homeland along the protected shores of South China and Taiwan, into the Philippines and Indonesia, and along the northern coast of New Guinea out toward the islands of the Pacific. Ultimately, these marvelous sea craft would take their descendants as far as Madagascar in the west and to Easter Island in the east.

The Austronesian expansion has been traced by archaeologists, such as Australia's Peter Bellwood, out of the South China–Taiwan homeland region southward into the islands of the Philippine archipelago, along a trail of sites containing earthenware pottery, stone adzes, and other characteristic artifacts. In the Cagayan Valley of northern Luzon, Austronesian pottery appears in sites such as Lal-lo and Magapit by about 2500 B.C. Soon after, similar sites with an Austronesian "signature" appear far to the east, in Sulawesi, Halmahera, and the small islands of the Moluccas. By about 1500 B.C., some of the Austronesians had skirted the northern shores of the large tropical island of New Guinea, arriving in the coral-rimmed islands of the Bismarck Archipelago, which includes Mussau. It was my task in 1985 to see where the Lapita finds first discovered by Egloff might fit into this emerging picture of the Austronesian diaspora.

· · ·

The Mussau Islands bake under a near-equatorial sun on the northern arc of the Bismarck Archipelago. The *Dick Smith Explorer*'s steel hull and deck were designed for Antarctic exploration rather than tropical cruising. It was unbearably hot belowdecks, so I opted to pass the night curled up on a hatch cover. We sailed through the night and made landfall the next morning. As we cruised slowly up the Malle Channel, a lookout was posted in the masthead to watch for coral heads, which abound in these poorly charted waters. Off Eloaua Island, a dugout canoe came offshore from one of the little hamlets of thatched huts. This was Johnny Saulo's home island, and he had alerted his uncle Ave Male by radio that our team would be arriving. Confirming that this was indeed Ave's village, I was relieved to find that we were expected. The ship's dinghy landed me and my two young Australian assistants along with our gear on Eloaua, one of several coral islets in the southwestern part of the Mussau group. It was here that construction of the mission airstrip had turned up Lapita pottery back in 1973. We would spend the next six weeks on Eloaua seeking to learn more about the Lapita past.

Ave greeted us warmly and made us feel at home in his tidy compound of sago-leaf-thatched houses nestled just behind the coral sand beach. Once we had unpacked our gear and gotten our bearings, our first task was to relocate Egloff's original find spot, situated in a yam garden to one side of the crushed coral runway, used periodically by the mission's light aircraft. Ave took us to the place. Opening up a new test excavation, we found more pottery sherds in a black soil overlying white coral beach sand. The sherds were disappointingly small and broken up, the result of centuries of turning the soil with digging sticks. After several days of digging, it began to look as if we might not get much further than Egloff had in answering our questions about the Lapita people who had once occupied this place.

Reconnoitering the island more extensively, I discovered that the airstrip where the pottery had first been exposed lay on a slightly elevated natural terrace a meter or two higher than the modern coastal plain. Knowing that sea level had been slightly higher about three to four thousand years ago, I reasoned that the Lapita settlements might have been positioned at the top of what was once the active beach, just as Ave's hamlet is at the top of today's. When sea level fell to its modern level, this would have left the former beach ridge slightly elevated, to be claimed later by the jungle as the shoreline advanced. To test my hypothesis, I laid out a series of test excavations extending across the old beach ridge and down onto the lower coral sand flat, at a place that Ave told me was called Talepakemalai. I expected that I would find the Lapita pottery and associated occupation debris distributed across the elevated terrace, and that these finds would cease as we crossed the slight slope that I thought marked the ancient beach line.

The first few test pits up on the terrace met my expectations. Lapita pottery, shellfish and fishbone remains, and obsidian flakes caught in our sifting screens all indicated that the Lapita peoples had lived on the higher beach ridge. Unfortunately, pit after pit showed the same shallow cultural layer, heavily disturbed by generations of gardening. Moving along our transect line, I then laid out a new excavation pit on the lower sandy flat, which I reasoned had once been the active beach. Expecting to find only a "sterile" deposit of beach sand, I was startled when Pru called out for me to come and have a look at what she was encountering at a depth of about half a meter in her test pit. Peering down into the neat meter-square pit, I watched as Pru's trowel exposed several large pieces of Lapita pottery. One sherd had a distinctive human face motif, whose almond-shaped eyes stared back at me. Unlike the pottery we had been finding up to that point, these sherds were well preserved and clearly had not been disturbed by later gardening. A smooth, shiny circular object proved to be a complete ring, exquisitely carved from the shell of a giant clam. Lapita people had used such rings as a form of shell "money" (not money, really, but valuables exchanged between social groups). Several large pieces of razor-sharp obsidian turned up next, followed by still more pottery sherds, and then an entire pig's tusk drilled for suspension, perhaps as a pendant.

The sand in the test pit had turned distinctly soggy; we were approaching the natural water table, or freshwater lens, that lies not far below the ground surface on coral islands. Yet pottery and other objects continued to appear. I took over digging from Pru and continued down even deeper, periodically bailing the water now rapidly seeping into the pit. A curious dark organic stain appeared in one corner. Mushy at first, it became firmer as I worked my trowel around the object to expose it. I soon discerned that this was a shaft of wood, extending downward into the sand, preserved all these thousands of years in the waterlogged, oxygen-deprived sediment. The shaft turned out to be the base of a post, sharpened to a point with a stone adz. I suddenly realized that a wooden structure of some kind had once stood on this spot.

Over the following weeks, we expanded our excavations out of the initial one-meter test pit, opening up a large area. This revealed not just one but numerous wooden post bases, set in two rows, or alignments, that had originally formed the sides of a stilt or pole house. This house had once stood over the shallow lagoon flat seaward of the beach ridge. Radiocarbon dating later indicated that the house had been occupied about 1300 B.C. My hypothesis about the beach ridge and location of the old shoreline had been correct. What I hadn't anticipated was finding the remains of a stilt-house village that had once extended out over the lagoon (fig-

FIGURE 2.
The author at the completed excavation of the Lapita stilt house remains at Talepakemalai, Mussau Islands, in 1986. The black, anaerobically preserved bases of the wooden house posts can be seen sticking up from the reef sands. The hoses were used to pump water, which continually flowed into the excavation. Photo by P. V. Kirch.

ure 2). Not that this should have been wholly unanticipated: other Austronesian societies in island Southeast Asia, along the New Guinea coasts, and as far east as the Solomon Islands were known to have built stilt houses over coastal waters. It was just that they had never been found in a Lapita context. After six weeks of intensive digging, Pru, Sally, and I returned to Kavieng with several crates full of precious Lapita pottery and a vast quantity of other specimens and samples. These materials would allow me to reconstruct in great detail the life of this early community of Austronesian settlers. I hand-carried the most spectacular finds with me, stopping off en route to show them to former colleagues at Honolulu's Bishop Museum. They were stunned at the beauty of the pottery with its fine decorations, and by the exquisitely carved shell and bone artifacts.

In 1986, and again in 1988, I returned to Talepakemalai to excavate and recover one of the largest collections of Lapita materials to date. What we subsequently learned from our detailed studies of the pottery, obsidian, food remains, and other materials that had been discarded by the Lapita occupants of that stilt-house village greatly expanded our knowledge of this early phase of Polynesian ancestors. Other field teams of the Lapita Homeland Project, including Chris Gosden in the Aware Islands off New Britain (Gosden also found the remains of stilt houses), Matt Spriggs on Nissan Island, and Ian Lilley in the Duke of York Islands, all added to the growing body of data about Lapita. Since then other investigators have joined in the quest to understand Lapita, including Glenn Summerhayes of New Zealand, Christophe Sand in New Caledonia, and my own student Scarlett Chiu, who has helped to break the code of the Lapita pottery design system. Archaeology is a science that depends on teamwork and the accumulation of knowledge. What we now know about this earliest phase of the movement of Polynesian ancestors into the Pacific owes much to the work of these colleagues.

· · ·

The Lapita Homeland Project showed that Lapita did not arise independently in the Bismarcks, as some had claimed. Instead, it was a key part of the expansion of the Austronesian peoples eastward out of the Sulawesi-Halmahera region of Southeast Asia into the Pacific proper. Plying their *wangka* sailing canoes through the equatorial waters north of New Guinea, perhaps in search of new trading opportunities, these Austronesian people moved into the Bismarck archipelago between 1500 and 1300 B.C. Pottery found in the earliest Lapita sites such as Talepakemalai closely resembles early Austronesian pottery in the Philippines, Sulawesi, and Halmahera.

However, the Austronesian-speaking Lapita people were not the first humans to settle this region. Evidence from limestone caves such as Panakiwuk excavated by Jim Allen on New Ireland showed that early hunting-and-gathering peoples settled the large islands of the Bismarcks as long ago as 36,000 B.C. Those early populations, doubtless small in number and probably highly mobile, subsisted from hunting marsupials, birds, and reptiles in the forest, gathering wild plant foods, and collecting shellfish and fish from the reefs and shallow lagoons. Their descendants followed this mode of existence for thousands of years, until the early Holocene. About 8,000 B.C., the indigenous occupants of the Bismarcks and New Guinea began to domesticate plants and gradually shifted to a horticultural way of life. The plants that they domesticated included tuber crops such as taro and yams, fruit crops including bananas and breadfruit, and nut-bearing trees like canarium almonds. These people spoke a diversity of mutually unintelligible languages that linguists group under the broad category of Papuan.

Between about 1500 and 1300 B.C., the Austronesians in their mat-sail-propelled *wangka* arrived in the waters of the Bismarck Archipelago, encountering the indigenous Papuan peoples already in residence. As nearly as can be ascertained, this contact of cultures was for the most part peaceful. Recent biological analysis of human DNA shows that these peoples exchanged marriage partners. Genetic traits that had evolved in the local Papuan populations, including a blood hemoglobin mutation that gave them resistance to the pervasive malaria of the Bismarcks, were taken up into the Austronesian gene pool. More than that, the distinctive Lapita culture was a kind of synthesis of the indigenous Papuan and the migrant Austronesian. From the former, the Lapita culture gained local knowledge of domesticated root, tuber, and tree crops, abandoning the older Austronesian reliance on rice. For their part, the Austronesians introduced the outrigger sailing canoe, and new fishing techniques such as trolling for pelagic fish, including bonito. They also brought to the Bismarcks the art of manufacturing pottery. Roger Green summed up the cultural synthesis that took place around the islands of the Bismarcks Sea in the mid second millennium B.C. with three words: *intrusion* (the arrival of the Austronesian people), *innovation* (the adoption of new technologies), and *integration* (the bringing together of indigenous Papuan and newcomer Austronesian cultural traits). We call this the Triple-I model of Lapita origins.

One of the most fascinating Lapita innovations involved the pottery vessels. Earlier Austronesian pottery in the Philippines or Sulawesi shows simple stamped or incised decoration. But Lapita pottery is unique in that many, if indeed not nearly all, of the designs represent human faces. Many of the beautifully decorated jars and bowls that we lifted from the waterlogged sands of Talepakemalai are covered in intricate, dentate-stamped motifs (figure 3). They portray faces with an elongated nose set between almond-shaped eyes, topped with a "flame" headdress. Sometimes there are upraised arms and fingers depicted to the sides of the face; other times there are geometric motifs that Matt Spriggs has likened to ear ornaments. We have yet to decipher this Lapita artistic code in full, but these were not mere utilitarian objects—they were valuable creations imbued with rich symbolic meaning. Indeed, the faces may have represented ancestors; the pottery (at least the decorated vessels) seems to have been used primarily in ceremonial contexts. In some Lapita sites such as Teouma in Vanuatu, people were buried in pots, or the pots themselves were buried after the end of a ceremony.

Tattooing is widely practiced among Austronesian peoples. The art was an early invention among the Proto-Austronesians in their South China–Taiwan world. Polynesians, like other Austronesians, tattooed profusely. In fact, our English

FIGURE 3.
An elaborately decorated ceramic cylinder stand as it was uncovered in the excavations at Talepakemalai, Mussau Islands. Note the finely executed dentate-stamped designs. Photo by P. V. Kirch.

word *tattoo* comes from the Tahitian word for this practice, *tatau*. (Sailors on Captain Cook's ships were among the first Europeans to be tattooed, and brought the word with them back to England, to be followed by generations of other seamen.) The intricate fine-toothed decoration that graces Lapita pottery mimics very closely the Austronesian method of tattooing. Both use a multitoothed comblike "needle," whether to insert the dye under the skin or to make the elaborate designs in the clay surface of the pots. In decorating their clay pots with the faces of deceased ancestors, the Lapita people were in essence tattooing the pots. The act of tattooing a human body was considered to be an important rite of passage, associated with coming of age or with assuming an important title or social role. In the same way, the act of tattooing a pot was likely to have been fraught with meaning. Scarlett Chiu has shown that some of the complex designs involving human faces combine two motifs found individually on other pots. She suggests that these new motifs represent the uniting of two lineages or family groups by marriage. There is much still to learn about the role that pottery played in the lives of the Lapita people, and much that we will never know. But the tattooed pots that emerged as a result of the Lapita cultural synthesis in the Bismarck archipelago in the mid second millennium B.C. are a key milestone on the long trail leading from the coasts of Asia to the shores of Hawai'i.

· · ·

About 1200 B.C., Lapita people from the Bismarcks began to expand eastward, as their ancestors had done, first through the main arc of the Solomon Islands, and then out past San Cristobal Island across more than 280 miles of open ocean to the Santa Cruz and Reef islands. This seemingly simple act, of exploring the open ocean eastward past the end of the main Solomons group to discover Nendö and the Reef Islands, marks a watershed in the history of human exploration. Up to this time, no human beings had gone farther east into the Pacific. Early hunting-and-gathering peoples had managed to cross narrow ocean channels separating New Britain and New Ireland from the large island of New Guinea as long ago as 36,000 B.C. They got to Bougainville in the Solomons by at least 26,000 B.C., and probably moved down into the other islands of the Solomons chain by the end of the Pleistocene (ca. 8000 B.C.). But at San Cristobal that slow, creeping expansion of the hunters-and-gatherers had come to an end. A vast ocean expanse lay before them, too formidable to cross and explore with simple bamboo rafts or dugout canoes. Whether some tried and failed we shall never know. But it is certain that until the arrival of the Lapita voyagers, the world beyond the end of the Solomons was an island universe known only to the birds, insects, tiny snails, and other creatures able to disperse by one means or another out into a profoundly oceanic world.

The Austronesian *wangka* was the first watercraft capable of exploring this formidable ocean, of carrying family groups and their precious cargo of seedling crops, domesticated pigs, dogs, and chickens, as well as other necessities for founding a new colony. The Lapita people had perfected the *wangka* since its initial invention in the calm waters of the South China Sea. For three centuries, Lapita canoes had been sailing among the islands of the Bismarcks and down into the Solomons chain, what my colleague Geoff Irwin calls a great "voyaging nursery."

About 1200 B.C., a Lapita *wangka* made that first crossing beyond San Cristobal Island, to discover Nendö and the adjacent Reef Islands. Its crew stepped ashore on beaches that only sea turtles had previously marked with their flippered crawl.

Some stayed to plant their taros and yams in the fertile earth, and to live off the rich resources of nesting seabirds, shellfish, and teeming fish in the lagoons. Others refitted their trusted outrigger canoes and returned with the news of their discovery: there were indeed more islands to the east. More canoes set out from the Lapita homeland to head eastward, bringing with them a precious cargo of sharp obsidian rock from the quarries at Talasea in New Britain. Flakes of this precious rock were excavated and identified by Roger Green in the sands of Nenumbo and other village sites of the Reef Islands thousands of years later, enabling science to trace this migration.

The Lapita dispersal picked up steam, possibly fueled by rapid population growth in islands that were free of malaria and other old-world diseases, and rich in resources. They may have been driven as well by the incentive to discover new islands in which one could be the unquestioned master (not having to dispute your claim with a prior indigenous population). Austronesian societies are characterized by what anthropologists call *ranking*, in which birth order determines the inheritance of land, houses, material property, titles, and even access to ritual or esoteric knowledge. Once the Lapita people discovered that new islands awaited them to the east, the incentive for a junior sibling to mount his own voyage of discovery—in which the chances of his being able to discover and claim a new island as exclusively his and for his own heirs seemed reasonably high—was compelling.

Within no more than two to three hundred years, or about eight to twelve human generations, Lapita people had used their *wangka* canoes to explore and colonize the previously uninhabited islands extending from the end of the Solomons down through the vast Vanuatu archipelago all the way to the Loyalty Islands and New Caledonia. Others moved directly eastward from the Reefs–Santa Cruz islands to cross a daunting five hundred miles of open sea and arrive in the vast Fiji archipelago. By 950 B.C. the most intrepid of them all had gone beyond Fiji—to Tonga, up through the Ha'apai Islands, and on to Samoa. In a mere 250 years, the Lapita people had expanded across roughly one-third of the Pacific Ocean!

· · ·

In 1976, I led a Bishop Museum expedition to Niuatoputapu Island, at the northern end of the kingdom of Tonga, approximately halfway between Tongatapu and Samoa. Over the course of seven months, our team discovered and investigated dozens of sites, tracing the island's history from initial settlement by a small group of Lapita voyagers until the period of contact with Europeans. The founding settlement, which we radiocarbon-dated to about 900 B.C., consisted of a small village or hamlet at the place called Lolokoka on the island's sheltered lagoon shore. Excavating at Lolokoka, we recovered pottery decorated in the classic dentate-stamped Lapita style, along with shell fishhooks, beads, rings, and other ornaments of shell, adzes of giant clam shell, and sharp flake tools of chert and volcanic glass.

Recently, Dave Burley of Simon Fraser University in Canada has conducted archaeological surveys throughout the Tongan archipelago, from Tongatapu in the south up through the Ha'apai Islands as far north as Vava'u. Dave has shown that small Lapita hamlets are scattered throughout these islands, all dated to about 950 to 900 B.C. Thus, once they discovered the Tongan Islands, the Lapita settlers quickly moved from island to island, establishing footholds on each one. The sites are small, of a size that might be expected from single households or extended families.

In the large and fertile Samoan archipelago, only a single Lapita site has yet been found, in a unique environmental setting. There, Lapita pottery was discovered beneath a reef flat in the course of dredging to make a small boat harbor. The Samoan Islands are rapidly subsiding because of their proximity to the edge of the Pacific tectonic plate. Thus, a Lapita village that was originally on the shoreline of 'Upolu Island (near Mulifanua) had, over the course of three thousand years, sunk beneath sea level and been covered over by a growing coral reef.

The people whom anthropologists call Polynesians all speak closely related languages and share many aspects of culture. They occupy islands and archipelagoes within a great triangle defined by New Zealand in the southwest, Easter Island in the southeast, and Hawai'i at the northern apex. This much has been known since Captain Cook first recognized the Polynesians as a "Nation." Thanks to the evidence from archaeological excavations in more than one hundred sites along the Lapita "trail of tattooed pots," we now know that the first people to enter the Polynesian Triangle were the *wangka*-sailing pottery makers who arrived on the islands of Tonga and Samoa about 950 B.C. They were the farthest-flung branch of the Austronesian diaspora that had started in Taiwan and Fujian some two thousand years earlier, the descendants of generations of voyager-colonists who had expanded out of the sheltered waters of Near Oceania to brave the open seas of the central Pacific.

Archaeological traces of Lapita have been found only at the western margins of the Polynesian Triangle, in the Tongan and Samoan archipelagoes and on the smaller western islands of Futuna and 'Uvea. Island groups farther east, south, and north were not settled until more than a thousand years later. By then, the Lapita voyagers had settled in to the Tonga-Samoa region, and their descendants had evolved a distinctive new Polynesian culture and language. This is a topic I will take up in the next chapter. The central point is that archaeology has now shown that the "gateway" to the Polynesian Triangle was the Tonga-Samoa region. It was here that people first arrived and here that they developed the traits we have come to recognize as distinctly Polynesian. The discovery of the Lapita "trail of tattooed pots," from the Bismarcks to Tonga and Samoa, is one of the great achievements of archaeology in the later half of the twentieth century.

· · ·

On a sunny August afternoon in 2002, representatives of the island nations of Papua New Guinea, the Solomon Islands, Vanuatu, Fiji, Tonga, and Samoa gathered on the grassy plain of Lapita, at Foué on New Caledonia. With the lagoon at our backs, gentle waves lapped at the beach of Lapita. The purple-gray nickel-rich mountains of La Grande Terre loomed ominously in the distance. A hundred or so of us—diplomats, archaeologists, and local Kanak inhabitants—gathered in a rough semicircle on the low grass to witness a unique ceremony. Fifty years earlier, Professor Edward Gifford of Berkeley had dug into the windswept plain where we now stood. We had come to remember and honor the event that had revolutionized our understanding of Pacific prehistory. Gifford had died in 1956, not long after making his discoveries. But standing there among us fifty years later was his former student Dick Shutler, now a distinguished emeritus professor of archaeology in Canada. The ground we were gathered on was where Shutler and Gifford had unearthed a key to the unwritten history of the Pacific Islands, including Hawai'i. Since their discoveries, the name Lapita has come to take on an almost reverent quality among Pacific scholars.

As the simple yet moving ceremony unfolded, I glanced over at Shutler, who could have little imagined fifty years earlier that the earthenware pottery sherds he was collecting from the sifting screen under the watchful eye of the balding and half-deaf Professor Gifford would have had such a lasting influence on Pacific affairs. Yet here the representatives of several island nations of the southwestern Pacific had come to acknowledge and bear witness to the common ancestry shared by their varied cultures. Dick's eyes were unmistakably tearing up.

The distant mountains framed the Kanak chiefs as they received their visitors. Tongan noble Tu'ivanuavou Vaea and his adjutant, Vili Vete, their hips girded with finely woven mats to mark the ceremonial occasion, strode forward. Between them they unfurled a magnificent tapa cloth with striking dark brown geometric patterns painted in *koka* bark pigment. The same designs had once graced the curved surfaces of Lapita pots, and the signaled an unbroken artistic legacy spanning three thousand years. Vaea and Vete laid their gift at the feet of the Kanak chiefs and then stepped back into the crowd. Chief Joseph Vita of Vanuatu presented shell necklaces, another link to Lapita times. Sepeti Matararaba of Fiji held a bundle of dried *yagona* (kava) root as he addressed the local chiefs. Kava was first domesticated by the Lapita people, later to become the sacred drink of many Pacific island cultures. Captain Cook had been offered ʻawa, as the Hawaiians called it, during his second voyage to the islands.

Finally, Taliaoa Pita Ulia from Samoa, the last representative to come forward, stripped off the Western dark suit coat and white shirt he was wearing. Baring his upper body, Ulia exposed an intricate set of dark blue tattoos, extending up from his buttocks over his waist and lower back. To our eyes he revealed a striking connection to the deep Oceanic past. The Lapita people had not only tattooed their bodies, but had "tattooed" their ceremonial vessels, in likenesses of their ancestors. The fine dentate-stamped decorations on the Lapita pots closely mimicked the tattooing of human bodies with similarly fine-toothed tattoo needles.

The trail of "tattooed pots" has given archaeologists clues to trace the origins of the Hawaiians and other Polynesians back more than four thousand years. The puzzle that Captain James Cook pondered in the great cabin of the *Resolution*, as she slipped away from Kauaʻi—"How shall we account for this Nation spreading it self so far over this Vast ocean?"—is finally being answered.

TWO · East from Hawaiki

The headlights of Chief Paopao's truck cast eerie shadows among the thick hibiscus and pandanus that threatened to engulf the rutted track hugging the eastern coast of Ta'u Island. As we bounced along in the predawn darkness, I reflected on what had brought me to this abandoned, wind- and wave-swept shore. I had arrived on Ta'u two days earlier, in June 1986, to begin an archaeological survey of the Manu'a group of islands that makes up the eastern extremity of the Samoan archipelago. Terry Hunt, a University of Washington grad student, was along to help out and gain experience. Little was known of the prehistory of these remote islands; we hoped to shed some light on the early history of Polynesians, especially in the period following the Lapita colonization.

Chief Paopao and his wife had graciously offered to lodge us in their house in Fitiuta village. The previous day we had discussed our research program with Paopao and other elders in the village guesthouse; later Paopao showed us stone structures in the Faga area. After explaining that our objective was to gain understanding of the deep past of Samoa, before the arrival of the *Papalangi* (Westerners), Paopao said that it would be important for him first to take us to Saua, a narrow coastal plain lying under the shadow of Ta'u's high mountain. But, Paopao stressed, we must go there before the first light of morning. I thought this was a bit curious but thought it best not to question the chief and follow his lead.

Paopao woke us from our slumbers about 4:00 A.M., offering mugs of hot coffee along with cabin biscuits and tinned butter for a quick breakfast. Shivering slightly in the damp air, Terry and I jumped into his 4x4 for the drive to Saua. It was essential, Paopao once again asserted, to arrive at Saua before the break of dawn.

Paopao pulled the truck into a narrow clearing off the main track, cutting the lights and engine. We stepped out into the salty air, guided down a sandy path to the beach by the roar of the surf crashing on the reef close offshore. The eastern sky was becoming distinctly lighter. Arriving at the beach crest, Paopao motioned for us to sit on the soft coral sand. He began to speak: "If you are going to work in Manu'a and want to understand our history, you must know about this place, Saua, and about our great creator god, Tangaroa. It was here, in Manu'a, that the sun first rose in the eastern sky. It was here that Tangaroa created the world, and the first man."

I already knew something about Tangaroa from my previous research into Polynesian history and culture. He was the first of the great Polynesian gods, the one who had with enormous effort pushed his parents—Sky Father and Earth Mother—apart from their eternal embrace, thus creating space for the world. In Samoa and elsewhere in Western Polynesia, Tangaroa was the supreme creator god, who had fashioned the first man from clay. In other Polynesian groups to the east, including Hawai'i, Tangaroa joined his brothers Tane, Tu, and Rongo as one of a pantheon of four great deities. The Hawaiians knew him as Kanaloa, god of the ocean. But here on the beach at Saua, with the east wind in my face and the surf rushing relentlessly toward us over the reef crest, I was not just reading about Tangaroa in some university library. Paopao, hereditary chief of Fitiuta, was sharing with me a tradition that extended back more than two thousand years, to the time of the original Hawaiki, the ancient Polynesian homeland.

By now the dawn's reddish-golden rays streamed skyward from behind a low-lying cloud bank. Paopao continued his account of how Tangaroa, having created the world, at first found only the vast, lonely expanse of waters, *Moana*. Taking the form of the migratory golden plover, known as *tuli* in Samoan (and to ornithologists as *Pluvialis fulva*), Tangaroa flew far and wide, searching for land but finding none. So Tangaroa caused the islands to emerge from the watery vastness, the islands of Samoa, the islands of Tonga, and those of Fiti (Fiji) to the west. And on the island of Savai'i (the Samoan spelling of Hawaiki), he caused a boy, Sava, and girl, I'i, to be born, Thus, here in Hawaiki, Tangaroa created the human race.

We sat for a long time quietly on the Saua beach as the golden dawn melted away to the humid warmth of a tropical morning. A solitary curlew waded in the foamy water at the foot of the beach. I was deeply touched, not only by Paopao's tale of Tangaroa, but by the simple fact that this proud chief wanted us to appreciate his

Polynesian point of view. Too often, I mused, Western scholars and scientists discounted Polynesian perspectives, their histories and traditions. Yes, some of these were "merely" mythology, but then what civilization is not founded on its own great mythic traditions? Just as the Greeks had Zeus and Ares, and the Romans Jupiter and Mars, so the Polynesians had Tangaroa and Tu. To get to the core of a culture, as anthropologists from Frazer to Lévi-Strauss have realized, one can do no better than to start with myth.

Hawaiki is a name deeply ingrained in Polynesian myth and tradition. Hawai'i itself is a variant spelling of this ancient name. So is Savai'i, the largest island of the Samoan group. (The old Polynesian *k* sound is often replaced with a glottal stop ['] in Eastern Polynesian dialects, and the Samoans further replace the *h* with an *s* sound.) The most widespread usage of the name, especially in the islands of central and eastern Polynesia, is as the ancestral homeland, the origin place of the Polynesians, from which their progenitors derived. It is to Hawaiki that their own souls will voyage after death, to join with the spirits of the ancestors.

Hawaiki is more than just a mythic place; it can be firmly situated in space and time. Decades of careful research in comparative ethnography, historical linguistics, and archaeology all point to the same conclusion: the ancestral Polynesian culture, along with its Proto-Polynesian language, emerged out of the small populations of Lapita voyagers who reached the Tonga-Samoa region about 950 B.C. Geographically, Hawaiki included the island chain stretching north-south from Tongatapu up through the Ha'apai and Vava'u island groups, to tiny Niuatoputapu, and on to the Samoan islands, as well as the Futuna and 'Uvea islands between Samoa and Fiji. The period during which the ancestral Polynesian culture took shape was from the end of the Lapita period, roughly 500 B.C., until a renewed period of exploratory voyaging took the Polynesians eastward once again, in the mid to late first millennium A.D. It was during this period of roughly a thousand years that the unique characteristics of Polynesian language and culture developed on the western margin of the vast Polynesian triangle. This was the true Polynesian homeland; this was Hawaiki, to be remembered forever after in myth and chant.

· · ·

Our 1986 reconnaissance in Manu'a, aided by the hospitality of chief Paopao, was more rewarding than we could have hoped for. Terry and I recorded numerous stone house platforms, star mounds, and other structures in the jungles of Ta'u and on the nearby islands of 'Olosega and 'Ofu. More important, we discovered the first sherds of ancient pottery in the Manu'a Islands, something that we knew belonged to the earliest period of Polynesian history. Inspired by our finds, we returned twice more to these rugged little islands at the eastern extreme of American Samoa, seeking clues that would help us to uncover the physical traces of ancient Hawaiki.

Having studied geomorphology at the University of Pennsylvania, I knew that the coastal zones of volcanic islands such as 'Ofu were likely to have gone through significant changes over the period that Polynesians had lived on these islands. Sea level had risen after the end of the last ice age, reaching a maximum about four thousand years ago, then falling back about one meter (roughly three feet) to the present level. The shorelines of islands would have responded accordingly, as we later saw in the Talepakemalai Lapita site on Mussau (see chapter 1). Thus early phases of human occupation on island coasts were likely to be buried under sand and coral rubble thrown up by storm surges, or under volcanic soil and debris eroding from the steep and unstable mountain slopes. An island archaeologist needs to be prepared to dig, sometimes quite deeply, to find what he is seeking.

When we returned to To'aga in 1987 to begin intensive archaeological fieldwork, I decided to use a method that I had already successfully applied on Niuatoputapu Island. We first cut a transect line through the thick vegetation from the shore running inland, perpendicular to the coastline, right up to the base of the steep mountain. A detailed elevation profile was measured along the transect, permitting us to detect subtle changes in the land surface. Then we began to dig test pits down into the coastal plain, at intervals of ten or twenty meters. The first few test pits near the coast revealed only fine white coral sand, with no artifacts. As we approached the center of the plain, halfway to the mountainside, our test pits began to turn up evidence of ancient human occupation and land use: stone flakes and a piece of an adz, shellfish and fishbone from ancient meals, and carefully laid gravel pavements of old house floors. But no pottery: clearly, this middle part of the plain had once supported a village, but one dating to a period after the Polynesians had given up the use of pottery.

Finally, our test pits reached the gentle slope below the steep mountain. Now the ground beneath our feet consisted of clay, not sand. We laid out a one-meter-square test unit and began to dig. The compact clay interspersed with rock and small boulders made the excavation difficult. At about a meter down, the clay abruptly ended, and we were back into coral sands. Our excavation had penetrated through a deposit of volcanic clay and rock that had eroded down from the mountain slopes, covering the original coastal plain. The sandy deposit being turned up by our trowels was a grayish color, stained with ash and flecked with small pieces of charcoal, the product of countless fires that had burned in ancient earth ovens. Peering into the sifting screens, my eye caught a glimpse of a flattish-sided object, reddish-brown in color. To the uninitiated, this might look like a piece of soft rock, but I knew otherwise. This was a potsherd, a fragment of a fired clay bowl made by one of the potters of ancient Hawaiki.

Soon our test pit had produced dozens of such pottery sherds, along with other artifacts including fishhooks made of turban shells and beautifully polished adzes of basalt. We spent many weeks, in 1987 and again in 1989, excavating more than twenty such test pits along the inland extent of the To'aga coastal plain, accumulating a sizable collection of artifacts. Thousands of bones of fish and other animals, and the shells of mollusks, all provided evidence for ancient Polynesian dietary habits. This evidence allowed us to reconstruct in some detail the life of the first inhabitants of 'Ofu Island. We sent fourteen samples of charcoal and shell from our excavations to the Beta Analytic laboratory for radiocarbon dating; the results gave us a precise chronology for the To'aga site. People had first settled on the narrow coastal plain of 'Ofu about 800 B.C., not long after the Lapita colonists first arrived in the Tonga-Samoa region. The layers containing pottery at To'aga begin about 800 B.C. and continue until about A.D. 500; more recent layers lack any pottery at all.

The cultural sequence contained in the deeply buried sands at To'aga tells us much about the material world of ancient Hawaiki, the Polynesian homeland and abode of Tangaroa. Perhaps the most surprising aspect of this story is the gradual loss of the art of making pottery, during the thousand years or so after the Lapita people first arrived in the region. Pottery was a hallmark of the Lapita culture, with its distinctive "tattooed" designs. But soon after the first Lapita voyagers settled the islands of Tonga and Samoa, their pottery began to change. First to go were the elaborately decorated vessels, so that the pottery became what we call plainware. This plainware continued to be fairly well made, with thin sides and well fired. But then, after several centuries, even the basic manufacturing of the pottery began to decline in quality. The simple bowls became thick, and coarse in texture. Not as much attention was paid to firing. The late Polynesian plainware that we found in the upper, more recent layers at To'aga was quite crude, an art form in the final stages of decay. Soon, pottery was no longer produced at all. By the time Europeans showed up in the Pacific (the Dutch explorer Le Maire arrived in northern Tonga in 1616), no Polynesian cultures produced pottery. Until archaeologists discovered that the early Polynesians of Hawaiki had once possessed ceramics, anthropologists thought that pottery had never been known in Polynesia.

Why did the early Polynesians in their Hawaiki homeland give up such a useful material art as ceramics? It is difficult to say with certainty, but the answer probably lies in gradual social changes, rather than in material necessity. Lapita pottery, with its elaborate decorations, played an important social and ceremonial role in that society. It is possible that in ancient Hawaiki, the Lapita descendants began to transfer that role to other kinds of objects, such as barkcloth and fine mats. Many of the same designs found on Lapita pottery also occur on Western Polynesian barkcloth. The pottery never played an important role in Polynesian cuisine, because the earth oven was the principal means of cooking food. And the function of holding water or cooked food could be served equally well by containers of coconut shell and carved wooden bowls as by fragile pottery. For all of these reasons, it seems, the Lapita descendants who became the first Polynesians gradually lost interest in making pottery, with the art form declining and eventually disappearing altogether.

· · ·

The excavations at To'aga, and at other early sites dating to the first millennium B.C. in both Samoa and Tonga, tell us much about the first Polynesians, descendants of the Lapita colonizers who settled this homeland region about 950 B.C. But the world of the ancestral Polynesians, of Hawaiki, can also be approached from detailed reconstructions of the Proto-Polynesian language. Historical linguists have reconstructed the speech of the ancestral Polynesians in remarkable detail, with more than four thousand Proto-Polynesian words, indicating how these people viewed their world and their society. (When a Proto-Polynesian word is used here, it is marked with an initial asterisk.) This linguistic perspective is especially useful because it provides evidence for social organization, religion, and ritual, topics that are difficult for the archaeologist to access. Together with my late colleague Roger Green, I developed a method that we call the triangulation approach, because it uses these different lines of evidence (archaeology, linguistics, ethnography) to take a bearing on a particular topic, just as a surveyor uses compass bearings from different base stations to locate precisely a point on a map. Let me briefly review some key aspects of this ancient world of Hawaiki as revealed by the triangulation method.

The ancestral Polynesians of Hawaiki, like their later Hawaiian descendants, were at home both at sea and on the land. Their economy was based on intensive cultivation of root, tuber, and tree crops, combined with a sophisticated knowledge of fishing and marine resources. They grew at least twenty-seven different kinds of crops, including taro, yams, bananas, breadfruit, sugarcane, and coconut. They also raised dogs, pigs, and chickens. A wide range of fishing techniques—including spearing, hook-and-line angling, trolling, netting of various types, and poisoning—allowed them to take a diversity of reef, benthic, and pelagic fish, along with sharks, turtles, and rays. Their cuisine was based on the starchy root and tuber crops along with breadfruit, supplemented by the fish, shellfish, and meat from the pigs and dogs they raised. The principal method of cooking was in an earth oven, called *umu in Proto-Polynesian (from which the Hawaiian word *imu* is derived). At To'aga we excavated one of these ovens, still filled with its volcanic cooking stones.

The early Polynesians of the first millennium B.C. lived in small villages or hamlets, mostly along the shorelines. As their populations grew some people moved into the inland valleys of the larger islands such as 'Upolu. The houses were of pole and thatch, often paved with gravel floors. Cookhouses were separate from the main dwelling houses. Within and around their houses one might have seen a variety of tools and implements. Pottery was still in use but becoming simpler and cruder in its manufacture. The main tool for all kinds of woodworking tasks was the adz, made of dense clamshell or, more commonly, hard basalt stone. Holes (necessary for lashing canoe parts together) could be drilled using a handheld pump drill with a stone or coral tip. Wooden vessels were finished and polished with abraders of pumice or coral. Coral and sea urchin files were also used to work turban shells into fishhooks, and to make other kinds of shell objects such as beads and decorative rings. Much of the early Polynesian material culture was fashioned from vegetable materials, for example, sturdy sennit cordage from coconut fibers. The tough, flexible leaves of the pandanus tree were processed and woven into a variety of mats, as well as baskets, fans, and even the sails of canoes. Coconut fronds provided material for weaving floor mats and baskets of various kinds. Their garments were made of barkcloth, fashioned from the inner bark of the paper mulberry as well as the breadfruit tree. Thus, even though the Polynesians lacked metals, they had a rich material culture, well adapted to their island environment.

To reconstruct the social world of these early Polynesians, we must rely on the evidence of linguistics, aided by ethnographic comparisons to determine the probable meaning of the Proto-Polynesian words. The two most important words indicating kinds of social groups in ancient Hawaiki sound similar, but are quite different: *kāinanga and *kāinga. Polynesians, like other Austronesian peoples, typically think of their social groups using a botanical metaphor. Instead of "descending" from their ancestors, as Westerners often put it, Polynesian family lines are seen to grow or "ascend" from a trunk or base of ancestors. Over time the expanding social group is like a tree, with the ancestors as the trunk, and the various branches of the family spreading out to the twigs, which are the current living members of the group. The *kāinanga was just such a tree. One belonged to a particular *kāinanga by virtue of genealogical continuity back to a prominent ancestor, after whom the group was usually named. Membership in a *kāinanga was an essential part of a person's identity, linking one back to ancestors and validating rights to land, privileges, and a shared history. At the same time, people lived in smaller residential groups, called *kāinga. These also had their own proper names, which were the names of the particular dwelling site and its associated garden lands.

Both kinds of social group, the *kāinanga and the *kāinga, had their own leaders. For the larger and more inclusive *kāinanga, the group leader was called the *ariki, a word that is found in almost every Polynesian language, usually translated as "chief." In Hawai'i, the word derived from old Proto-Polynesian *ariki is ali'i, meaning a member of the chiefly class. As best we can infer, in ancient Hawaiki the *ariki played a role that was part secular leader, part priest. He had the responsibility for conducting most if not all of the sacred rituals of the group, including supplications to the ancestors. The individual *kāinga household groups were headed by senior family members, probably male in most cases, who were called *fatu. One other word in the ancient Proto-Polynesian speech is of interest here, *tufunga, meaning an expert of some kind. The original *tufunga probably included experts in fishing, canoe making, medicine, or other areas of specialized knowledge.

Our studies of Proto-Polynesian language have also revealed much about the spiritual life ancient Hawaiki. As with all later Polynesians, the concepts of *mana (supernatural power) and *tapu (sanctity) were essential to the people of Hawaiki. *Mana, the life-giving power that emanates from gods and ancestors, had to be carefully protected and channeled, which was one of the roles of the *ariki, the leader of the major social group. The protection of *mana required *tapu, various kinds of prohibitions, for example against touching the head of the *ariki and thus polluting him.

The ancestral Polynesians in Hawaiki had not yet developed the pantheon of four great gods whom we find later in Hawai'i and elsewhere in Eastern Polynesian societies. Only the name for the god *Tangaroa can be reconstructed to Proto-Polynesian. *Tangaroa was regarded as having been born from the "primordial pair" of Earth Mother and Sky Father. (We see this cosmogony in the later

Hawaiian creation myth of Papa and Wakea.) But the early Polynesians certainly also venerated their ancestors, their *tupunga, who were the source of growth (*tupu) of the social unit as well as the givers of *mana.

We can glimpse only very dimly the ritual world of ancient Hawaiki. We know that the main ritual space was an open assembly area (the *marae), probably in most cases on the seaward side of a cult house (*fare-qatua), which often may have been the dwelling of the *ariki, with his own ancestors buried beneath its floor. Rituals involved the preparation of *kava, a psychoactive drink made from the root of a kind of pepper plant (Piper methysticum), which was first domesticated by the Lapita people. Prayers or incantations, *pule, were offered by the group leader and elders, presumably seeking *mana from the ancestors and spirits, to ensure that the crops would be abundant, the fish would be plentiful, and the people would be protected and would multiply.

Essential to the ritual life of ancient Hawaiki was its calendar, based, like those of many other early horticultural peoples, on the lunar cycle. The year was divided into two main seasons, in accordance with the wet and dry periods that prevail in the Tonga-Samoa region. There were twelve lunar months, each with its own name; most names indicated such things as the seasonal rains or planting times for yams. A short thirteenth month was required to keep the lunar month sequence synchronized with the solar year. The early Polynesians were adroit observers of the heavens. They knew that it was necessary to recalibrate their lunar calendar every year, to keep the months corresponding properly to the environmental and horticultural phenomena they indexed. To do this, they used the rising and setting of the star cluster Pleiades, which they named *Mata-liki (the "little eyes"). This practice of closely observing Pleiades, and using it to reset the annual calendar, was carried down into later Polynesian cultures, including that of Hawai'i. It was the rising of Pleiades that determined the onset of the Hawaiian new year's season, when the god Lono returned from Kahiki. That Captain Cook happened to return to Hawai'i at precisely this moment in November 1779 was one of the reasons he was associated with Lono.

This then was the world of ancient Hawaiki, the homeland to which all of the later Polynesian societies and cultures trace their ancestry. It was out of this homeland—this Hawaiki—that Polynesian voyagers expanded later in the first millennium A.D., in a great diaspora that led them to discover, explore, and settle virtually every speck of land in the far-flung eastern Pacific, including Hawai'i.

· · ·

In 1956, a young American archaeologist named Robert Suggs arrived on Nuku Hiva Island in the Marquesas archipelago, part of French Polynesia. Not since 1920 had an archaeologist explored the Marquesas. Back then Ralph Linton of the Bishop Museum's Bayard Dominick Expedition had decided that there was nothing in the Marquesan earth to dig up, only stone monuments and *tiki* statues on the surface. Suggs represented a new generation, schooled in stratigraphic methods of excavation. He had read about Gifford's exciting finds in Fiji and New Caledonia; he knew about the fishhooks and other artifacts that Kenneth Emory was digging out of rock shelters in Hawai'i. Suggs was determined to have another look at what might lie buried in the Marquesan earth.

Hearing a rumor that large quantities of "pig bones" littered the sand dunes in a remote, uninhabited valley called Ha'atuatua, on the northeastern corner of Nuku Hiva, Suggs and his guide, Tunui, rode out to the valley on horseback to see what they might find. Arriving on the beach, Suggs later wrote that he "nearly fell out of the saddle" when he realized that the bones eroding out of the wave-cut sand bank were human ribs, vertebrae, thigh bones, and skulls. More significantly, he could see a dark layer with charcoal and beds of ash, evidence of former occupation. Eroding out of this layer were pieces of carved pearl shell, fragments of fishhooks, stone and coral tools, and large stones that seemed to be part of a pavement.

The next year, Suggs returned for a full-scale excavation at Ha'atuatua, which turned out to be one of the earliest Polynesian settlements not just in the Marquesas, but in all of Eastern Polynesia. The site was rich in shell fishhooks, stone adzes, coral files, and other artifacts, but what was most startling was a handful of small, reddish-brown pottery sherds. Suggs immediately recognized that with these pieces of pottery he had a direct link between the Marquesas and the ancestral Polynesian homeland in Western Polynesia. A few years later, Yosihiko Sinoto of the Bishop Museum excavated at another early site, in the Hane Valley on the nearby island of Ua Huka. He too recovered a small quantity of potsherds. The first Polynesians to depart from Hawaiki, the Western Polynesian homeland, to explore the vast eastern Pacific did so just as the art of pottery manufacture was in its dying throes. The use of pottery goes out completely in Samoa and Tonga by about A.D. 600–1000. These few small, crumbling potsherds from Ha'atuatua and Hane are the only evidence of ceramics that have ever been found in Eastern Polynesia.

Other archaeologists have followed the path first blazed by Bob Suggs when he dug into the sands of Ha'atuatua more than a half-century ago. Barry Rolett and Eric Conte have refined the sequence of Marquesan settlement. In the Cook

Islands, Richard Walter, Melinda Allen, and I have all dug early sites dating to the period of Polynesian expansion. My former student Marshall Weisler explored the remote Pitcairn and Henderson islands of southeastern Polynesia. David Steadman and Terry Hunt have independently both probed the sands of Anakena on Easter Island. In the Mangareva Islands, Eric Conte and I recently excavated a colonization-period site at Onemea, where the sands are full of the bones of seabirds that densely populated the islands before humans arrived. The combined evidence from all of these excavations gives a broader context to the Polynesian discovery and first settlement of Hawai'i.

What has emerged is a picture of an amazing dispersal, a diaspora, of Polynesian peoples out of the Western Polynesian homeland, out of Hawaiki, around the close of the first millennium A.D. We can only guess at the motives that propelled this exodus. It is possible that after more than a thousand years in the relatively small archipelagoes of Tonga and Samoa the pressure of growing populations made renewed explorations to the east inviting. Continued interarchipelago voyaging between the islands of Hawaiki also encouraged the development of Polynesian canoe technology and seafaring skills. In particular, the double-hulled (as opposed to outrigger) canoe was probably first invented in Western Polynesia by the beginning of the first millennium A.D. The large twin hulls were capable of holding substantial amounts of cargo in addition to pigs, dogs, and chickens. Decking positioned over the cross-booms could support a thatched house to shelter the human passengers. Such craft could have transported as many as forty people, along with the means needed to establish a new colony, for as long as a month at a time. With favorable winds and currents, these canoes could take such a group over thousands of miles in search of new islands.

Just when the first double-hulled canoes began to depart Hawaiki for the east is still a matter of debate and ongoing research. The centrally located and resource-rich islands of the Society group, including Tahiti, are likely to have been among the first islands of Eastern Polynesia to be discovered and occupied. Unfortunately, our knowledge of early settlements in the Society Islands lags behind that of other island groups. Domesticated coconuts found by my former student Dana Lepofsky, in a buried deposit on the island of Mo'orea, were dated to the seventh century A.D. More work on Mo'orea and elsewhere in the Societies will hopefully resolve whether it was one of the first East Polynesian islands to be settled.

Meanwhile, the earliest occupation sites in the Cook Islands, Marquesas, Austral Islands, Mangareva, Pitcairn-Henderson, and remote Easter Island are all beginning to show a pattern of settlement dating to between A.D. 900 and 1100. New

Zealand—Aotearoa as the Polynesians called it—was not found and settled until later, about A.D. 1250. As we will shortly see, the evidence from Hawai'i increasingly supports an initial settlement date about the close of the first millennium. Thus the settlement of Hawai'i is part of a much larger Polynesian expansion out of Hawaiki into the eastern Pacific. It was truly one of the greatest diasporas of human history.

Did the Polynesian explorers halt when they reached remote Easter Island, or Rapa Nui as they named it? There is no reason to think that they should have, and indeed, it is almost certain that one or more canoes reached the coast of South America. The strongest evidence for this lies in the sweet potato, a crop of South American origin that was an important food plant of the Polynesians as well. In Hawai'i, the sweet potato became the principal staple in the drier, leeward regions of the archipelago. Some scholars speculated that sweet potatoes were brought into Polynesia by early Spanish voyagers. That theory has now been falsified by the excavation of pieces of carbonized sweet-potato tuber from pre-Columbian contexts in archaeological sites on several Polynesian islands. In 1989, I excavated well-preserved pieces of sweet potato from the Tangatatau rock shelter on Mangaia Island in the southern Cooks. We radiocarbon-dated these to between A.D. 1000 and 1200.

Thus, about the same time that Polynesians were exploring and settling the easternmost archipelagoes of the Pacific, they went all the way to South America, returning with a cargo of sweet potato tubers. Contact may have been fleeting, and whether the Polynesians had any impact on the cultures of South America (or North America) remains a matter of heated debate. But by reaching the coast of what is now Chile, the Polynesians completed the final leg of the Austronesian expansion—one that began along the shores of the South China Sea and ultimately reached one-third of the way across the globe.

THREE · Follow the Golden Plover

Resting high atop the fine black sands of a Marquesan beach toward the close of the first millennium A.D., a *vaka moana* received its finishing touches in preparation for a voyage of exploration into the unknown. *Vaka*, the Polynesian word for canoe, comes from the older Austronesian word **wangka* with the *ng* sound dropped. *Moana* was what the Polynesians called the great ocean engulfing their islands. *Vaka moana*—not just an ordinary canoe, but an oceangoing craft capable of transporting entire families and their cargo on extended voyages of a month or more, in search of undiscovered islands. The Polynesians invented the double-hulled canoe, the original catamaran, long before Western sailors thought of the concept. Such double-hulled craft had carried the Polynesians out of Hawaiki, beyond their ancestral homeland in Tonga and Samoa, to find and settle the islands of Eastern Polynesia, including the Marquesas Islands, just south of the equator. This *vaka moana* and her crew were destined to discover Hawai'i, the vast archipelago that they would name in honor of Hawaiki, the homeland.

It had taken more than two years of tedious preparations to arrive at this point. First, the great canoe itself had to be built, beginning with cutting and hewing the massive *tamanu* trees that provided the keel, washstrakes, and gunnels. *Tamanu*, known as *Callophyllum inophyllum* to modern botanists, is one of the most prized island woods for carving. Durable and resistant to rot, *tamanu* takes the cut of an adz blade cleanly. Under the watchful eye of the *tufunga vaka*, the expert canoe-maker, the keels and planks had been skillfully carved using stone adzes. The adz blades themselves were works of art, fashioned by experts from dense basalt, painstakingly ground to a fine edge. To give the hulls sufficient depth to ride out the open ocean swells, and to add cargo capacity, they were raised by attaching planks to the dugout keel. The individual wooden pieces had to be expertly fitted, and then joined together by "sewing" with line made from coconut fiber. The holes through which the line passed were bored using basalt awls and pump drills. After the planks were fitted and sewn, the seams were caulked with strips of barkcloth and sticky breadfruit sap. The end product was nearly watertight yet flexible. The finished hulls measured fourteen fathoms, or a bit more than eighty feet, from stem to stern.

While the younger men carried out the painstaking work of hewing, fitting, sewing, and caulking the large hulls, under the partial shade provided by lean-to sheds thatched with coconut fronds, others contributed to the group effort in different ways. Hundreds of yards of coconut sennit were needed, not only to lash the hull pieces together, but also for the canoe's rigging. The coconut husk must first be soaked and beaten, and the individual fibers teased apart. The rolling together of fiber strands was the work of the old men, whose thighs became smoothed from the countless strokes of palm against skin, endlessly twisting together the fibers into short segments that were then braided into line and rope. They sat chatting in the shade of their houses, telling tales of voyages past, of Hawaiki, endlessly rolling, twisting, and braiding the ropes that would help transport the next generation to islands as yet undiscovered.

The village women too had their role to play in building the *vaka moana*. Her sails would be great sheets of tightly woven pandanus matting, sewn together in the shape of a crab's claw. The women knew this work well; it too was tedious with many steps. First the girls and younger women had to pick the long, thorny leaves from the pandanus forests that clung to the rocky headlands, and carry the heavy bundles back to the village. Then the thorny ridges would be stripped off, and the leaves softened by rolling around a clenched fist. They had to be dried in the sun, and wound up into large disk-shaped wheels until needed for weaving. Now the older, expert women took charge, carefully splitting the prepared leaves into long, narrow strips of uniform width, and tightly weaving these into mats. When enough long mats were assembled, after months of labor, they were joined by sewing them together with line, until the sail—the *ra*—was complete.

With the twin hulls ready, the *tufunga vaka* turned his attention to the vessel's superstructure. First the hulls, laid parallel and about ten feet apart, were joined together by a series of stout timber cross-beams, *kiato* as they were called. These too were fixed to the hulls by lashing with sennit, again strong but flexible. The *kiato* were then decked over with split bamboo, so that the space between the hulls became a platform spanning nearly the entire length of the vessel. Twin masts of long, straight *mara* trees were stepped into specially prepared *kiato* fore and aft; these would hold aloft the twin crab-claw sails of pandanus matting. Between the two masts, the craftsmen erected a low thatched house and lashed it securely to the *kiato* and the bamboo decking. This house would shelter the crew and their most prized possessions during the long voyage. A sort of wooden box, built into the deck and filled with sand, would serve as a hearth for warmth and to roast fish or seabirds.

The valley's young *ariki*, hereditary leader of his *mata-kainanga* (lineage), had decided to make this voyage in search of new lands to settle permanently. Not that life here was unkind. But theirs was one of the smaller valleys on the island, with limited garden land. The valley's stream sometimes dried up when the rains stopped. He and his clan were descended from a younger brother of the voyager-chief who had first left Hawaiki to arrive here in the Marquesas, several generations earlier. The most desirable valleys had therefore gone to the senior siblings; his ancestor had had to be content with the leftovers. This was the way things worked in a ranked society—the senior lines always had preferred access to the best lands, the richest fishing grounds, the highest-ranked women. There were two ways to overcome the lesser status conferred by birth. The first was to directly challenge and usurp the power of the more senior line. But this entailed considerable risk, especially if one's resources were limited. The second strategy was to discover and settle a new island where your own line would be paramount, second to none. It was a strategy that had been enacted time and again throughout the Austronesian and Lapita expansion into the Pacific. Indeed, it had helped to drive that expansion. Now the young Marquesan *ariki* would repeat the history of his ancestors, well known to him in their family traditions. He too would seek a new land.

The *ariki* had long pondered in which direction to pursue his quest, consulting the elders, the *tupunga*, of the island. They were by no means completely isolated, these early Marquesans. Occasional voyages between here and the Tuamotus to the southwest, and beyond to Tahiti and Ra'iatea, brought news of related families who had settled into the other islands of the central eastern Pacific. The *tupunga* told him that many islands to the south and east had already been discovered and claimed. Indeed, they had heard stories of one *vaka moana* that voyaged so far to the east that it had come upon a great land so vast that no end to it could be discerned. That endless desert land was already inhabited by people whose language was completely unintelligible. The crew of the *vaka moana* had traded fishhooks and adzes for a supply of edible tubers, picking up from the strangers the word for this new plant: *kumara*. Following the winds and currents, they had sailed back to Polynesia, bringing the new crop of sweet potatoes with them. They had found the limits of the Pacific world.

Based on what the *tupunga* had told him, the young *ariki* decided that there would be no point in heading eastward on his quest for a new island where he and his descendants would be masters of the land. He had a different idea. For many years now, he had observed the seasonal behavior of a particular kind of migratory bird that his people called *tōrea*. European ornithologists, when they explored the Pacific at the end of the eighteenth century, would give the name Pacific golden plover to this species. Every year, flocks of *tōrea* would arrive in the Marquesas and in other islands of the central Pacific, individual birds always returning to the same patch of grass or beach. After mating and rearing their young for six months, the birds would suddenly depart, as if on a signal given by an invisible hand. The *ariki* surmised that these flocks of *tōrea* must be flying to another place where they lived for the other half of the year before they returned again to his island. Today we know that the golden plover spends the northern hemisphere's summer months in Alaska and Siberia, making the long over-water flight south again when the winter cold begins to set in. The *ariki* did not have that knowledge. But he was willing to take the risk that following the path of the departing *tōrea* might lead him and his people to new, undiscovered islands. He had observed that the birds always departed in a northerly direction, a direction the Marquesans called *tokerau*. It was to the *tokerau*, then, that he would steer his double-hulled canoe.

The *vaka moana* was now nearly ready. But other preparations were still under way. They knew that they could be weeks, a month or more perhaps, alone on the vast *moana*. The deep hulls needed to be fully laden with enough supplies to keep the crew fed. Baskets filled with sun-dried fish and leaf-wrapped parcels of fermented breadfruit would provide food, while hundreds of green coconuts offered drinking water as well as coconut meat rich in nutrients and vitamins. Fish, of course, could be caught while at sea, and some seabirds might also be procured. Water from rain squalls funneled with mats into gourds would augment the drinking supply.

But the canoe had to carry far more than her crew and its supply of food and water. They were setting forth on a voyage of colonization, to islands that would have none of the crops or other useful plants upon which their culture was based.

Thus in the final round of preparations the valley's best gardeners were at work carefully wrapping up young living shoots of taro in bundles with protective moist earth, to keep them from perishing on the salty voyage. The first task upon arriving at a new island would be to transplant these precious shoots. Seed yams, cuttings of the kava plant, stalks of sugarcane, and young banana seedlings were all lovingly packed away. The people were fond of eating pigs and dogs, so the best-looking breeding pairs of these animals were also selected to make the trip. Chickens, too, were readied for the voyage by securing them in wickerwork cages. Other passengers would go along uninvited and undetected as stowaways—the diminutive Pacific rat, which frequented the thatch of people's houses, probably sneaked on board while the canoe was laden with food on the eve of its launching. The house gecko, *moko*, may have hidden in the rolls of barkcloth or mats.

After years and months of planning and intensive preparation, all was at last ready. The *vaka moana* was hauled off the beach ridge, her twin hulls gliding on coconut log rollers as the sweating villagers heaved on sennit ropes. Her launching was not without appropriate ritual and ceremony, for it was essential that the ancestors and the gods be called on to witness this momentous event, to bestow their *mana* on the *vaka*, the *ariki*, and his people. A name had been given to the great *vaka* by the *tufunga*, for all canoes had proper names, and were thought to have souls, much like people. (I well recall how—on Anuta Island in 1971—the people wailed and cried for a canoe that had been damaged on a reef, caressing its hull as they might caress an injured child.) We can never know the name of this canoe, but for the sake of my narrative let me call her *Mahina-i-te-Pua* (Moon-in-the-Flower). *Mahina-i-te-Pua* is a canoe name recorded in ancient Polynesian chants, a metaphorical name evoking the white, moonlike light that bursts forth from the sea foam as the canoe's bow cuts through a wave.

The white-haired *tufunga* chanted as the people tugged at the ropes, and the great hulls of *Mahina-i-te-Pua* slowly heaved, and then gently slid down the black sand beach to taste for the first time the waters of the bay. His chant may have gone something like the words of an ancient Tuamotuan song for the seagoing canoes, recorded by Kenneth Emory on remote Hao atoll in the 1930s:

> Set up the canoe on the land
> Open the door of the canoe shed
> Haul the *vaka* straight to the ocean
> Tane-with-Ruanuku "drink of the sea"
> Set upright the mast and the yards

> To carry the sail Moon-in-the-Clouds...
> The *vaka* drives through sparkling spray
> Her starboard is white with foam
> A short-crested wave slaps
> A long-crested wave strikes
> Banked in clouds is the face of Hawaiki
> The outline of the land rises...

Lightly tethered by ropes as she floated gracefully on the calm waters of the bay, the *vaka moana* was now quickly loaded with the lovingly prepared stores and provisions. There was no time to waste, as the precious seedlings and sprouts had only a finite time before they would begin to wither and die. Not everyone in the valley would make the voyage. There was room on board for perhaps forty persons. No infants or young children, who would simply get in the way; no feeble elders, who could not work the sails or hold the steering paddles. Only those who were past puberty, and in their reproductive prime, would make the voyage. Perhaps in a few years, if the voyage was successful and the new colony established, they would return with the news that others could now follow and join them. For this pioneering trip the crew was limited to about ten couples, all young adults. Led by their chief, their *ariki*, they would carry on the proud tradition that had led their ancestors east from Hawaiki.

A final, precious item of cargo was lifted onto the canoe to be safely stored away in the deck house: a slab of basalt dikestone that had stood upright on the *ariki*'s family shrine, representation of the *atua* (ancestors). This sacred stone, for the moment carefully wrapped in barkcloth dyed with red turmeric, would provide the foundation for a new temple, a new place of sacrifice, should they be fortunate enough to make landfall successfully. It was a physical link to their past, to remind them of their roots, their origins, their *tupunga*.

The time for farewells had come. In the way that Polynesians over the generations have expressed their *aroha*, their love for their kinsfolk, a great wailing rose up from the beach. *Aue! Aue! Aue!* Grandmothers and grandfathers watched their offspring begin to fade as the *vaka* rode the tide out into the bay; small children cried for their mothers and fathers who were leaving them, perhaps forever. First one sail, then the next, pointed their crablike tips into the breeze. *Mahina-i-te-Pua* began to make headway, approaching the basalt pinnacles that guarded the head of the bay. The wailing could no longer be heard by those on board. Salt spray stinging their eyes, everyone focused on the tasks at hand, securing cargo, lashing lines taut, holding the large steering paddle. The *ariki* stood near the bow, grasping the foremast. Follow the *tōrea*, follow the *tōrea* to a new Hawaiki.

· · ·

The first Europeans to enter the Pacific were often amazed at the seafaring abilities of the island peoples they encountered. Captain Cook, without doubt the greatest European navigator of all time, never ceased to be impressed, both at Polynesian watercraft and at their navigational knowledge and skills. Cook harbored no doubts that the ancestors of the Polynesians had sailed eastward through the Pacific, discovering island after island on their way. In 1769, while at Tahiti on his first voyage, he wrote the following lines in his journal:

> In these Proes or Pahee's [canoes] as the[y] call them . . . these people sail in those seas from Island to Island for several hundred Leagues, the Sun serving them for a compass by day and the Moon and Stars by night. When this comes to be prov'd we Shall be no longer at a loss to know how the Islands lying in those Seas came to be people'd, for . . . it cannot be doubted but that the inhabitants of those western Islands may have been at others as far to westward of them and so we may trace them from Island to Island quite to the East Indias.

Cook already understood that the Polynesians used astronomical phenomena for their navigation, especially "the Moon and Stars by night" as he put it. No doubt this was due to the close acquaintance he had made with a remarkable priest and navigator from the island of Ra'iatea. Tupaia, a priest of the cult of the war god 'Oro, was then living on Tahiti engaged in supporting the political aims of a high-ranking chiefess, Purea.

Tupaia was possessed of an extraordinary intelligence. He quickly became friends with the expedition's naturalist, Joseph Banks. Banks and Cook were products of the Enlightenment (Banks would later become famous as president of the Royal Society of London), predisposed to take the Polynesians on their own terms and try to understand, not judge them. A most unusual cross-cultural exchange began between the Polynesian priest and the English gentlemen, a sort of shipboard anthropology seminar. Tupaia and the English taught each other the rudiments of their respective languages. To the degree that their linguistic facility allowed, each tried to learn and understand the ways of the Other. Tupaia, of course, did not write down what he learned of British naval culture. But Banks, Cook, and others did learn a great deal from Tupaia, even though what they gleaned of his navigational knowledge was surely the tip of the iceberg.

Tupaia told Banks and Cook that he had personally sailed to dozens of islands, many of them a considerable distance from Tahiti. He named Manu'a in the Samoan group and in all likelihood had been there. Tupaia decided to join Cook's expedition; the captain quickly agreed to take this knowledgeable guide along. Leaving Tahiti, where they had been anchored for three months, the *Endeavor* sailed westward through the Society Islands, Tupaia giving directions. They landed first at Huahine, and then proceeded to Ra'iatea, center of the 'Oro cult and location of the great *marae* (temple) of Taputapuātea, where Tupaia performed his priestly duties. After Ra'iatea, Tupaia guided the *Endeavor* to Borabora, near the western end of the archipelago. From here, he proposed that they sail farther to the west where, Tupaia claimed, there were many islands about ten to twelve days' sail away. Cook, however, made the decision to turn south, in a search for the fabled *Terra Australis*, the great southern continent.

Tupaia had been fascinated by the illustrations drawn by Banks's assistants. Banks gave Tupaia a set of pencils and watercolors so the priest could sketch scenes of his own people and landscapes. Some of Tupaia's drawings, in a sort of "naïve" art style, survive to this day. While at Ra'iatea, Tupaia drew a kind of map of the islands that he had voyaged to or knew about, which in all numbered about 130. Cook, the consummate cartographer, was fascinated and redrew the chart in pen and ink, labeling each of the islands with the names given to him by Tupaia. Many of these are clearly recognizable. That the map does not exactly match a modern hydrographic chart is probably due more to the problems of translating Tupaia's cognitive model of his island world—which he thought of in terms of star paths, wind directions, and sailing times in days—into a grid of latitude and longitude. Cook's drafted version of Tupaia's map, still preserved in the archives of the British Admiralty, stands as a lasting testament to the incredible geographic and seafaring knowledge of a Polynesian master navigator.

Tupaia never returned to Ra'iatea to pass on his knowledge to the next generation. He caught a fever and died in the colonial port of Batavia, Java, on the *Endeavor*'s return voyage to England. Over the next several decades, the islands of Polynesia were beset with European newcomers: explorers, beachcombers, missionaries, whalers, merchants, and colonial officials. The introduction of diseases to which the islanders had little or no resistance took a devastating toll on the population. Some demographers have estimated that the cumulative death toll may have approached 90 percent. Among those who passed away of dysentery, measles, or smallpox were the priests and elite who, like Tupaia, had carried the cumulative knowledge and wisdom of their culture. Those skilled in carving the double canoes also perished, and in any case the younger generation was more enamored of the new European sailing ships. Many Polynesians went to sea on European trading and whaling vessels, learning to sail and handle these different kinds of ships. The last of the great voyaging canoes rotted in their sheds, with no one interested enough to keep them repaired and refitted.

By the late nineteenth century, the Polynesian tradition of interisland voyaging, which had been so vibrant in Cook and Tupaia's day, was now a fading memory. No master navigators remained who could recount the star paths to distant islands. But some of the older people still knew the legends and traditions of famous voyages from generations earlier. In Hawai'i, a Swedish immigrant named Abraham Fornander, who had become a citizen of the Hawaiian Kingdom and learned the Hawaiian language, began assiduously to collect these traditions. About the same time, a land surveyor named S. Percy Smith did the same thing in New Zealand. Fornander and Smith published books in which they drew on the Polynesian voyaging traditions to advance historical interpretations of Polynesian origins and migrations. Although their theories differed slightly, both Fornander and Smith believed that the Polynesians had originated in Asia, and that their ancestors had been great seafarers who had purposefully voyaged from archipelago to archipelago until they had explored and settled the breadth of the eastern Pacific.

The field of anthropology, the scientific study of human cultures and their history, did not really get going in the Pacific until the early twentieth century. As I will recount in more detail in the next chapter, the great push in the 1920s and 1930s was to capture a sort of "memory culture" from those older Polynesians who still knew something about the traditional lifeways. The dean of these early twentieth-century anthropologists was himself a half-Polynesian, who went alternatively by his Scottish father's family name as Peter Buck or by the name given to him by his Maori mother, Te Rangi Hiroa. In *Vikings of the Sunrise*, published in 1938, Hiroa summed up the prevailing anthropological interpretation of Polynesian origins. His very title conjures up a romantic vision of great seafarers who set out fearlessly into the dawn in search of new islands to settle. Like Fornander and Smith before him—indeed, harking back to Cook's view that the Polynesians had come from the west—Hiroa traced the origins of his mother's people back to the shores of Asia. In his eyes, their conquest of the Pacific had been a triumph of seamanship and navigational skill—perhaps the greatest seafaring saga in the history of humankind.

In 1936, as Hiroa was writing *Vikings of the Sunrise* in his study at Honolulu's Bishop Museum, a young Norwegian student named Thor Heyerdahl, accompanied by his bride, Liv, arrived to spend a year on isolated Fatu Hiva Island in the Marquesas. Heyerdahl had come to the Marquesas to study the islands' flora and fauna. But during the course of his sojourn he became increasingly intrigued with Marquesan culture, with the question of where the Polynesian ancestors had come from. An old man named Tei, who lived alone in a remote valley, told Thor and Liv that his ancestors had been led across the ocean by a leader named Tiki, whom Tei now venerated as a god. When Heyerdahl asked the aged Tei where his ancestor had come from, Tei responded "From *Te-Fiti*." According to Heyerdahl's account, the old man "nodded toward that part of the horizon where the sun rose, the direction in which there was no other land except South America." Not understanding that Te Fiti is actually an old Polynesian name for islands to the west (Fiti is the Polynesian version of Fiji), Heyerdahl focused on the old man's nod to the east. That simple nod set Thor Heyerdahl off on a lifelong quest—indeed, one could say an obsession—to prove that the Polynesians had not come from the west, as Cook, Fornander, Smith, Hiroa, and all the others had asserted. Heyerdahl became convinced that the Polynesians had originated in the east, in South America.

Returning to Norway, Heyerdahl plunged into the university library to dig up as much evidence as he could possibly muster to support his theory of a South American origin for the Polynesians. Unlike the Polynesians, no South American cultures were known to have had deep-water sailing craft (a problem for Heyerdahl's theory), but the Inca had possessed crude balsa rafts on which they sailed hugging the continental coastline. Heyerdahl also thought he detected similarities between Inca and other South American stonework and carved images, and the statues and temples of Easter Island. He became convinced that Tiki had been an Inca (or pre-Inca) lord who had sailed to Polynesia on a balsa raft.

The idea was not entirely new. It had been proposed by a Spanish priest, Joaquin Martinez de Zúñiga, and by an English missionary, William Ellis, in the early nineteenth century. But little evidence supported this view, and it had never been accepted. Heyerdahl compiled his research into a ponderous manuscript, sending it around to possible publishers. No one was interested. Then he hit on a plan that must surely rank as one of the cleverest public-relations schemes ever devised. He would build a replica Inca raft of balsa logs and sail it across the Pacific, thus "proving" his theory. In 1947 the *Kon-Tiki* raft, with Heyerdahl and five other Norwegian and Swedish adventurers aboard, crashed on the reef of Raro'ia atoll in the Tuamotu Islands, after drifting for 101 days from Peru. The public, weary after years of war, was enthralled with the story. A black-and-white film made from footage shot during the voyage was a hit in movie theaters. Heyerdahl's account of the trip, *Kon-Tiki*, became a bestseller, translated into more than twenty languages. And he finally found a publisher for the lengthy tome describing his theory, *American Indians in the Pacific*.

Heyerdahl was not a trained anthropologist or archaeologist; the scholarly world viewed his raft voyage as more of a stunt. But he had achieved so much fame that his theory could not be ignored out of hand. Heyerdahl had directly challenged the prevailing wisdom that the Polynesians had voyaged into the Pacific from Asia. Moreover, he was soon joined by another contrarian, a New Zealander named Andrew Sharp. Like Heyerdahl, Sharp was not a trained anthropologist, but a civil servant who had turned amateur historian in his retirement. Sharp didn't buy the idea that the Polynesians were South Americans, but he did question the view of Hiroa and others that the Polynesian ancestors had been great seafarers who had purposefully navigated their canoes eastward to discover new lands.

In *Ancient Voyagers in the Pacific*, published in 1956, Sharp compiled historical accounts of "drift voyages" that had been documented throughout the Pacific. Canoes or boats traveling between nearby islands had sometimes been caught up in storms, eventually to be cast up on some distant island hundreds or even thousands of miles away. To Sharp's mind, this was how the Polynesians had eventually ended up on all of the far-flung islands of the Pacific. There had simply been enough cumulative "drift voyages" over the centuries that castaways had ended up on all of the islands.

Heyerdahl and Sharp had a major influence on the thinking of anthropologists and archaeologists in the 1950s and 1960s. Scholars began to question whether Fornander, Smith, and especially Hiroa had not overly romanticized the Polynesian past, attributing to them greater seafaring skills than they had, in reality, commanded. Perhaps the many legends of great long-distance voyages were just myths, created after the fact. Perhaps the Polynesian canoes had not been able to sail upwind, against the prevailing east-to-west winds and current that sweep across the Pacific. Heyerdahl had pointed to these winds and currents as major support for his theory of South American origins. By the early 1970s, some archaeologists were prepared to discount the Polynesian traditions of voyaging. Tupaia's map had been forgotten.

• • •

The great Harvard biologist Ernst Mayr claimed that science is a "self-correcting system." Scientists and scholars sometimes get off course in their research, lose their bearings, and follow a trail that is wrong or becomes a dead end. Fortunately, another scientist will come along and recognize the error, disprove the erroneous hypothesis, correct the theory. Just as Heyerdahl and Sharp were rejecting the idea that the Polynesians had purposefully navigated their canoes upwind to discover their islands, a young anthropologist named Ben Finney became interested in the sailing abilities of traditional Hawaiian canoes. At the University of California, Santa Barbara, where he taught, Finney and his students built a forty-foot double-hulled canoe, christening her *Nālehia* (The Skilled Ones). The sight of a double-hulled canoe with an inverted crab-claw sail must have startled a few surfers in the Santa Barbara Channel! From *Nālehia*, Finney learned that such a canoe could, in fact, sail to windward, and that she handled quite well even in heavy seas. But *Nālehia* was a small canoe, modeled after one the Hawaiian king Kamehameha III had used for inshore sailing. It would not have gotten a colonizing party of Polynesians, with all of their supplies and animals, from the Marquesas to Hawai'i.

In 1970, Finney moved to Honolulu to join the faculty at the University of Hawai'i. There he met up with a racing-canoe paddler named Tommy Holmes and a Native Hawaiian artist, Herb Kawainui Kane, both of whom shared Finney's passion for Polynesian canoes. Together, the three of them founded the Polynesian Voyaging Society, with the aim of building a full-scale oceangoing double-hulled canoe and sailing this craft along the legendary seaway between Hawai'i and Tahiti. In 1975, *Hōkūle'a*, her twin hulls measuring sixty-one feet from stem to stern, was launched at Kualoa on O'ahu's windward shore. *Hōkūle'a* is the Hawaiian name for Arcturus, the zenith star of Hawai'i, which Polynesian navigators may have used to determine the latitude of that island on return voyages from central eastern Polynesia.

Though constructed of Western materials—fiberglass hulls, canvas sails, nylon rigging—*Hōkūle'a*'s lines and design are strictly Polynesian (see color plate 1). Over the course of the next year, Finney and *Hōkūle'a*'s crew, made up mostly of Native Hawaiians who had become enthusiastic about re-creating the voyages of their ancestors, took the craft through a series of sea trials. Since no one had sailed a vessel like this in hundreds of years, there was much hands-on learning to do. But, in Finney's words, "*Hōkūle'a* proved to be a seakindly vessel, remarkably smooth and stable even in high winds and seas."

A more challenging task than building the canoe and learning to handle her at sea was to find someone who could navigate *Hōkūle'a* from Hawai'i across 2,700 miles of open ocean to Tahiti, using only traditional, noninstrumental navigational methods. Tupaia and the other great navigators of Polynesia were long gone. They had died without passing along their knowledge to later generations. But in the islands of Micronesia, to the northwest of Polynesia, a tradition of inter-island navigation was still alive. The Micronesians were distant "cousins" of the Polynesians, both descendants of the Lapita people. About the same time that the early Polynesians were developing their distinctive culture in Hawaiki (Tonga and Samoa), the first Micronesians were settling the islands of the Caroline and Marshall archipelagoes. Finney and his colleagues reasoned that if they could find a traditional navigator in one of the Micronesian islands, he might be able to apply methods that were similar to those that the great Polynesian navigators would have used. These would include such devices as the use of star paths, wind directions, ocean swell patterns, and similar natural signs to determine one's position and course while at sea. Because these methods rely entirely on an ability to read the natural world, without the aid of instruments such as a sextant or a GPS device, some prefer to call this the art of "way-finding," to distinguish it from Western "navigation."

Hōkūle'a found her master way-finder on the Micronesian atoll of Satawal. Mau Piailug had piloted his own sailing outrigger canoe among the islands of the Carolines. He was widely recognized as an expert in star lore and other way-finding skills. Mau agreed to join the historic voyage from Hawai'i to Tahiti. To prepare for this, Mau spent hours in the planetarium at the Bishop Museum in Honolulu, studying the night sky and star positions that he would see as *Hōkūle'a* proceeded southward on her voyage. An ancient Polynesian navigator would have had this knowledge in his head, just as Mau knew intimately the skies over his home islands in Micronesia.

On May 1, 1976, *Hōkūle'a* departed Maui Island with Kawika Kapahulehua as captain and Mau as way-finder. For the first part of the voyage, they steered toward the southeast, using the northeast trades to propel them. When they entered the infamous doldrums along the equator, overcast skies frequently denied Mau a crucial view of the heavens, while inconsistent winds made steering difficult. Some of the crew began to doubt Mau's ability. But after a few anxious days they were south of the doldrums. A brisk wind from the southeast steadily drove *Hōkūle'a* forward. Polaris, the north star, disappeared under the horizon while the Southern Cross rose higher each night, marking *Hōkūle'a*'s southward progress. Thirty days out from Maui, Mau predicted that they were about to enter the low coral atolls of the Tuamotus, the "Dangerous Isles" as early Europeans had called them. A short time later, the crew spotted white fairy terns, seabirds that nest on land. Mau had been correct in his prediction.

On June 4, 1976, *Hōkūleʻa* sailed through the pass off Tahiti, where Cook and Tupaia had once stood together on the deck of the *Endeavor*. Anticipating their arrival, the Tahitian governor of French Polynesia declared a public holiday. A crowd of thousands mobbed the black sand beach to greet joyously their Hawaiian cousins who had reopened the ancient seaway between these distant archipelagoes. Nothing could have more dramatically disproved Thor Heyerdahl or Andrew Sharp than this experimental voyage of a re-created Polynesian *vaka moana*. Tupaia's legacy had been resurrected. The Polynesians had not floated passively across the sea on balsa rafts. They had not drifted at the whims of storms and currents to their islands. No, their ancestors had invented and built great seafaring canoes with an excellent upwind sailing capacity. They had perfected the arts of way-finding using the stars and swells. Hiroa had been right after all—his ancestors had been the true "Vikings of the sunrise." Science, as Mayr correctly noted, is a self-correcting system.

In the years since *Hōkūleʻa* made her first voyage from Hawaiʻi to Tahiti, she has gone on to make many other experimental voyages across the Pacific, tracing the ancient routes to Samoa, the Cook Islands, Aotearoa (New Zealand), and isolated Rapa Nui (Easter Island). A young Hawaiian man, Nainoa Thompson, became Mau's protégé and taught himself the way-finding art, working with Ben Finney and others to understand more fully and document how the Polynesians accomplished their great navigational feats. Thousands of schoolchildren have visited and sailed on *Hōkūleʻa*, helping to fuel a cultural renaissance in Hawaiʻi and elsewhere in Polynesia. Tupaia, no doubt, would be pleased to see that the voyaging skills of his people are once again being resurrected.

· · ·

Mahina-i-te-Pua had now been at sea for five days. She had handled beautifully as the southeast trades filled her woven mat sails, making steady progress toward the northeast. The steersman kept her pointed upwind in this direction to counter the steady current from windward, so that they would continue to follow the flyways of the *tōrea* toward what they hoped would be new lands. Her people had settled into a regular routine on the ocean, taking their turns at holding the great steering paddle; tending to the pigs, dogs, and chickens; trying their luck at catching bonito with pearl-shell lures trolled on lines from the stern. They had become accustomed to the steady rhythm of the twin hulls rising up the crest of an oncoming swell, slicing through the inky blue water, and then gliding gracefully back down into the trough. An infinity of swells lay before them, to the horizon and beyond.

As the sun went down that evening, the wind unexpectedly dropped. Mat sails went slack, hanging on their booms. The equatorial heat became oppressive, even out on the vast ocean. A low haze had been pulled, like a shroud, over the arc of the heavens. The *ariki* could no longer see the "Great Fishhook of Maui" (Scorpio) hanging in the eastern sky, or the other stars he had been using to check their course night after night.

Dawn broke. As the morning progressed, the sun's rays bore down on the drifting canoe, no wind or breeze to cool them. A brief rain squall gave some relief toward the late afternoon. But still there was no steady wind to drive them on toward their goal. For four days they drifted like this, only an occasional breeze allowing them to make a little headway toward *tokerau*, the north. Languid pigs and dogs panted in the hulls. The people began to grumble among themselves. Perhaps the *ariki* did not know what he was doing. Why had he taken them on this *tokerau* route, when others had always sailed eastward, toward the rising sun?

Mahina-i-te-Pua had arrived in the doldrums, something unknown to the *ariki*. The knowledge he had absorbed from the sage *tufunga* was of a world of regular southeast trade winds, interrupted by predictable westerly reversals. He had no idea that he and his crew were about to cross over from the great South Pacific gyre of counterclockwise circulating winds and currents to the North Pacific gyre, with its clockwise circulation system. This was a region into which no Polynesian canoe had ever before ventured.

Finally, on the tenth day out from their island, the hint of a steady breeze began to rise, not from the southeast as during the first leg of their voyage, but now from the northeast. At first tentative, the breeze steadily gained in strength. By the following morning *Mahina-i-te-Pua*'s sails were billowing out once again, the sennit stays straining to keep the booms in check. The *ariki* ordered his steersmen to keep her bow pointed up into the wind, toward the northeast. If they were to follow the true path of the *tōrea*, the canoe must not veer too far to leeward. Long and steady swells now swept down from *tokerau*, from the north. The *ariki* took them as a good omen from the *atua*.

For another ten days *Mahina-i-te-Pua* continued on her new northeastern tack, twin wakes hissing as she made steady progress into the unknown. Each night the *ariki* stood on the bamboo deck, one hand grasping the foremast, his eyes scanning the heavens. The stars so familiar to him, whose names he had been taught by the old *tufunga*, were now progressively lower in that part of the sky they called *tonga*, south. Fētū-teka, the Southern Cross, hung lower and lower in the sky each night off the canoe's stern. Meanwhile, new stars had begun to rise in the *tokerau* region off the bow, stars he had never seen before. One especially bright star, rising each night in the east, was getting progressively higher and higher overhead. It was now nearly directly over *Mahina-i-te-Pua*'s masts. *I must give this* fētū, *this star, a name,* the *ariki* thought. Another new star had recently begun to appear, out of its "pit" *(rua)* on the horizon. (The Polynesians thought that each star had its own pit out of which it rose each night.) This one was dead-on *tokerau*, and thus did not move; all the other stars appeared to rotate around it. It is what Europeans call Polaris, the North Star. (Hawaiians would later call Polaris Hōkū-Paʻa, the "fixed star.")

The northeast trade winds increased in force, causing the rolling swells to crash over the starboard hull's bow. The crew was working day and night with wooden bailers to keep the hulls from swamping. Three weeks out from the Marquesas, time was running out. The tender taro and banana shoots would not last much longer. They needed to be planted in moist, nurturing earth or they would shrivel and die. The pigs and dogs had lost much of their weight as well; they would not survive much more than another week or two at sea.

The *ariki* stood alone at his accustomed position next to the foremast. The bright, as yet nameless star shone down on *Mahina-i-te-Pua*. Under his breath, he began to chant the names of his ancestors, invoking the generations before him, back to Hawaiki and beyond. He chanted to his ancestors to send *mana*, the unseen power, the strength to persevere. As he prayed, the twin hulls shuddered with each crash of the ongoing swells. Salt spray drenched his sunburned face and soaked his long, matted hair. The time had come to make a decision. They could not continue endlessly on this course. He would take the bright star overhead as his sign from the *tupunga*, the ancestors. Calling out to the steersmen, he gave the order to bring *Mahina-i-te-Pua* about and let her run full with the wind at her stern.

Through the night they sailed, the twin crab-claws straining with the northeast trades. They no longer had to bail, a great relief to the weary crew. But no one slept much nonetheless. All knew they were at a turning point. Either they found new land soon or they would have to turn back. A return trip at this point, as supplies were dwindling, was not an inviting prospect.

Mahina-i-te-Pua was running virtually due west. As the dawn broke only long, low cloud banks marked the horizon. As the day wore on, the *ariki* became quiet and sullen. Had the spirit of his ancestors, his *tupunga*, deserted him? Perhaps they did not know how to find him, out here on the vast *moana,* so far from his homeland. His thoughts were interrupted by a shout from the steersman. He had spied a *kaveka*, a sooty tern, flying low across the water. Soon several more birds were spotted off the port bow. They knew that *kaveka* were land-nesting birds, going out to fish at sea during the day but always returning to land at night. Land must be near. Just at sunset, one of the young men with an especially sharp eye thought that he had seen the sun dip behind a long, broad slope, not of a cloud, but of land. The morning would tell.

Mahina-i-te-Pua continued her steady run to the west throughout the night. As the dawn rays cast a golden light on her mat sails, all eyes strained to the southwest, where the young man thought he had seen the slope of a mountain. At first clouds obscured the horizon. Then a gap opened up, revealing a sight none of them had ever witnessed. It was indeed a mountain, *maunga* in their language, but one so high and so vast that it now loomed up into the clouds. More amazing still, its summit reflected the morning light, glistening white, completely unlike the forest-clad summits of the Marquesan mountains they had left behind. *"Maunga tea!"* they exclaimed, *"Maunga tea* (White mountain)!" Only much later, after the little colony had established itself, would several of their strongest young men venture to climb the summit of that great mountain, the first Polynesians to encounter snow. Mauna Kea, as this 13,796-foot-high volcanic giant is known today, is the highest mountain in the Hawaiian Islands. Before the day was out another peak had appeared to the west. This they would later name Haleakalā (House of the Sun). At 10,023 feet it dominates the island that bears the name of the great Polynesian-culture hero Maui, he who snared the sun to slow its path across the sky.

We do not know exactly where the *ariki* and his people landed *Mahina-i-te-Pua*. Perhaps they guided her through the fringing reef off Waiheʻe on northwestern Maui, where archaeologists later found early forms of pearl-shell fishhooks eroding out of a sand bank. Or they may have continued on, passing the towering cliffs of windward Molokaʻi, to land along the inviting Waimānalo or Kāneʻohe shores of Oʻahu. Other early archaeological deposits have been found there. The traditions are silent on such matters.

Mahina-i-te-Pua, as I have called her, was the first *vaka moana* to succeed in voyaging across 2,400 miles of open ocean from the Marquesas to Hawaiʻi. Her people quickly established a foothold in this vast archipelago, known previously only to seabirds and other creatures whose progenitors had dispersed there on the wind and waves. Her cargo of precious taro and banana shoots flourished in the rich, humid soil, providing the basis for a new economic foundation. The pigs and dogs, too, survived the voyage, to reproduce and provide sources of food and sacrifice for future generations. But there were also failures. The delicate breadfruit saplings did not survive the salt spray, and withered before they could be established in the new land. Not until a later voyage, by the famous Kahaʻi-a-Hoʻokamaliʻi, did the nourishing breadfruit tree get to Hawaiʻi. But for the most part the voyage had been a success. Like his ancestors before him, the *ariki* had a found a new island home in which he and his people would be their own lords, their own *fatu*. Once the colony was securely established, he would make the return voyage to send word that a vast new group of islands had been discovered. Other canoes would make the long voyage from the south, guided by the bright star that passes directly over Hawaiʻi (Arcturus). He had given that zenith star a name now, in his early Proto–Eastern Polynesian dialect: *Fētū-leka* (Star of Rejoicing). With later changes to the sounds in their dialect, Hawaiians of Cook's day would call this star *Hōkūleʻa*.

FOUR · Voyages into the Past

In the nineteenth century, many Hawaiians could recount the ancient legends and chants chronicling the great Polynesian voyages. Native scholars like David Malo, John Papa 'Ī'ī, and Samuel Kamakau hurried to collect these *mo'olelo* (traditions). They feared the inevitable day when the last of these knowledgeable elders, *kūpuna*, would be gone. Abraham Fornander, a Swedish immigrant who arrived in Hawai'i in 1838, was one of the first to study thoroughly the Hawaiian traditions gathered by Kamakau and others. Fornander married 'Ālanakapu, a Moloka'i chiefess, learned her Hawaiian language, and mingled in the same social circle as King David Kalākaua, himself a devoted student of the *mo'olelo*. In his three-volume work *An Account of the Polynesian Race*, published in 1878, Fornander laid out his theory tracing the Polynesians back to a distant homeland in South Asia. Many of Fornander's ideas were based on far-fetched comparisons, with no real scientific basis. Nonetheless, his compilation of the Hawaiian legends and traditions remains a classic source of traditional information. I will draw on Fornander's collection, as well as on those of Kamakau, 'Ī'ī, and Malo, to tell the tales of Hawaiian history in later chapters.

With such direct indigenous knowledge about the Hawaiian past, it did not occur to anyone in Fornander's day to excavate in the ground—to use the methods of archaeology—to trace Polynesian history. In the late nineteenth century archaeology was only just emerging as a serious scholarly field, casting aside its roots in antiquarianism. In North America, archaeology became linked with ethnography, the study of living non-Western cultures, in the combined new field of anthropology. Harvard offered the first doctoral degree in this new subject in 1894. At the same time, several museums were established in the United States and Europe that incorporated the new field of anthropology in their scope. Among these were the American Museum of Natural History in New York, the Smithsonian's Natural History Museum in Washington, D.C., and Chicago's Field Museum.

In far-off Honolulu, the idea of a natural history and anthropology museum focused on Hawai'i and the Polynesian islands caught the fancy of Charles Reed Bishop. Like Fornander, Bishop was a Haole immigrant to the Kingdom of Hawai'i. He had married the royal princess Pauahi, heiress to the vast estates of her great-grandfather King Kamehameha I. Pauahi died in 1883, leaving her lands to establish the Kamehameha Schools for Hawaiian children. Her husband, Charles, retained possession of her priceless collection of Hawaiian antiquities, including precious feather cloaks, capes, helmets, *kāhili* (feather standards), finely carved bowls, and many other objects that had passed down through the Kamehameha family to Pauahi. Bishop fixed on the idea of building a memorial to his beloved Pauahi, a "treasure house" of the Kamehameha dynasty. Thus the Bernice Pauahi Bishop Museum of Polynesian Ethnology and Natural History was founded in 1884. By the turn of the century under the inspired leadership of William T. Brigham, its first director, the Bishop Museum was housed in an imposing complex of cut-basalt stone buildings in the Romanesque style.

Brigham was a Victorian intellectual who dabbled in everything from volcanoes to Hawaiian feather work. Brigham especially wanted to test one part of Fornander's theory of Hawaiian history, based on the traditions of a fourteenth-century Tahitian priest named Pā'ao. According to Fornander, in the earliest period after Polynesians arrived in Hawai'i, the people constructed "truncated pyramids" or platforms of stone, upon which they conducted rites in honor of the creator god, Kāne. When Pā'ao arrived around the fourteenth century, bringing the chief Pilika'aiea to rule Hawai'i Island, he introduced the cult of the war god Kū, along with the practice of human sacrifice and a new kind of walled temple. Fornander believed that the construction of walled temples went along with a hardening of Hawaiian society into rigid classes; had the lowly commoners looked upon the sacrificial bodies, they would have defiled the sacred rites. Thereafter, the common people had to remain outside the stone walls, silent and obeisant.

Brigham reasoned that this postulated change in temple architecture from pyramid to enclosure might be verified through archaeological study of the ruins of *heiau*. Many temples were associated with the names of the chiefs who had built them, and the chiefs in turn could be put in chronological order through Fornander's compilation of genealogies. If temples of the platform type were linked to the earliest generations of chiefs, and walled temples with later generations, Fornander's theory would be confirmed. To undertake this project Brigham enlisted a young Australian, John F. G. Stokes, who was appointed in 1903 to be the Bishop Museum's curator of ethnology. Stokes was the first true archaeologist of Hawai'i.

In August 1906 Stokes sailed by interisland steamer from Honolulu to the then-sleepy backwater village of Kona, on Hawai'i Island. Along with his surveying equipment and camping gear, Stokes was armed with a list of ancient temple sites drawn up by Thomas Thrum, a Honolulu editor and amateur archaeologist. Stokes threw himself into the task of surveying *heiau* ruins in the vicinity of Kailua, of which there were many. Quickly realizing that this was going to take much longer than Brigham had led him to believe, Stokes wrote back to his museum patron, requesting additional funds and asking that his wife be allowed to join him. Assisted now by Margaret, Stokes spent the remainder of 1906 traveling on horseback to more than one hundred *heiau* from Puna in the southeast, through Ka'ū and Kona in the south, and up to Kohala at the island's northern tip. Among the temples Stokes mapped were Waha'ula and Mo'okini, both reputed to have been built by the voyaging priest Pā'ao.

Stokes had learned to speak some Hawaiian. He interviewed local *kūpuna* who knew the names and sometimes the traditions associated with each temple. Stokes used his transit and measuring tapes to make detailed plans of the stone platforms and walled enclosures. He combined empirical archaeological observations with the existing emphasis on oral traditions. His carefully drawn plans and notes, accompanied by his glass-plate photographic negatives, remain an irreplaceable legacy in the Bishop Museum archives.

Stokes's work failed to support a clear-cut division of Hawaiian temples into platform and walled types, a disappointment for Brigham. Rather, there was an almost infinite variation in architectural forms: walls, platforms, and terraces combined in endless ways. Additional fieldwork on Moloka'i Island in 1909, where Stokes again obtained a priceless record of that island's *heiau*, confirmed the messiness of the Hawai'i Island data. Fornander's interpretation of Hawaiian traditional history could not be verified or rejected. But archaeology had for the first time made an independent contribution to knowledge regarding the Hawaiian past.

So far, Stokes's temple surveys were all focused on above-ground ruins. No one had yet hit upon the idea that archaeological *excavation* might yield new insights about the past. Then, in 1913, Stokes—ever the pioneer—made a field trip to arid Kahoʻolawe Island, lying in the rain shadow of Maui. In an overhanging rock shelter called Kamohio that had served as a fisherman's camp and shrine to the sea god Kanaloa, he laid out an excavation trench and began carefully to peel away the layers of dusty earth. Stokes's notes, written in his fine pencil script, show that he correctly recognized a succession of three different strata. Stratigraphy, the principle that successive *strata* (layers) laid down one over the other correspond to a temporal sequence, had been accepted by geologists starting with British scholar Charles Lyell in the early nineteenth century. But it was still a new idea for archaeologists, and by no means widely accepted. Had John Stokes continued this line of archaeological investigation at Kamohio and other sites in the islands, he might have truly put Hawaiian archaeology at the forefront of the field. But Stokes was a cautious scholar—a perfectionist, some would later say. He was slow to write up his results and hesitant to publish until he was certain that his interpretations were watertight. The fascinating Kahoʻolawe materials, organized by their distinctive strata, languished in the museum's specimen cabinets.

By 1919 Brigham had wearied of the ceaseless task of building the Bishop Museum; photographs show that his long beard had turned snowy white. Stokes's patron decided to retire. The museum's board of trustees wanted a new director with top-notch credentials who would bring the latest scientific methods to Honolulu. All of them missionary descendants, the trustees turned to their Yankee roots and chose Herbert E. Gregory, then Silliman Professor of Geology at Yale University. Gregory's goal was to turn Brigham's Victorian-era "treasure house of Kamehameha" into the premier center of scientific research in Polynesia. Gregory convened an international conference of scholars in Honolulu in 1920, urging the delegates to proclaim "the problem of Polynesian origins" as the foremost issue facing Pacific science. He promptly convinced Yale to hand over forty thousand dollars, not an inconsequential sum for the time, from alumnus Bayard Dominick to underwrite a series of expeditions to the far-flung islands of Polynesia. Gregory brought in other new staff members from elite eastern universities, including young Edward S. Craighill Handy of Harvard.

Handy headed up the Marquesas Islands team of the Bayard Dominick Expedition, working with young archaeologist Ralph Linton. Where Stokes had had the temerity to dig into the earth of Kamohio rock shelter, Linton never bothered to put a spade into Marquesan soil. Simply assuming that nothing of interest would be preserved in the dank tropical earth, Linton confined himself to drawing up plans of Marquesan megalithic feasting platforms, and recording stone *tiki* and petroglyphs. Handy, for his part, interviewed the elders and recorded Marquesan legends and genealogies, which he saw as the keys to understanding the past. Young Edward Handy quickly stole the spotlight as the emerging guru of Polynesian studies.

Unfortunately, Handy was caught up in the prevailing anthropological theory, promulgated by German scholars such as Nazi anthropologist Fritz Graebner, seeking to prove that all "high civilizations" ultimately had an Aryan origin. The Polynesians had migrated into the Pacific, Handy wrote, in two great waves, the first of which he traced back to "Brahmanical" cultures in India, and the second to later "Buddhistic" civilizations in East Asia. Handy's theory, concocted of patches and shreds of ethnographic evidence, fitted well with the racial anthropology of the time, yet it hadn't a single bit of archaeological evidence to back it. In his willingness to embrace the racist *Kulturkreise* theories, Handy set Polynesian anthropology back. At the same time that Handy's star was rising, Stokes had run afoul of museum director Gregory. A man of decisive action, Gregory had no time for anyone who lingered over results before publishing. Gregory and Stokes clashed repeatedly. In 1929, Stokes was relieved of his position as curator of ethnology. It was a deep blow to Hawaiian archaeology.

By the 1930s the initial promise of archaeology, as a way to obtain independent evidence regarding the Polynesian past, had been forgotten. To be sure, the Bishop Museum continued to support archaeological field surveys on all the main Hawaiian Islands. It was deemed important, for posterity, that the temple sites be recorded. But this work focused almost entirely on mapping and drawing plans of the larger stone temples, and occasionally a fishpond or house site. No digging was done. Stokes's discovery of stratification in the little Kamohio rock shelter on Kahoʻolawe had been forgotten.

Gregory retired in 1935, having seen his famous Bayard Dominick Expeditions through to completion. Gregory recommended that the museum's trustees appoint the brilliant young part-Maori ethnologist Te Rangi Hiroa to succeed him at the museum's helm. A tall, handsome figure who unfortunately suffered from the pervasive racial discrimination against "colored" peoples that dominated his time, Hiroa thought that the key to his Polynesian past was to be found in the traditions and legends, as well as in detailed comparisons between the "material culture," the artifacts, of various island groups. Hiroa disdained archaeology. Calling it "a dry subject," he only grudgingly admitted that the detailed plans of ruined temple sites ought to be recorded "for subsequent comparative study."

In 1938, Te Rangi Hiroa published *Vikings of the Sunrise*. His book was intended to be the definitive statement on the history of his Polynesian ancestors. It was based heavily on the work of the Bishop Museum during the preceding three decades. John Stokes is briefly mentioned, but archaeology is almost completely ignored. Stokes's painstaking maps of the Hawai'i and Moloka'i island temples, his testing of Fornander's theory, along with the pioneering excavations on Kaho'olawe Island, all mattered naught. Hiroa used the Polynesian traditions, along with his own meticulous studies of "material culture," to weave a story of a "Europoid people" (his term), who were neither "Negroid" nor "Mongoloid" and who had conquered the vast Pacific. The Polynesian story as told by Hiroa is one of heroic proportions.

Hiroa's grand synthesis marked the end of an era in Polynesian and Hawaiian scholarship. It had begun a century earlier with the efforts of Kamakau and Malo to collect and write down the traditions of their people. Fornander had brought order and intelligent inquiry to the subject, using the traditions to outline the indigenous history of a proud civilization. Brigham and Stokes tried to build on that foundation, without rejecting the traditions. They sought to bring an entirely new approach—that of archaeology—to bear on the questions. Unfortunately, they lacked the one essential tool needed to exploit fully the evidence from the ruined temples or the earthen floor of Kaho'olawe's fisherman's shelter. There was no means then available to directly *date* these material vestiges of the past. Although Stokes understood the principles of stratigraphy, and therefore knew that the fishhooks and coral files from the deeper layers of his rock shelter were older than those at the top, he had no way of knowing whether the whole sequence represented one hundred or one thousand years of occupation. The only way to get at *time*, the backbone of all history, seemed to be through the genealogies of the Hawaiian and Polynesian chiefs. And this led one back again to the oral traditions. The traditions reigned supreme. Archaeology's time was not right. It was a "dry subject" to Hiroa and most of his contemporaries, and told them nothing new about the Polynesian past.

· · ·

The tool that John Stokes lacked, that would free Polynesian archaeology from its inferior status—the tool that would in fact revolutionize archaeological research around the world—finally arrived as a by-product of World War II's Manhattan Project and the new nuclear age. Willard Libby, a young physicist at the University of Chicago, had been a member of the top-secret Manhattan Project during World

War II. In the postwar era of increased support for nuclear physics, Libby was exploring the properties of radioactive carbon-14, or ^{14}C. Most carbon in our atmosphere has six neutrons (and hence is designated ^{12}C), which is the stable form of the element. However, when carbon in the upper layers of the atmosphere is hit by cosmic rays, this can bump up the number of neutrons to eight. Being radioactive or unstable, this ^{14}C isotope of carbon will, over time, shed its extra neutrons and return to the stable ^{12}C form. Libby's crucial discovery was that the process of "decay," by which the radioactive forms of carbon shed their neutrons, occurs at a predictable exponential rate, with a half-life of 5,730 years. That is, after 5,730 years, half of the radioactive carbon isotopes in a given quantity of carbon atoms would have shed their extra neutrons and returned to the stable ^{12}C form; after another 5,730 years, half of the still-remaining radioactive isotopes would again have returned to the stable state; and so on.

Carbon is an essential building block of life. Every living thing on Earth takes in carbon from the atmosphere, incorporating carbon into its organic molecules. Most of the carbon taken in by a plant or an animal, or a human being for that matter, is the stable ^{12}C form, but some is of the radioactive ^{13}C and ^{14}C isotopes. During life, carbon is continually cycled through living organisms, through their metabolic pathways. At death, however, the total amount of carbon is set, with a fixed ratio between the three isotopes of ^{12}C, ^{13}C, and ^{14}C. Because ^{12}C is stable, its quantity will not change in the bones or tissues of the dead organism. But the radioactive carbon will continue to decay, and its relative frequency diminishes following the constant rate that Libby had determined.

Libby realized that these fundamental laws of physics could give him an "atomic clock," a method for dating accurately the age of any material that had once been part of a living organism. If he could measure the amount of ^{14}C left in the organic matter relative to the stable amount of ^{12}C, he could calculate the elapsed time since death, since the rate of decay of ^{14}C was constant. The trick was to measure precisely the quantities of the two isotopes of carbon. Libby determined the rate at which ^{14}C atoms were decaying over a period of time by assembling a circular array of Geiger counters surrounding the sample of organic material to be dated. The apparatus was protected from extraneous decaying carbon by a thick shield of lead bricks. The organic matter of unknown age, after being burned down to pure carbon, was placed in the Geiger counter array and left for a period of a month or more. Each time one of its radioactive ^{14}C atoms decayed to a ^{12}C atom, loosing its extra neutrons, the energy released would send a "Beta particle" flying outward, to be caught by the Geiger counter's detector screen. Beta particle by Beta particle, tick by tick on the Geiger counter, the amount of remaining ^{14}C would be recorded by means of the rate of decay. If a sample was very young, the counter would click away rapidly. If it was old, the clicks generated by Beta particles would be few and far between. The method was primitive, but it worked.

Libby's discovery was so revolutionary that it won him the Nobel Prize for Chemistry in 1960. Libby tested his new method by applying it to wooden door lintels from the tombs of Egyptian pharaohs, whose dates were already thought to be known from their hieroglyphic inscriptions. When the ^{14}C results matched those of the hieroglyphic dating, the world of archaeology came knocking at Libby's door. Everyone wanted to have their latest excavated wood, charcoal, or bone dated. The Wenner-Gren Foundation for Anthropological Research in New York provided a special fund to help cover the cost of dating these archaeological samples. Kenneth Emory of the Bishop Museum was ready to take them up on the offer.

· · ·

Kenneth Emory had joined the Bishop Museum in 1920, when Herbert Gregory first assembled his dream team of young researchers to tackle the "problem of Polynesian origins." Emory's first project was to investigate enigmatic stone ruins reported to lie inside the great volcanic crater of Haleakalā on Maui. Next, Kenneth carried out a "survey of the native culture" of Lāna'i, an arid island in the lee of Maui. In 1924, he spent a year mapping the ancient temples of Tahiti and the Society Islands. Emory next joined the museum's Tanager Expedition, which was exploring the great chain of northwestern Hawaiian islands that runs for nearly one thousand miles from Kaua'i up to Kure and Midway atolls. On the tiny volcanic islets of Nihoa and Necker, uninhabited in historic times, Emory surveyed mysterious temples and house sites, mute testimony to the prior presence of Polynesian settlers. These ruins closely resembled *marae* he had explored in the upland valleys of Tahiti and Mo'orea, offering clues to contact between Hawai'i and the Society Islands at some time in the past. After his Nihoa and Necker expeditions, Emory's attention turned back to Tahiti and French Polynesia, with several expeditions through the Tuamotu Islands. By the late 1930s, Imperial Japan had cast a long shadow over the Pacific. With the bombing of Pearl Harbor, no research could be done in the Pacific until Japan was defeated. Emory played his part in the war effort by training thousands of U.S. Pacific Theater troops in jungle survival skills, using his knowledge of atoll culture.

In the spring of 1950, Emory was asked by the University of Hawai'i, which had begun a fledgling program in anthropology, if he would conduct a field course in archaeological methods. Bishop Museum salaries were notoriously poor (especially under the miserly Hiroa). Emory needed the extra compensation he would be paid for teaching the course. A young student, Jack Porteus, had been poking around some rock shelters at Kuliʻouʻou Valley on Oʻahu's southeastern coast and had found one that had a considerable buildup of ancient habitation debris in its floor. Emory decided that this would be a good location for the university's new field school in archaeological methods.

Emory gridded out the Kuliʻouʻou shelter floor in 3 × 3 foot squares for horizontal control, removing the sediments in uniform six-inch levels. This method thus did not follow real stratigraphy, but it did allow for vertical control of artifact distribution. At Kuliʻouʻou, Emory and students alike were delighted to discover, as they passed the organically rich sediment through their sifting screens, artifacts of various kinds appearing in considerable numbers. Fishhooks carved of bone and pearl shell, adzes of basalt and chips that had flaked off adzes during use, awls of dog and bird bone, files of coral and sea urchin spine, and, perhaps the most exciting, exquisitely carved bone tattooing needles. Emory could also begin to see that there were subtle changes in the stylistic details of the fishhooks and other artifacts. Nearly forty years after John Stokes had first dug in the floor of Kamohio fishing shrine, the potential of stratigraphic archaeology was being rediscovered in Kuliʻouʻou rock shelter.

But now Emory had the ace up his sleeve that Stokes had lacked nearly four decades earlier. Libby's revolutionary method of dating charcoal and other organic materials had been announced to the world. Emory was keen to see if he could use it to date the occupation of Kuliʻouʻou. A sample of charcoal from a fire hearth near the base of the Kuliʻouʻou cultural strata was sent off to Libby's laboratory. On February 19, 1951, Hiroa told his secretary to call Emory into the spacious Bishop Museum director's office on the second floor of Paki Hall. Sitting across the imposing *koa* desk that had once been Gregory's, Hiroa, ill with cancer, read aloud to Emory the letter he had just received from Willard Libby. The charcoal from the hearth at Kuliʻouʻou, after sitting for a month or more in Libby's array of Geiger counters at Chicago, ticking off its decaying Beta particles, had returned a calculated age of A.D. 1004, plus or minus 180 years. The plus or minus factor was due to statistical uncertainty; after all, the method was still a bit crude. That didn't matter to Emory. As he later recalled, "Boy, was I excited. Immediately it opened a whole new vista of possibilities." Archaeology had finally come into its own in Hawaiʻi and the Pacific.

. . .

When Te Rangi Hiroa succumbed to cancer in 1952, the museum's trustees appointed a young anthropologist named Alex Spoehr to fill the position of director. Spoehr himself had excavated archaeological sites in the Mariana Islands of Micronesia, in 1949. Emory did not have to convince his new boss that the museum should launch a program of archaeological excavations throughout the Hawaiian Islands. He began searching for sites that might produce abundant artifacts from the Hawaiian past. Emory had become especially interested in finding bone and shell fishhooks, with their fascinating diversity of shapes. Elsewhere in the world, archaeologists had long focused on pottery because the subtle style differences in shape and decoration could be used to put archaeological finds in an ordered sequence. This method of ordering materials according to gradual changes over time is called seriation. The ancient Hawaiians had never made pottery, but the fishhooks that Emory was digging up could be ordered by seriation.

Most of the sites Emory tested were rock shelters, the former haunts of ancient fishermen who had camped out in them. Sitting around their stone-lined hearths (which conveniently provided charcoal samples for radiocarbon dating), those ancient fishermen had painstakingly carved their delicate bone and shell fishhooks with files and abraders of coral and sea urchin spine. But at Ka Lae, the southernmost point of Hawai'i Island, Emory found a different kind of site, richer than any other in fishhooks and the tools used to carve them. This was a low undulating sand hill called Pu'u Ali'i (Hill of the Chief). The upper part of the sand dune had been used as a burial ground. Below the burial layer, and hence predating it, was a layer of compacted earth, stained dark gray from the charcoal and ash of countless hearth fires. This deposit had accumulated over a long period of time, the remains of an ancient camp of fishermen. South Point is famed to this day among island fishermen as a prized area for catching *'ahi* (yellowfin tuna), *aku* (bonito), and *mahimahi*, among other big game fish.

When Emory and his crew started digging into the dark gray earth of the ancient fishing camp at Ka Lae, their sifting screens quickly filled with archaeological treasure. Bone fishhooks, and some of shell, of several styles appeared in numbers they had never seen before. Some 1,443 hooks had been found by the time the excavations were completed. The files and abraders of coral, pumice, and sea urchin spines used to carve the fishhooks numbered ten times that.

Emory now turned up the pace of his Hawaiian archaeological program. Assisted by a young Japanese archaeologist named Yosihiko Sinoto, and by his

University of Hawai'i student William Bonk, Emory tested every promising rock shelter or coastal sand dune from Hawai'i to Kaua'i to see if it could yield the prized bone and shell fishhooks and other artifacts. With the South Point fishhook sequence as their key, Emory, Sinoto, and Bonk thought that they could bring the artifact assemblages from all of the different sites across the archipelago into a systematic correlation. South Point anchored the sequence. A hearth at the bottom of the thick dark gray fishermen's camp at Pu'u Ali'i had produced a radiocarbon date of A.D. 124, plus or minus 60 years. This, Emory believed, was a true approximation of the date that Polynesians first arrived in Hawai'i. In 1959 the Bishop Museum published Emory, Bonk, and Sinoto's definitive study, *Hawaiian Archaeology: Fishhooks*. Stokes, no doubt, would have found it gratifying. The stratified deposits of Hawaiian archaeological sites were telling a new story of the Polynesian past.

. . .

Such was the state of play in Polynesian archaeology and anthropology when I entered the scene as a young student volunteer at the Bishop Museum in the early 1960s. Emory was a towering figure, not just literally tall and lanky (and firmly set in his opinions), but the acknowledged leader of his field. By 1959, with the publication of his *Fishhooks* monograph, Emory felt that the Hawaiian archaeological scene was more or less played out. He had found and excavated the best sites, skimmed off the cream, put Hawaiian history on a new firm, scientific basis. The task of mopping up, sorting out the details, could be left to junior colleagues, such as Lloyd Soehren of the museum's staff, under whose watchful eye I first learned to excavate. Emory moved on to his first love, Tahiti, where he and Sinoto were now scouting for fishhook-rich sites that would allow them to link their Hawaiian sequence back to the Polynesian homeland.

The late 1960s were a time of radical rethinking in the field of archaeology. New methods and approaches were being discussed at North American universities. At Harvard, Gordon Willey—a specialist in the ancient Mayan cultures of Central America—had developed what he called the "settlement pattern" approach. Willey encouraged a young Harvard graduate student, Roger Green, to apply the settlement pattern approach in the Society Islands. In 1960 Roger and his wife, Kaye, began work in the deep valley of 'Opunohu, on Mo'orea Island a short distance from Tahiti. Their project would change the way archaeology was done throughout Polynesia.

Whereas Emory and Sinoto had concentrated on artifact-rich archaeological sites (those containing lots of fishhooks, adzes, and other items that could be used to create a chronology), Green now set out to record entire archaeological *landscapes*. The idea was to look at the full range of sites across a landscape, as a means to reconstruct the social organization and even the ecological adaptations of pre-European societies. By carefully mapping out and recording *all* of the sites in a valley, ranging from simple shrines to the largest temples, from house platforms to irrigated agricultural complexes, Green showed that it was possible to gain new insights into how these ancient societies and populations had been organized. Changes in settlement patterns over time likewise gave important clues to how society itself had evolved.

The Bishop Museum hired Roger Green in 1965. He promptly began organizing a new Polynesia-wide research program that would incorporate the settlement pattern approach. The museum also hired Douglas Yen, an ethnobotanist from New Zealand who was carrying out a Pacific-wide study of the sweet potato plant. Roger Green and Doug Yen became my mentors. Following their lead, I wanted to explore the ways in which Hawaiians had not only adapted to, but transformed, their island ecosystems. I was inspired by the new theoretical movements within the field of archaeology. Some called it the New Archaeology, in opposition to Culture History, the kind of sequencing of fishhooks that Emory and Sinoto emphasized. Archaeology had the potential to be a much more broadly encompassing science of the human past. I wanted to be a part of this, to apply the New Archaeology to Hawai'i.

In the early fall of 1968 I boarded a Pan American Airlines 707 jetliner at Honolulu Airport, bound for Los Angeles and then on to Philadelphia. I was enrolling at the University of Pennsylvania, a renowned center of anthropological studies based out of its famous University Museum. Since I had never been farther east than Seattle and San Francisco, my first few months in Philadelphia were ones of culture shock. But I buried myself in my courses, absorbing everything I could about the New Archaeology. One winter's day a letter arrived bearing the Bishop Museum's logo. It was from Roger Green, asking if I would like to join two graduate students in a new project on the island of Moloka'i. Roger knew that I had already mapped ancient stone remains of house sites, temples, and irrigation systems in the island's Hālawa Valley. He proposed that my reconnaissance form the basis for an expanded project. Better still, he had funds to help finance the work. What a gift to a young undergraduate anthropology student! In May 1969 I was back in the Hālawa Valley, working with Gil Hendren of Harvard and Tom Riley of the University of Hawai'i. That summer, and again in 1970, we explored and excavated the valley's archaeological remains, from a rich coastal village site to small stone habitations along the interior slopes, overlooking hundreds of ancient stone-faced irrigation terraces. It was there in Hālawa that I began my lifelong quest to understand how ancient Hawaiian civilization had developed.

It is now more than forty years since I had the privilege of conducting my first archaeological project in Hālawa Valley. The field of archaeology has continued to change over these four decades, with new theoretical approaches and many new methods of extracting evidence about the past. It is not just the methods and techniques of archaeology that have dramatically changed since I dug into the sands of Hālawa. The practice of archaeology has also undergone radical transformations. In the late 1960s, most archaeology was still carried out by academics, university professors (who typically did their fieldwork during the summer break) or museum scholars like Emory. Today, the majority of archaeological work in Hawai'i is carried out not by university or museum people, but by private archaeological contractors who operate consulting practices. We call this kind of archaeology cultural resource management (CRM, for short). CRM developed in response to federal and state laws requiring that archaeological (or "cultural") resources be identified and evaluated prior to land development, whether for public works such as highways, or for commercial projects such as hotels, subdivisions, and golf courses.

After statehood in 1959, new capital infusion from the mainland United States, Japan, and elsewhere led to a huge expansion of Hawaiian tourism and the hotel industry, which today is the economic mainstay of the fiftieth state. As hotels and resorts began to be constructed throughout the islands, along with the public infrastructure of highways and airports, the demand for CRM archaeology expanded enormously. There are now more than two dozen archaeological consulting firms operating in Hawai'i. They conduct the vast majority of archaeological work in the islands. In contrast, the Bishop Museum has just a single young Polynesian archaeologist on its staff.

As a senior scholar, I have the luxury of reflecting back over the years at the many changes that have transformed my field. I think back to Kenneth Emory, who when I first met him was the reigning guru of Polynesian archaeology. As a young man, Emory had experienced the paternalistic views of Edward Handy and Te Rangi Hiroa, who dismissed archaeology as no more than a minor source of evidence for the Polynesian past. Then Libby's revolutionary method of radiocarbon dating had opened up undreamed-of possibilities. Emory had been at the forefront of revitalized archaeological research in Polynesia. In the late 1960s,

Emory thought that he and his colleagues had resolved all of the important questions that archaeology in Polynesia could answer. How wrong he was! Entire new areas of inquiry and methods to address these have arisen in the past forty years. Emory, were he still alive today, would barely recognize the field of archaeology, with its use of GPS and GIS technologies, its laboratory analysis of pollen grains and ancient bird bones and fishbones, its new uranium-thorium dating methods.

The practice of archaeology in Hawai'i and elsewhere in Polynesia is also changing in other ways. As critical historians of science have rightly pointed out, anthropology and archaeology had their origins in a colonial world. Until fairly recently, most archaeologists were white (Haole in Hawaiian) and typically male. In our contemporary postcolonial world, the demographic composition of the archaeological community is changing rapidly. Not only are many more women joining the archaeological profession, but ethnic diversity is also increasing. Two Native Hawaiian women, Kehau Abad and Kathy Kawelu, now have doctoral degrees in archaeology (Abad from the University of Hawai'i, Kawelu from my own department at Berkeley). Increasing numbers of young archaeologists of Native Hawaiian ancestry are actively engaged in CRM consulting on the islands, or work for state or other government agencies. Inevitably, this expansion of the archaeological community to include Native Hawaiians as well as members of other ethnic groups will change the ways in which archaeology is practiced. Indigenous scholars bring new questions to the table, and they have new ideas about how archaeological sites should be studied and about how archaeological data should be collected, analyzed, and preserved.

Archaeology, in its theory, its methods, and its practices, will continue to evolve. The young scholars who are emerging today will carry the field forward. They will build on the knowledge already accumulated, but they will ask new questions, apply new methods, and take the field in directions that I cannot imagine. The quest to discover who we are—and the deep history that makes us who we are—is deeply ingrained in humankind. It is part of what makes us human.

FIVE · The Sands of Waimānalo

No one knows for certain how long there have been islands in the geographic space now occupied by Hawai'i. Oceanic islands, like people, are born, live out their lives, and eventually die. This is, of course, just a metaphor. The timescale is geologic, the lifespan of an individual island covering millions of years. Each island in the Hawaiian archipelago begins its life cycle with the eruption of countless flows of lava from a fracture, a "hot spot" on the ocean floor. This hot spot, where a gigantic magma plume rises from the Earth's mantle, has been stationary for a very long time, while the thin ocean crust moves glacially but relentlessly over it. The hot spot spews out enormous quantities of molten basaltic rock, at a rate fast enough to build up a mountain mass that rises above the sea surface. The Big Island of Hawai'i, with its great volcanoes of Mauna Kea and Mauna Loa, was the most recent to emerge. Because it is so youthful, its lava slopes are barely weathered or eroded. Only in the northern tip of Kohala district, which was the first part of the island to emerge, about one million years ago, has sufficient time passed to form several deep valleys, including the famous Waipi'o.

As each island passes off the hot spot and its volcanic fires cool, nature's forces begin to shape the smooth volcanic slopes, waves eroding cliffs along the shore, rainstorms cutting gullies that gradually turn into valleys. Sometimes change is catastrophic, as when huge chunks of unstable volcanic mass collapse in gigantic landslides. The northern half of eastern Moloka'i fell into the ocean about one million years ago, leaving behind some of the tallest sea cliffs in the world (see color plate 2). By the time an island is about two to three million years old, its landscape has reached "middle age," dissected by deep valleys separated by razor-backed mountain ridges. Oʻahu and Kauaʻi in the western part of the main Hawaiian chain, some three million and five million years old, respectively, are typical of this phase in an island's life. Eventually, the ceaseless work of wind and water, combined with the gradual subsidence of the Pacific Plate on which the island masses rest, wears the island down until it sinks beneath the ocean. Tiny Nihoa and Necker Islands, 155 and 333 miles to the northwest of Kauaʻi, respectively, are the remnants of islands that were once much larger. Stretching even beyond them is a string of reefs and coral atolls that have formed atop the sunken volcanic peaks of even older, now-submerged islands, extending back seventy-five million years in time.

From the perspective of a canoe-borne Polynesian exploring expedition in search of new islands to settle, it was the middle-aged islands that would have been the most attractive. Islands such as Oʻahu and Kauaʻi offered the best combination of resources to the newcomers. Their broad valley slopes provided fertile soils in which to plant root and tuber crops, while the freshwater streams coursing down from the mountains could irrigate taro fields. These islands were also old enough that extensive coral reefs had formed around their coastlines. On the younger islands such as Hawaiʻi and eastern Maui, much of the coastline consists of sea cliffs where the ocean swells pound dangerously. But Oʻahu and Kauaʻi, along with the southern coast of Molokaʻi and parts of western Maui, are ringed with well-developed coral reefs that protect their shores. These reefs support a diverse ecosystem of marine life. The algae and seaweeds that grow on the reefs provide food for herbivorous parrotfish and wrasses, which in turn are the prey of larger carnivorous species such as jacks and barracuda. At the top of this food chain are the sharks. The reefs of Oʻahu and Kauaʻi abounded in fish, sea urchins, octopuses, limpets and other mollusks, and seaweeds, all foods that the Polynesians prized and knew how to prepare in a variety of ways.

The age progression from youthful to older islands, beginning over the hot spot in the southeast and extending gradually to the northwest, influenced the ways in which Hawaiian civilization developed. Not surprisingly, both Hawaiian oral traditions and the evidence of archaeology suggest that during the first few centuries after settlement, Hawaiian culture flourished primarily on the westerly, older islands. Only later, after the population expanded and the pressures on land increased, did the vast lava slopes of eastern Maui and Hawaiʻi attract people from the increasingly crowded salubrious valleys. That is a story I will recount in subsequent chapters. For the moment, let us pick up the thread of initial discovery and arrival of the first voyagers from the Marquesas.

. . .

On the windward side of Oʻahu, near the island's eastern tip, the graceful arc of Waimānalo Bay dazzles the eye with its four-mile-long white coral sand beach buffeted by the relentless surf of the North Pacific. Whenever I round the point of Makapuʻu and Waimānalo comes into view, backdropped by the sculpted crest of the Koʻolau Mountains while the blue Pacific extends to the horizon, it never fails to take my breath away. The beach at Waimānalo is longer than the more famous Waikīkī on the leeward side. In my opinion Waimānalo is more beautiful, especially now that Waikīkī is hemmed in by high-rise hotels, its sands crowded with pale tourists from Tokyo and Chicago. Waimānalo, thankfully, remains largely unscarred by development. You can still stand on the promontory at Makapuʻu, looking out over the broad sweep of Waimānalo Bay with its lush valley stretching away inland, and get a sense of what the landscape looked like when the first Polynesians arrived.

Early Polynesians—such as the crew of *Mahina-i-te-Pua* or other voyagers from the south—would have been looking for a place much like this to establish a new settlement. Colonizing voyagers were seeking a particular combination of natural resources and topographic features in order to reestablish their economy and lifestyle. First, they wanted access to fertile alluvial soils in which to plant their seedling crops, especially taro, banana, yams, and other useful plants such as paper mulberry and kava. Waimānalo Valley possesses good, gently sloping alluvial terrain, where the annual rainfall measures between forty and sixty inches, virtually ideal for growing most Polynesian crops. Second, the settlers wanted a permanent source of fresh water for drinking and cooking, as well as bathing. Swampy or marshy areas adjacent to a stream were ideal for planting taro, which grows best when its roots can tap flowing water. Waimānalo Stream begins in innumerable small rivulets descending from the crenellated amphitheater walls of the Koʻolau Mountains, joining to form a single watercourse that enters the sea near the northwestern end of Waimānalo Bay. Third, the colonizers wanted to live where they could go fishing and gathering on an extensive coral reef. This, too, was readily available at Waimānalo. Waimānalo offered just the combination of topography and resources that would have been highly attractive to early Polynesian voyagers.

During World War II, the U.S. military seized much of Waimānalo to use as an airfield for fighter planes to stave off a possible Japanese invasion. After the war, Bellows Air Force Base became a communications center and recreational area for servicemen who, with their families, enjoyed the use of beach bungalows. In the winter of 1967, an air force officer telephoned Lloyd Soehren at the Bishop Museum to say that human skeletons had been seen in sand dunes just north of Waimānalo Stream.

Soehren drove over to Bellows to have a look. Human burials were indeed present in several spots. Some had yellow and red glass trade beads with them, indicating that they probably were the remains of Native Hawaiians who had perished during the "great dying" after European contact. Thousands had died in various epidemics of influenza, smallpox, measles, and other diseases to which the Hawaiians had no resistance.

What interested Soehren much more than the burials was what he saw in a little sand hill across the road from the row of recreational bungalows, near the stream mouth. Where the grass- and ironwood-covered mound had been cut by a backhoe that was mining sand, distinct layers of black-colored sand alternated between zones of clean white beach sand. Occasionally, the black sand layers were interrupted by oval-shaped pits, their bottoms filled with chunks of charcoal and firecharred basalt rocks. Soehren immediately recognized these pits as Polynesian earth ovens, *imu* in Hawaiian. To his archaeologically trained eye, the black-stained layers were surely old land surfaces where people had once lived. The sand had become stained with charcoal and ash from their ovens and fireplaces, combined with organic refuse. In earlier times the surface of the sand hill must have been slightly lower and had been covered with a small settlement, perhaps a little hamlet or village. Indeed, there had been several periods of habitation atop the dune, each interrupted by a period of abandonment when the drifting white sands blowing inland from the beach covered the stained sands of the former village. These phases of habitation, followed by abandonment, and then by habitation again, made the interior of the sand hill—when cut through by the backhoe—resemble a cake with alternating layers of chocolate and vanilla. The oldest habitation layer, of course, was the deepest.

Soehren scrambled up the eroding front of the sand hill, where the backhoe's bucket had exposed the black layers. Here and there bits of shell and bone caught his eye. Picking up one pearl-shell fragment that was sparkling in the bright sunlight, Soehren turned it over in his fingers. It was a piece of a fishhook, with a well-shaped shank and part of the bend, but missing the point. He also found fragments of files or abraders made from coral, which the Hawaiians had used to grind down and shape pearl shell to fashion their fishhooks. He decided that it would be worthwhile to begin a more systematic excavation into the sand hill, to see what might be found within the telltale dark layers.

At the time, Soehren lacked the resources to mount a major excavation. At the Bishop Museum, he was deeply engaged in working up the results of a multi-year excavation in the Nuʻalolo rock shelters of Kauaʻi Island. The thousands of well-preserved artifacts from Nuʻalolo were demanding nearly all of Soehren's time to study and describe. For the dig at Waimānalo he would need some assistance. Soehren rang up Richard Pearson, across town at the University of Hawaiʻi in Mānoa Valley. Pearson, a recently hired young faculty member in the Anthropology Department, was keen to find local archaeological projects that he might use as opportunities for training students. When Soehren described the Waimānalo sand hill, Pearson was immediately interested.

For several weekends in February and March, a small band of volunteers and students drove out to the Bellows Air Force Base to join Soehren and Pearson. They cleaned off the face where the backhoe had sliced into the sand hill, getting a better look at the strata. The team laid out a grid of stakes on a north-south axis, so that excavation could proceed systematically, by carefully gridded out squares each one meter on a side. As the digging with trowels and other small hand tools proceeded, the sand from each square and stratum was placed in buckets, which were passed along to the screeners. The fine sand passed easily through the wire mesh sieves, leaving behind bits and scraps of ancient debris that would tell the tale of daily life on the sand hill. Pieces of fishbone along with mollusk shells that had been discarded after a meal; larger fragments of pig and dog bone, some showing signs of having been cut with a coral saw in order to work them into fishhooks or other objects; coral saws and files; flakes of hard basalt; and pieces of worked shell, including more fragments of fishhooks like the one that Lloyd had picked up the first day. Everything was bagged by square and layer number, to be sorted out later and catalogued in the university laboratory.

After several weekends of work, Pearson and Soehren were convinced that the site deserved a more comprehensive investigation. Some of the fishhooks recovered from the sifting screens resembled early styles that had been found at only one or two other sites in the islands. Indeed, a few resembled hooks that the Bishop Museum's Yosihiko Sinoto had dug up on Ua Huka Island in the Marquesas. Pearson proposed that he excavate further into the sand hill during the coming summer, as a University of Hawaiʻi field school or archaeology training project. The field school students would provide essential labor while receiving training on what seemed to be an interesting and possibly important ancient site.

A student at Honolulu's Punahou School, I had already spent a couple of summers volunteering with Lloyd Soehren at the Bishop Museum. Two years earlier I had accompanied him to the Big Island of Hawai'i; our team had excavated two rock shelters used by ancient fishermen at Kahakahakea in the lava flows north of South Point. At Kahakahakea, Soehren taught me the methods of archaeological excavation. The next summer, in 1966, I had assisted another Bishop Museum field team mapping hundreds of house sites and garden enclosures on Maui. By this time I was thoroughly hooked on archaeology and knew that it would be my life's passion.

Richard Pearson asked if I would like to help out with the excavation at the Waimānalo sand hill later that summer. Aware that Soehren had taught me surveying, Pearson suggested that I might make a topographic map of the site and also be responsible for maintaining the daily level records (the depths at which various objects were being recovered or plotted, as measured from a fixed datum on top of the sand hill). I was thrilled at yet another opportunity to expand my repertoire of archaeological experience. While Soehren was an excellent teacher, he was gruff and formal, a man of few words as he puffed away on his ubiquitous pipe. Pearson was just the opposite—young, lively, talkative, and willing to share his enthusiasm for archaeology. He had recently received his doctorate from Yale University, under the tutelage of Kwang-Chih Chang, the famous Chinese archaeologist (with whom I would also later study). From Pearson I heard of the intellectual debates then beginning to swirl around anthropology departments at mainland U.S. universities. This was heady stuff for a young student. I eagerly accepted Pearson's offer to join in the Waimānalo excavation.

Throughout August 1967, we dug into the side of the sand hill next to the stream at Waimānalo (see figure 4). Early each morning, the team—undergraduate students, a couple of grad student assistants, Pearson, and this eager young Punahou School student—assembled at the rickety old wooden shack on Maile Way that served as the University of Hawai'i's archaeology laboratory. Everyone would pile into the back of a university motor-pool pickup, Pearson at the steering wheel as we headed off on Kalaniana'ole Highway for the roughly forty-five-minute drive to Waimānalo. In the back of the pickup, we passed the time talking about the previous day's finds and what we might discover in the hours ahead. The sea breeze filled our nostrils as the truck rounded the narrow highway in the shadow of Makapu'u's sheer cliffs, and that never-to-be-forgotten view of Waimānalo Bay stretched out before us.

Pearson had the idea of opening some larger areas of the site so that we could

FIGURE 4.
University of Hawai'i excavations in progress at the Bellows Dune site, Waimānalo, O'ahu Island, in 1967. Note the charcoal-stained dark layers, resulting from periods of habitation, separated by white dune sands. The author is kneeling, second from the left. Photo courtesy of the Oceanic Archaeology Laboratory archives, University of California, Berkeley.

try to discern the spatial patterns that might be evident from the postholes and pits. There had been two main phases of occupation on the sand dune, each marked by its own distinctive layer of charcoal-stained black sand. Layer II, slightly higher, was the younger; it had a number of postholes as well as several burials in pits. One of these burials, in a pit that had cut down through Layer II from higher up, was of a young female (estimated to have been about nine years old when she died) on whose chest had been placed a hook-shaped pendant made of rock oyster shell. This kind of ornament, called *lei niho palaoa*, was a symbol of the Hawaiian chiefs. The young chiefess also wore an anklet of carved and perforated pig tusks. We could only wonder what tragic event had caused her untimely departure from the world.

After Layer II had been peeled away, we proceeded to remove the clean sand separating this from the underlying and older Layer III. As the darkened sand began to appear, we could once again discern postholes where the wooden timbers that once supported pole-and-thatch structures had been planted in the dune. In the southwestern part of the larger area now opened up, a pavement of waterworn pebbles was exposed by our trowels and brushes. Polynesians frequently paved their house floors with such gravel, which they called *'ili'ili*.

When the gravel was scraped away, the top of a pit, about one meter in diameter, appeared. This proved to be another burial, containing the tightly flexed skeleton of an old woman, with her lower legs doubled back under the thighs. Close examination of the bones revealed that she had lived a long and eventful life. She had suffered two fractures during her life, one of her right thigh bone, the other of her right arm. Yet she had survived this trauma, for the fractures had healed. But in her old age she suffered from severe arthritis in both her legs and her lower back. This old woman, who likely was a revered elder, had been buried beneath the floor of the gravel-paved house. Polynesians in some islands of the South Pacific, such as Tikopia, still bury their dead beneath their house floors, so that when they sleep at night, their heads are "pillowed on the bones of the ancestors."

By the end of August we had removed Layer III with its gravel pavement. Beneath this there was only clean beach sand. We closed up the excavation, and spent the last few days of the summer term making sure that all of the artifacts and midden samples had been properly bagged and catalogued in the University of Hawai'i archaeology lab. I was sad about having to end the excavation. There was still more of the site that we hadn't probed, and I felt it had held back some secrets from us. Meanwhile Pearson would send several samples of the charcoal we had collected from hearths and earth ovens to the Gakushuin Laboratory in Japan to be dated by the radiocarbon method. We were all eager to see whether the lab would report dates that might be as old as Emory's sites at South Point, or as the Kuli'ou'ou rock shelter.

· · ·

Mahina-i-te-Pua had been running with the wind for several days along the windward coasts of the newly discovered large islands. At first the shorelines appeared wild and forbidding—sheer cliffs of lava rock upon which the North Pacific swells exploded ferociously. To attempt to land the canoe on such a dangerous shore would have been an act of suicide. So they continued onward, keeping a watchful eye, especially during the night, lest they run aground on the rocks. Now a different kind of shoreline came into view. A thin white line of surf divided the inky blue of the deep ocean from the azure hue of a shallow lagoon. Beyond the lagoon a coral sand beach beckoned. Inland of the beach a broad, fertile valley could be discerned. The trick now was to find a pass, an *ava*, through the reef. Soon a gap in the white line of breakers appeared, and the steersman guided *Mahina-i-te-Pua* straight into the pass. Where there was such a pass, there would also be abundant fresh water, for they knew that coral does not grow in the outflow from a stream.

The twin keels of *Mahina-i-te-Pua* crunched on the sharp coral of the pristine reef flat. No canoe, no human footsteps, had preceded her. It might very well have been on the fringing reef of Waimānalo Bay that *Mahina-i-te-Pua* came ashore to end her long journey from the Marquesas. We will never know for certain. Perhaps it was the reef off Waiheʻe on Maui. Or at Wailua on Kauaʻi. Or in the bay of Hālawa on eastern Molokaʻi. All of these places possess the same combination of desirable environmental features that the first Polynesians to arrive in the Hawaiian Islands were seeking. Wherever the exact place, what matters is that they had survived that crucial first voyage, following the *tōrea*, the golden plover. Now what was foremost in everyone's mind was to get the precious cargo ashore, to plant the tender shoots and seedlings that would give life to the land, to make sure the pigs and dogs and chickens got their land legs back so they could reproduce.

The people waded ashore, following their *ariki*. In his arms he held the basalt slab wrapped in red turmeric–dyed barkcloth, taken from the shrine of his ancestors in their home valley, sheltered throughout the long voyage in the thatched hut amidships. This stone was a physical connection to their homeland. When the *ariki* offered his prayers, his *pule*, to the ancestors, their spirits would descend and occupy the stone. The ancestors had not deserted him on the long and risky voyage. Now he must offer a *pule* of thanksgiving for bringing them to these magnificent new islands. Cresting the broad slope of the beach, he set the precious bundle down where coral sand gave way to herb-covered earth. Unwrapping the layers of reddish-yellow-dyed barkcloth, he exposed the prismatic column to the bright sunlight, setting it upright. Seated cross-legged before the ancestral stone, he removed a dried stem of *kava* from the folds of his loincloth and set it before the stone. Then the *ariki* began to chant his genealogy. He invoked Tangaroa, who with heroic effort had pushed apart Sky Father and Earth Mother in the vastness of Te Pō, The Darkness, to beget the world. Then the long generations in Hawaiki, down to those more recently departed, who had led other *vaka moana* eastward into the dawn, to discover and populate the islands whence they had sailed in *Mahina-i-te-Pua*. He gave thanks for the successful voyage, in which no life had been lost. And he beseeched the ancestors to send *mana* that they would flourish in their new home.

Now the difficult task of establishing a beachhead in the new land would begin in earnest. The work would not be easy, and the first year of their colony would be precarious. For the first few months, until the fledgling crops matured, they would have to subsist on whatever the land and sea provided. Fish, shellfish, seaweeds, and birds there would be in abundance. But it is difficult to live on an exclusively protein diet. The Hawaiian Islands, like most of those in the Pacific, did not naturally have many plants yielding edible fruits or berries. They might have to cut and boil the pith of tree ferns to survive; indeed, they probably did.

The people began to unload the precious cargo contained within the sturdy hulls of *Mahina-i-te-Pua*. The breeding pairs of pigs and dogs could scarcely stand up after weeks in the bilges, and had to be carried ashore. As some men constructed wooden pens to hold them, others went off onto the reef with nets to catch fish with which to feed the animals, so they could regain their strength. In some ways it was more important to feed these domestic animals than to feed the people. Another group of men, having reconnoitered along the stream bank for suitable garden land, began to clear the undergrowth with their stone adzes. Here they would plant the precious shoots of taro, the stalks of banana and sugarcane, and cuttings of kava and paper mulberry.

There was yet other work to be done. After weeks at sea, they wanted shelter, a roof to protect them from the rain squalls that periodically came in with the trade winds, and to shelter their belongings. Unlike on other islands of the South Pacific, there were no natural stands of coconut to offer ready thatching material. They had brought coconuts with them, and would plant these, but it would be years before the trees would mature and bear fruit. Fortunately, the *hala* (pandanus tree) grew naturally on these islands in abundance. A party of the women set out to collect *hala* leaves. Returning to the encampment, they folded the long, tough leaves over fathom-long sticks, pinning the leaves in place with twigs. These sheets of pandanus leaves, when stacked in an overlapping fashion and supported by sturdy posts, made convenient lean-to shelters. Later, when there was more time, they would construct permanent houses with *hala*-leaf thatch, paving the floors with clean gravel from the stream.

Someone had dug a pit in the soft sand, filling it with dry driftwood. Quickly rubbing a hardwood stick against a softer piece of hibiscus wood, he ignited some kindling, and then set the driftwood in the pit on fire. Volcanic stones from the stream were placed on top of the burning logs. When the fire had burned down the rocks would be white-hot, ready to receive the leaf-wrapped bundles of fish and freshly caught turtle meat that had been brought back by the fishermen. For the first time, a Polynesian earth oven wafted its perfumed smoke into the air of a Hawaiian evening. Long shadows began to fall over the beach as the sooty terns came in after a day out at sea. The sound of the breakers on the reef drowned out the seabirds' cries as they settled into their treetop nests. After so many long nights on the restless ocean, they would sleep the sleep of exhausted voyagers. Exhausted, yes, but also exhilarated in the excitement of their newly discovered land. The *tōrea* had not disappointed them.

· · ·

Richard Pearson opened the blue aerogram envelope from Professor Kigoshi of Japan's Gakushuin University radiocarbon-dating laboratory. He was eager to read the results of the dating of five samples of charcoal from Layers II and III of the Bellows sand dune site at Waimānalo, which he had sent off to Kigoshi some months earlier. Scanning the typewritten page, his eyes glanced at the numbers, reported in years "B.P." or "before present," standardized to 1950, the year in which Libby had perfected the method of radiocarbon dating. One sample from the deep Layer III had an age of 1030 plus or minus 110 years. This would put the calendar age of Layer III at the end of the tenth century. Then Pearson's eyes fixed on an even larger number, 1600 plus or minus 90 years. That date could mean that the site had been settled back in the fourth century. That would make the site almost as old as Emory's Puʻu Aliʻi sand dune site at South Point. But then Pearson noticed a problem. The older date came from Layer II, which was stratigraphically *above* Layer III and therefore by the law of superposition had to be younger. The radiocarbon dates were inverted. Kigoshi's letter reported two other dates: one of 1110 plus or minus 120 years from the upper Layer II, and 700 plus or minus 125 years from a fire pit in Layer III. A final sample, also from the deep Layer III, had been determined to be younger than 380 years; it was probably contaminated in some way.

Pearson wasn't sure how to deal with these dates. They were all relatively old. Even the youngest sample indicated that Polynesians had occupied the site no later than the mid thirteenth century. The other samples suggested that a date before A.D. 1000 was highly likely. But he was troubled by the fact that the oldest date did not come from the deepest layer, that there were inversions in the dates relative to the stratigraphy. He put the aerogram in a file folder on his desk; he would need to think about this further.

In the spring of 1970, three years after the excavations at the Bellows dune site had begun, I was now studying anthropology and archaeology at the University of Pennsylvania. The Philadelphia winters were a harsh experience for someone raised in Hawaiʻi. I decided to spend the spring term at the University of Hawaiʻi, taking courses by several leading experts on Polynesian anthropology.

Knocking on Pearson's office door on the Mānoa campus, I inquired about the results of our work in Waimānalo. He showed me Kigoshi's letter, and we talked about the interesting but somewhat confusing results. I had spent the previous summer digging a sand dune site near the mouth of the Hālawa Valley on Molokaʻi. There I had found bone fishhooks and basalt adzes that resembled those we had recovered from the Bellows sand hill. I had also compared these with artifacts that Yosi Sinoto had excavated from Hane and other sites in the Marquesas Islands. I told Pearson that I that I thought these artifacts were of early types, quite different from those typically found in later Hawaiian archaeological sites. Thus in spite of the somewhat conflicting radiocarbon dates, I felt that the sand hill at Bellows dated to an early phase of Polynesian occupation in Hawaiʻi.

Pearson's expertise was in the archaeology and prehistory of eastern Asia, especially the Ryukyu Islands, Japan, and Korea. Realizing that with my experience at the Bishop Museum I had gained considerable knowledge of Polynesian artifacts, he invited me to help write up the Waimānalo finds. I jumped at the offer to coauthor an article on the site, agreeing that I would analyze the artifacts as well as the "faunal remains" of bone and shellfish. Pearson would write up the excavations and one of his colleagues at the University of Hawaiʻi, Mike Pietrusewsky, a specialist in physical anthropology (the study of human skeletal remains), would analyze and describe the burials.

Pearson handed me the bags of artifacts and specimens that we had carefully catalogued back in 1967. I took these to the Bishop Museum, where I had the use of a small table in the archaeology laboratory. Carefully examining the various fishhooks of pearl shell and bone, adzes of basalt, coral files and abraders, pendants, and other objects, I described their forms, took their measurements, made drawings, and took photographs. I then compared these specimens with similar artifacts that Kenneth Emory, Yosi Sinoto, Lloyd Soehren, and other museum archaeologists had recovered from sites throughout the Hawaiian Islands, as well as with Sinoto's recent finds from the Marquesas and Tahiti.

The fishhooks were especially interesting. They lacked the protruding lashing knob at the head of the shank that was characteristic of later Hawaiian hooks, and instead had a simple angled notch, or in one case, a double notch. Also, the shanks tended to curve gracefully inward, giving the hook a "bent shank" appearance. These were traits typical of fishhooks from early sites in the Marquesas Islands. There were also two small "two-piece" fishhook points of bone, with simple notched bases, similar in form to some points that I had recently excavated in the sand dune at Hālawa on Molokaʻi. Once again, these did not match classic Hawaiian forms of two-piece fishhooks, which were more massive and had protruding knobs at the base.

The stone adzes, though few in number, were also unlike the typical quadrangular Hawaiian form. One small adz had a "reversed triangular" cross section and tapered toward the butt end. Another was long and quite thin, well ground with an "incipient tang." These were forms that were known from early sites in southeastern Polynesia but were very rare in Hawaiʻi.

An especially intriguing artifact, found deep in Layer III, was a beautifully carved section of a large cone shell, one end of which had fourteen serrated "teeth" defined by parallel incisions. This was immediately recognizable as the blade of a coconut grater, which would have been lashed to a wooden seat or stool. A half coconut is rasped over the serrations, as shreds of the coconut meat fall into a waiting basket. These shreds are then squeezed through hibiscus bark to render coconut cream, a key ingredient in Polynesian cooking. Coconut graters (usually made of pearl shell) were common at early sites in Tahiti and the Marquesas, but none had previously been found in a Hawaiian excavation. Here, too, was another link with islands to the south.

I wrote up the results of my analysis and comparisons and gave my manuscript to Pearson before heading back to Philadelphia to continue my studies at Penn. Pearson combined this with Pietrusewsky's analysis of the burials and skeletal remains, and with his own description of the excavations and stratigraphy. Our article, "An Early Prehistoric Site at Bellows Beach, Waimanalo, Oahu, Hawaiian Islands," was published in October 1971.

In the years since the Bellows site was published, other archaeologists have attempted to reexcavate at the sand hill, or to find other early sites in the

Waimānalo region. Unfortunately, after we finished our dig in August 1967, the sand dune was allowed to undergo further erosion. More sand was mined, and by the time others attempted to locate the early Layers II and III again, these were completely gone. Other thin cultural strata were found, but these dated to a later time period. Later studies made by archaeologists under contract to the Air Force, after the U.S. military had begun to take seriously its obligation to protect cultural sites, found scattered archaeological deposits in various places on the base. But none has yielded the same kinds of early fishhooks or adzes that came out of the 1967 excavations.

For three decades, archaeologists have debated the significance of the Bellows sand dune site. While the initial radiocarbon dates included some that were very early, there was always the nagging problem of the inversions. Radiocarbon dating has undergone huge improvements in recent years. Back in 1967 most dating was done using an older method of gas proportional counting, which required large quantities of charcoal. Moreover, we lacked the means to identify what kind of wood was being dated. In some cases dates were obtained from the charcoal of large trees that had grown for many years before they were burned, or even from driftwood that had floated around the North Pacific for decades or more before being cast up by the surf on a Hawaiian beach, providing ready fuel for Hawaiian earth ovens. Thus the dates might be giving ages older than the actual earth-oven burning event. Today, we routinely identify wood that is to be dated by comparing a microscopic thin-section with reference materials of known species. Charcoal from long-lived trees, and from driftwood with a non-Hawaiian origin is rejected. And the dating itself is performed using the accelerator mass-spectrometry (AMS) method, which counts individual atoms of ^{14}C and is therefore more precise and accurate.

Tom Dye, a consulting archaeologist who had done some of the contract work for the U.S. Air Force at Bellows, decided to see if he could apply these new dating methods to the older materials originally excavated from the Bellows sand hill in 1967. He obtained permission from the Air Force, which now holds the 1967 collection, to date several pieces of charcoal as well as small worked fragments of pearl shell, from Layers II and III. Analyzing the results, Tom came to the conclusion that the oldest date Richard Pearson had received from Dr. Kigoshi back in 1967, the one suggesting settlement as early as the fourth century, was erroneous. The date itself might be valid, but it was probably from old wood, or driftwood. The new AMS results give a high probability that the initial Polynesian settlement at Bellows was established sometime between A.D. 1040 and 1219. The statistical nature of radiocarbon dating makes it impossible to tie down this range more precisely. However, this age range agrees well with dates for the earliest Polynesian sites in the Marquesas, in Mangareva, in the Austral Islands, in the Cook Islands, and on remote Rapa Nui.

It is impossible to say whether the charcoal-stained black-sand Layer III in the sand hill next to Waimānalo Stream, with its early forms of pearl-shell fishhooks, adzes, and coconut grater, derived from the first Polynesian settlement in Hawai'i. The old woman, hunched over with debilitating arthritis, who on her death was lovingly placed in her grave beneath the floor of the gravel-paved house, may or may not have sailed to Hawai'i on a double-hulled canoe like *Mahina-i-te-Pua*. Trying to find the very first place where ancient Polynesians set foot on Hawaiian shores is far more difficult than trying to find the proverbial needle in the haystack. In any case, we will never know for certain. But it is clear that this site dates to the same general time frame, the eleventh to thirteenth centuries, when Polynesians were discovering and establishing colonies across the vast eastern Pacific. The Bellows site allows us to draw back the curtain of time ever so slightly. I am grateful that fate allowed me to participate in the excavation at Bellows in 1967. And I owe much to Richard Pearson, who encouraged a precocious young student. It was a gift and a privilege to sift the sands of Waimānalo and help to open a window into the distant past of Hawai'i.

PART TWO · IN PELE'S ISLANDS

SIX · Flightless Ducks and Palm Forests

On Oʻahu's north shore, not far from Waimea Bay where the world's most daring surfers come in winter months to ride fifty-foot breakers, Haleʻiwa town nestles between sheltered Waialua Bay and the Koʻolau foothills. Ramshackle board-and-batten storefronts date back to the sugar plantation era, now a fading memory. Two centuries ago, however, the area was a thriving center of Hawaiian population, who farmed taro on the rich alluvial bottomlands and raised mullet and milkfish in the royal fishponds. Driving northeast out of Haleiwa, crossing the double-arched concrete bridge spanning the Anahulu River, you can catch a glimpse—beyond the jumble of surf shops and T-shirt stands—of Lokoʻea and ʻUkoʻa ponds. Visitors and locals alike drive past these vestiges of a bygone era in a few seconds. Yet these prosaic and ancient ponds are a treasure trove of deep Hawaiian history.

ʻUkoʻa Pond, by far the larger of the two Waialua fishponds, extends for half a mile between the base of the steep volcanic slopes and the gently undulating sand ridges over which the modern highway runs. The pond is a kind of swampy lagoon formed behind the dunes, themselves heaped up by the surf of countless winter storms. ʻUkoʻa is fed by freshwater springs that seep out of the layers of basalt lava; the subterranean waters originate in the rains that soak the distant Koʻolau Mountains. The pond first formed about seven thousand years ago, when sea level began to rise to its modern stand, allowing the coastal sand dunes to stabilize. The long oval pond became what geologists call a "depositional basin," a sink into which sediment and debris from the slopes surrounding the pond gradually accumulated. Year after year, grain by grain, silt and clay were washed into 'Uko'a, along with bits and pieces of vegetation, to settle down through the still waters, forming a layer of mud and peat cloaking the pond's floor. The sediments continued to be laid down over thousands of years, eventually building up to a thickness of more than twenty feet. Following the principle of stratification, the deepest sediments are the oldest, while those at the top are the most recent.

Silt and clay, along with odd bits of plant matter, were not the only materials washed into 'Uko'a Pond. Entombed along with them were literally billions of microscopic pollen grains that came from the trees, shrubs, and grasses that grew on the hillsides and coastal plains surrounding 'Uko'a. Each species of flowering plant produces its own distinctive pollen. These pollen grains can be identified by botanists based on their attributes of shape, size, and surface sculpturing, readily visible through a microscope. Pollen is, moreover, remarkably resistant to decomposition, especially when trapped in waterlogged, oxygen-deprived sediment.

The accumulated silts and clays at the bottom of 'Uko'a Pond encapsulate a priceless record of environmental change on northern O'ahu. The record begins seven thousand years ago when the first mud and peat began to accumulate at the bottom of the pond, continues through the arrival of Polynesians at the close of the first millennium A.D., and comes right up to modern times. Each centimeter of the pond's muddy sediment is a page in this record, a record in the form of billions of pollen grains—an indelible witness to the trees and plants that once grew in this part of the island. When one kind of tree or shrub grew in profusion on the lands surrounding 'Uko'a Pond, its pollen would accumulate in greater quantities than that of less common species. Similarly, when a particular kind of plant declined in frequency, owing to natural causes such as climate change, or after the arrival of people who cleared land for agriculture, its pollen declined in the pond's sediments.

Palynology, the science of identifying and interpreting pollen histories, allows us to read this record, unlocking the past environmental history contained in the mud of 'Uko'a, and other ponds and swamps like it. The first step is to extract a continuous core of sediment, usually done from a raft floating on the pond surface. A metal coring device is pushed down into the pond sediments one meter at a time. Each meter-long core segment is carefully labeled and taken back to the laboratory. There the core is sliced open and studied for changes in the nature of sedimentation (such as layers of gravel that might signal major storm events and flooding). The age of the core is determined by radiocarbon-dating vegetative debris or other organic matter at various points down the core column. Then the core is systematically sampled throughout its length for the all-important pollen content.

In order to see and identify the pollen, it must be "unmasked" from its clay and peat context, using a process known as acetolysis. The obscuring clay and organic matter are chemically burned away through a succession of caustic acid and base washes, with intervening rinses; finally, the pollen is separated out by centrifuging. When the acetolysis is complete, the palynologist deposits a drop of the purified liquid containing concentrated pollen on a microscope slide. He then begins the tedious work of identifying and counting a representative sample of the grains for each sample down the length of the core. It can take months in the lab to analyze a deep core, but the end product is a marvelous record of environmental history. This is typically presented as a graph showing the changing frequencies of various plant types over time, with the timescale provided by the radiocarbon dates from key samples up and down the core. Such pollen diagrams have been instrumental in tracing the history of vegetation change around the world, providing key evidence for climate change as well.

Steve Athens is an archaeologist who has used the science of palynology to unlock the story of Oʻahu's changing environment before and after the arrival of the Polynesians. Tall and lanky, with a Chaplinesque moustache, Athens first came to the islands in the early 1980s, working with the Bishop Museum on several archaeological projects. He later founded a nonprofit archaeological consulting firm, International Archaeological Research Institute, which has carried out some of the most innovative archaeological research in Hawaiʻi. In the early 1990s, Athens began to drill cores into the sediments of ʻUkoʻa Pond and other swampy places around Oʻahu Island, such as Kawainui Marsh inland of Kailua town. Athens realized that both ʻUkoʻa and Kawainui offered excellent potential for pollen analysis. An archaeologist by training, he could carry out the physical coring operations in these ponds and swamps, but enlisted the help of palynologist Jerome Ward to identify the pollen grains extracted from the cores. What they found revolutionized our ideas about Oʻahu's environment at the time the first Polynesians stepped ashore on the beaches of Waimānalo and elsewhere.

. . .

Among the rarer trees found only occasionally in the remaining native forests of Hawaiʻi is a majestic palm, reaching up thirty feet or more, topped by a great crown of fan-shaped leaves. The Hawaiians called these palms *loulu*. They used their leaves for thatch, and to weave hats and fans. The seeds can be eaten raw, and supposedly taste much like coconut. Botanists classify the *loulu* as members of the genus *Pritchardia*, widely distributed in the Pacific islands. There are as many as fourteen species in the Hawaiian archipelago, each typically confined to a single island. Most *loulu* species in Hawai'i are now extinct in the wild, or very nearly so.

When Jerome Ward began to identify and count the pollen grains in the samples that Steve Athens had pulled up from the muddy depths of 'Uko'a and Kawainui, he was surprised to find an unusually high percentage of *Pritchardia* pollen. This was especially the case for the deeper, older sediments. In the core from Kawainui Marsh, for example, sediments below about twelve feet deep had anywhere between 20 percent and 40 percent of their pollen grains coming from *Pritchardia* palms. The 'Uko'a Pond core showed a similar pattern, although in this case there was essentially zero palm pollen in the upper four feet or so of the twenty-foot-deep core; in the samples taken from below four feet, *Pritchardia* pollen was abundant, right to the base of the swamp.

Athens and Ward were surprised by these results. Since *loulu* palms are so rare in surviving Hawaiian forests, they expected that their pollen would be a minor component of the core samples. Instead, it appeared that for most of the time that pollen was being deposited into the 'Uko'a and Kawainui sediments, these majestic palms had been very common in the lowland terrain surrounding these depositional basins. At some point, however, *Pritchardia* pollen became increasingly scarce, declining rapidly, to become virtually nonexistent in the pollen record. These changes occurred in both the Kawainui and the 'Uko'a sediments, though at slightly different depths, reflecting different rates of sedimentation. When Athens and Ward looked at the radiocarbon dates they had obtained from the peat samples, it became clear that the sudden decrease in palm pollen in both sets of cores occurred at the same time. The radiocarbon date for the change in 'Uko'a was between A.D. 922 and 1152, while at Kawainui the change began slightly below a radiocarbon sample dated between 1219 and 1403. The actual timing of palm decline was essentially the same in these two areas of O'ahu, some twenty-seven miles apart, between about A.D. 1000 and 1200. Something major had happened on O'ahu during the eleventh to thirteenth centuries that had caused a vast lowland palm forest to decline.

Declining *Pritchardia* pollen was not the only dramatic change in the upper parts of the cores. At 'Uko'a, several other kinds of native trees and shrubs, such as the *koa* tree *(Acacia koa)*, prized for its beautiful hardwood, and a wiry shrub called *'a'ali'i* in Hawaiian *(Dodonaea viscosa)*, also showed declines in their pollen frequencies. More puzzling to palynologist Ward was a steep drop-off in yet another kind of pollen grain, one that did not match any of the pollens in Ward's extensive reference collection. From its shape and features, Ward knew that it had to be a member of the legume family, but it didn't quite fit any of the known Hawaiian species of legumes. In their first reports, Ward and Athens just called it "Unknown Tricolporate Type 1," after its distinctive features. Given that the pollen grains of Unknown Tricolporate Type 1 accounted for as much as 30 percent of the total pollen content in the deeper layers of Kawainui Marsh and was also common at ʻUkoʻa, it must have once been prolific. And yet the mystery legume could not be identified, even though generations of botanists had collected plants all over Oʻahu for nearly two hundred years. This was indeed puzzling.

In 1992, just as Athens and Ward were publishing the first results of their coring project, a field team from the National Tropical Botanical Garden landed on the uninhabited island of Kahoʻolawe to carry out a survey of its plant life. Using climbing gear, they scaled the steep cliffs of a narrow rock pinnacle at the base of the island's sea cliffs. This pinnacle was so sheer that not even the introduced goats that had decimated so much of Kahoʻolawe's native vegetation could scale them. Reaching the small summit of the pinnacle, the botanists found two individuals of a small shrub never before seen in the Hawaiian Islands. Specimens including dried flowers from the leaf litter under the plants were collected. Shortly thereafter this was described in the scientific literature as not just a new species, but also a genus new to the botanical world, *Kanaloa kahoolaweensis*. The two little shrubs that had escaped the marauding goats on Kahoʻolawe were the last surviving members of this species. Jerome Ward, hearing of this exciting new discovery and knowing that the newly named plant was a legume, requested a sample of the pollen taken from the anthers of its dried flowers. He was amazed to find that the features of *Kanaloa kahoolaweensis* pollen were indistinguishable from those of Unknown Tricolporate Type 1. The mystery of the unknown pollen grains had been solved, thanks to those two surviving plants atop that rocky pinnacle. Sadly, one of the plants has since died, but several clones are being cultivated in the National Tropical Botanical Garden on Kauaʻi.

Out of Ward's painstaking work of counting thousands of pollen grains under his microscope, a picture of the vegetation that had once surrounded the lowlands of ʻUkoʻa and Kawainui on Oʻahu has begun to emerge. Dryland forests towered over by tall *loulu* palms, interspersed with *koa* and other hardwood trees, had shaded lower spreading shrubs such as *ʻaʻaliʻi* and the now nearly extinct *Kanaloa*. Between 1000 and 1200, give or take a century or two, all this had changed dramatically, so dramatically that the *loulu* palms became rarities found only in the upland forests. The *Kanaloa* itself was completely wiped off Oʻahu Island. What had happened to precipitate such drastic changes?

The timing of these changes was provocative, since it correlated closely with archaeological evidence of the Polynesians' arrival on Oʻahu. Humans are often the precipitators of major disruptions to island ecologies, especially islands that have been isolated for millions of years, whose plants and animals are vulnerable thanks to the gradual evolutionary loss of defense mechanisms. But the coincidence of timing was only what scientists call a "correlation." It did not demonstrate causation. To link the decline in the palms and other native plants to the arrival of humans, more direct evidence would be required.

Such evidence came in the form of another microscopic component of the cores, tiny specks of charcoal derived from the smoke and ash of ancient fires. When a fire burns in the vicinity of a swamp or pond, microscopic ash and charcoal get wafted up by the wind and then settle out over the water, or are washed in with the clay and silts after rainstorms. Either way, these carbon particles settle to the pond's bottom along with the pollen grains, to become part of the site's environmental record. When palynologists plot their results on a pollen diagram, they usually also plot the frequency of charcoal particles, as a proxy measure of the degree of fire and burning in the area over time.

In the ʻUkoʻa core, there is a total absence of charcoal particles below about two meters in depth. This is not surprising, because natural fires would be extremely rare on an island like Oʻahu. With a subtropical, humid climate and lacking active volcanoes, Oʻahu has no natural sources of ignition to set the forest on fire. But in the samples higher than two meters, coinciding precisely with the steep declines in *loulu* palm, *Kanaloa*, and other pollens, microscopic charcoal becomes abundant. Whereas there had been no fires on the landscape surrounding ʻUkoʻa for some six thousand years, suddenly about A.D. 1000 fires became frequent. What had changed? The answer was clear: Polynesians, farmers who used fire to clear the land for their gardens, had arrived on the island.

The destruction of the lowland palm forests at ʻUkoʻa and Kawainui, and probably across much of Oʻahu, did not occur all at once. Although the declines in pollen frequency are steep, they probably occurred over two to three centuries. We are not talking about wholesale burning of the landscape. The first Hawaiians were not eco-vandals intent on destroying the forests of their new island homes. But they were farmers, and farmers everywhere must clear land if they are to sustain themselves. As their population grew and their descendants spread out over the fertile lowland slopes and valleys of Oʻahu, more and more land was cleared of native palm forest and converted to productive gardens. The transformation of O'ahu's lowland ecology was inevitable, given a growing human population with an agricultural subsistence base. And along with the palm forests, other unique creatures disappeared. These creatures were birds, large flightless birds the likes of which have not been seen for close to a thousand years. They were products of millions of years of island evolution, vulnerable once the curtain of isolation was broken. Their bones first turned up in large numbers in archaeological excavations for a deep-draft harbor on the 'Ewa Plain of leeward O'ahu, and startled the world of science.

·   ·   ·

The first hint that the Hawaiian Islands had once been the habitat of large flightless birds came in 1971, as an accidental discovery by an amateur naturalist named Joan Aidem. Walking along the sand dunes at Mo'omomi on Moloka'i Island's desertlike western end, Aidem saw fossil bones eroding out of a layer of cemented, or "lithified," sand. Aidem collected a skull and some loose bones and took them to Alan Ziegler, zoologist at Honolulu's Bishop Museum. Ziegler was so surprised by Aidem's bones that he immediately returned with her to Mo'omomi to see if he could find the remainder of the skeleton. Digging into the hardened sand, which had been blown inland during the late Pleistocene (the last ice age), Ziegler found most of the articulated skeleton that belonged to the skull. Extinct *Amastra* land snail shells found along with the bones were later dated to about 25,150 years ago, fixing the time of the bird's death.

Ziegler's office and workroom were next to my archaeology lab on the second floor of the museum's Bishop Hall, a stately old edifice of hewn basalt that had originally been the home of the Kamehameha Schools. I well remember Ziegler, with his crew cut and deeply lined face, talking excitedly about this new fossil find from Moloka'i. "Pat," he said, "this bird appears to be a goose," drawing out the word "goose" in his distinctive southern drawl. "But it just doesn't look like any other goose we've ever seen. The darn things got tiny wing bones, do ya see? No way this thing could have gotten off the ground. But just look at those massive leg bones!"

Ziegler was a first-rate zoologist, but his specialty was mammals, not birds. He knew that this remarkable new find needed to go to an appropriate specialist in the taxonomy (classification) of birds. The specimens were carefully packed and mailed off to the most respected expert on birds in the United States, Alexander Wetmore, at the Smithsonian Institution in Washington, D.C. Wetmore, assisted by his protégé Storrs Olson (Olson and his collaborator, Helen James, have become the world's experts on Hawaiian fossil birds), examined the Mo'omomi bones and realized that this was not just a new species, but something so distinctive that it deserved to be in its own new genus. They gave the name *Thambetochen* to the new genus, which translates as "astonishing goose." And indeed, it was astonishing: a stocky, flightless gooselike bird, standing three feet tall (and weighing as much as sixteen pounds) as it lumbered about through the dryland forests of Moloka'i during the last ice age. The bird's jawbones (technically, its premaxillae and dentaries) were massive, with a series of blunt projections that resembled teeth. Obviously such jaws had evolved to crush seeds and fruits; the bird must have been a grazer or browser on the forest floor.

The discovery of *Thambetochen chauliodous* in the Mo'omomi dunes set off a flurry of interest in the possibility that other fossils of extinct Hawaiian birdlife might be discovered. Storrs Olson led a field trip to Kaua'i Island, geologically the oldest of the main islands, where he hoped that Pleistocene sand dunes of similar age to those on Moloka'i might yield bird bones. His hunch paid off with more fossil bones at the Makaweli dunes along the island's southwestern shore. Then, in 1976, as Olson was just getting his program of fossil-hunting organized, Bishop Museum archaeologists hit a veritable gold mine of bird bones at Barber's Point, on O'ahu's 'Ewa Plain.

At least twice during the Pleistocene, first about 300,000 years ago and then again about 120,000 years ago, sea levels rose higher than they are at present, during periods when the Earth's glacial ice caps melted. The higher seas drowned the coastal plains of leeward O'ahu, allowing extensive coral reefs to grow in the shallow water. When the waters retreated again, they left a vast plain of dead coral limestone. Over time, the leaching action of fresh rainwater on the porous limestone left pockets, sinkholes, and even underground caverns. These sinkholes and caverns were perfect depositional basins and traps, accumulating sediment and debris much as 'Uko'a Pond had accumulated mud, only in this case the sediments were mostly dry. Consisting of calcium carbonate, the gravelly sediments in the sinkholes provided a protective environment for any bones that happened to accumulate in them.

In the mid 1970s, the U.S. Army Corps of Engineers began to draw up plans for a deep-draft boat harbor in the vicinity of Barber's Point, to relieve overcrowding in Honolulu harbor. The corps contacted the Bishop Museum, requesting an archaeological survey of the area proposed for the new harbor. The survey at Barber's Point was directed by Aki Sinoto, son of the museum's senior Polynesian scholar, Yosihiko Sinoto. Aki Sinoto uncovered rich fossil deposits in the Barber's Point sinkholes. Some of the sinkholes had been used by the ancient Hawaiians for habitation or temporary shelter, while others had simple stone structures built on the karst plain near their entrances. Sinoto and his crew dug test pits in some of these sinks, finding Hawaiian artifacts in the upper sediments; one site had a charcoal-filled hearth. But they also found large quantities of bird bones, including new kinds of *Thambetochen*, along with seabirds such as the dark-rumped petrel *(Pterodroma phaeopygia)* and Newell's shearwater *(Puffinus newelli)*. Whereas the *Thambetochen* geese were long extinct, the seabirds still nest on Oʻahu's small offshore islets, and in the protected northwestern Hawaiian islands. Large numbers of such seabirds were historically unknown on the main islands, yet here was the evidence that they had once been so plentiful that their bones accumulated in dense numbers in these crevices and sinks.

Many of the bones were from sediments that underlay and predated the Hawaiian use of the sinkholes. Some of the bones, however, were associated with the period of Hawaiian occupation. In fact, the densest concentrations of bones seemed to be just prior to the use of the sinkholes as habitations. The question posed itself: did the arrival of Polynesians on the ʻEwa Plain have something to do with the decline in bird populations, and the extinction of the "astonishing goose"?

Storrs Olson and his wife, Helen James, visited the Bishop Museum's zoology department, working with Alan Ziegler to sort and classify the finds in a preliminary way. We were all sitting on the second-floor balcony of Bishop Hall late one afternoon after the official workday was over, sipping some beers from Ziegler's cooler and mulling over the significance of these remarkable new finds. My interest was piqued, and I wanted to explore the Barber's Point sites myself—not so much for the bird fossils, about which I had no expertise and which were best left to Olson and James—rather, I was fascinated by the fact that the sediments were loaded with thousands and thousands of tiny shells of native land snails. Hawaiian snails were something I did know a bit about, having apprenticed early in my budding scientific career with Yoshio Kondo, the Bishop Museum's malacologist (specialist in mollusks). Sorting through Sinoto's samples, I realized that many of the snail shells were from species that were long extinct on Oʻahu. I wondered whether further study of these little snails might give us some clues about what happened on the ʻEwa Plain. I invited Carl Christensen, another of Kondo's acolytes who had gone to get his doctorate in malacology and now worked at the museum, to join me in this endeavor.

By now the corps was proceeding rapidly with its plans for the deep-draft harbor. For the next phase of the archaeological "salvage" work they chose a private consulting firm, Archaeological Research Center Hawai'i, headed by Hal Hammatt, a young archaeologist who had come to the islands from Washington State. Fortunately, Hammatt was amenable to having us take sediment samples for our snail research from his new excavations in the Barber's Point sinkholes. The methods we used were similar in many respects to those used in pollen analysis. Since the Barber's Point sinks were dry and shallow, we didn't need to take cores, but simply dug out carefully controlled samples in vertical columns from the sidewalls of Hammatt's larger excavations. Because the Hawaiian snail shells can be quite small (less than 1 millimeter, or 0.04 of an inch in diameter), we then ran the samples through extremely fine mesh, to get rid of the dirt and collect the snails. Christensen then sorted these into the various genera and species. We plotted the frequencies on a graph for each site, much like the frequencies of changing pollen.

Our snail frequency diagrams, especially the one from a particularly large and deep sinkhole labeled Site B6–78, revealed a fascinating pattern. The deeper deposits contained the shells of thousands of native species of snails, little creatures that had flourished under a canopy of native dryland vegetation. As one traced the frequencies higher up the column, approaching the uppermost levels, the native snails began to decline precipitously, much as the *Pritchardia* pollen had declined in 'Uko'a Pond. At the same time, a new kind of snail appeared, with an elongated little shell, called *Lamellaxis gracilis*. This snail we knew to have been carried as a stowaway by the Polynesians on their colonizing voyages, probably adhering to soil wrapped up with seedling crop plants. The same upper deposits yielded two other unmistakable signs of human arrival: the bones of the Polynesian-introduced rat, and the tiny toothed mandibles of geckos and skinks, other stowaways on voyaging canoes.

The combined evidence from the Barber's Point sinkholes, the abundant bird bones along with the snail shells, rat bones, and gecko jaws, supported the interpretation that the rich biodiversity that had once characterized the 'Ewa Plain had persisted until the arrival of Polynesians. Initial radiocarbon dates from the Barber's Point sites pointed to perhaps A.D. 1200 as the time Polynesians began to make use of this dry, leeward zone. This richly diverse ecosystem had collapsed quite rapidly, evidenced both by the precipitous decline in native land snails and by the accumulation of the bird bones, including the last of the *Thambetochen*.

Since the original discoveries on the 'Ewa Plain, a great deal of additional research has taken place in the vicinity of Barber's Point. Olson and James have found that the *Thambetochen* is not actually a true goose, but rather more closely related to ducks. (Geese and ducks are part of the same Anatidae family of birds.) In addition to *Thambetochen,* Olson and James have named two more genera of these flightless ducks, called *Ptaiochen* and *Chelychenlynechen.* The latter genus, with its tongue-twisting name meaning "turtle-jawed goose," was a Kaua'i Island relative of *Thambetochen.* Its name derives from the shape of its jaw, which looks like that of a turtle! These scientific names, essential for science as they may be, are hard for even specialists to remember and pronounce, so Olson and James coined a Hawaiian name for the family of flightless ducks: *moa nalo. Moa* is the pan-Polynesian name for chicken, also applied by the Maori of New Zealand to their own great flightless birds. *Nalo* means lost, forgotten. Hence *moa nalo,* the forgotten great flightless birds that once roamed Hawai'i's dryland forests.

Steve Athens has continued field research in the Barber's Point area. Picking up where Carl Christensen and I left off with our discovery of rat bones in the upper levels of the sinkholes, Athens applied the advanced method of AMS radiocarbon dating to directly date the little rat bones, a technique not available to us back in 1976. With AMS, it is possible to date minute quantities of bone and to get dates with tighter ranges than with the older gas-counting radiocarbon method. Athens found that the rat bones, which appear in significant numbers in the upper levels of the sinkholes, date to between A.D. 900 and 1200. This not only helps to tie down the chronology of human arrival on O'ahu (since the rats were transported on Polynesian voyaging canoes), but also raises a new hypothesis about the causes underlying the rapid collapse of the native dryland ecosystem on the 'Ewa Plain.

Rats are prolific breeders. It is likely that once the first rats were ashore on O'ahu their population exploded exponentially. The seeds of *loulu* palms are edible, and other native Hawaiian trees, shrubs, and herbs probably offered a tasty range of seeds, fruits, and young shoots and seedlings to the hungry rats. It is a sobering thought that the delicate and vulnerable Hawaiian lowland forests may have been subject to a tidal wave of exploding rat populations, hundreds of thousands of little jaws munching away at the defenseless vegetation. While it is plausible that rats contributed to the collapse of the lowland ecosystems on O'ahu and other islands, it would be unwise to jump to the conclusion that the rats provide the exclusive explanation. After all, the sudden appearance of all those charcoal particles in the 'Uko'a and Kawainui cores cannot be ignored. They remind us that human-ignited fires and forest clearance for gardens also had to be a big part of the story. No doubt, like most ecological phenomena, the reality of environmental change on O'ahu in the centuries following Polynesian arrival was complex, with several causal factors interacting. Nonetheless, one thing seems indisputable.

Had Polynesians never arrived in their double-hulled canoes, carrying not only the precious seedling crops to establish their new gardens but also the stowaway rats, Captain Cook would have been greeted only by flocks of large, lumbering flightless ducks!

· · ·

The world of ancient Hawai'i that has emerged from the collaborative, interdisciplinary work of archaeologists and natural scientists is quite astounding. It is an island world entirely different from the Hawai'i recorded in historical times. That ancient world was populated by stocky, flightless *moa nalo*, crunching palm seeds in their toothed beaks. It was a world in which the skies teemed with shearwaters and petrels, nesting on the cliffs high above Waimānalo. *Loulu* palm forests carpeted the lowlands from Waikīkī to 'Ewa. Under their shade the now-extinct *Kanaloa* shrubs and countless other native herbs and grasses provided a richly textured habitat for thousands species of tiny insects and land snails. A wholly unique, living world had evolved over millions of years on Earth's most remote archipelago.

Into this world, which had remained so isolated, so remote, and hence so fragile and vulnerable, stepped the first people. Hauling their weather-beaten double-hulled canoe up onto the beach, whether it was at Waimānalo, or Kualoa, or perhaps the much-later-to-be-famed Waikīkī, these explorers from islands to the south had decisively pierced the curtain of isolation wrapped around Hawai'i for at least sixty million years.

We can only speculate about how the voyagers reacted, wide-eyed, to the new landscape that greeted them. Much of what they saw was familiar: the seabirds were well known, as were most of the kinds of fish on the reefs and in the lagoons. Many of the plants looked familiar, and were at least related to plants they already knew and had names for in their southern island homelands. But others were new and untested. Were their fruits edible, or poisonous? Might their leaves or bark offer as yet unknown medicinal properties? Was the timber good for building houses, or for the hulls of canoes? All this must have run through the minds of the *tufunga* (experts). This was a world they would have to explore carefully, and test, and come to know intimately.

And then there were the lumbering, flightless *moa nalo*, who had never before known a predator. To the Polynesians, weary after a long voyage and in need of immediate food sources while they established their gardens and guarded the precious founding stocks of pigs, dogs, and fowls, the *moa nalo* were an irresistible resource. No doubt the defenseless *moa nalo* lacked any innate prey response, their ancestors having never faced the challenge of being hunted. Most likely they just stood there, their avian brains unable to process this new two-legged creature that walked up to them with spear or club in hand. The bones of *moa nalo* are so scarce in archaeological deposits that it would seem they disappeared very fast, shockingly fast.

Other changes took much longer. The *Pritchardia* forests of the Oʻahu lowlands did not disappear overnight. But two to three centuries after the first double-hulled canoes had arrived not only with their human crews but also with their portmanteau cargo of stowaway rats, dogs, pigs, chickens, geckos, weeds, crop plants, and seedling trees, the landscape of Oʻahu—and the other islands of Hawaiʻi—had gradually been transformed. This was especially so in the lowlands, where the human population was concentrated, where they made their gardens and their homes. The forests had been cleared and pushed back to the lower mountain slopes. The landscape became one of managed organization, with fields and boundaries, paths and houses, temples and fishponds. *Homo sapiens* had made its mark on the land, as people have done on every other piece of real estate across the planet. The pristine world of prehuman Hawaiʻi was gone forever. Only modern science can resurrect for us a memory of it, in fossil fragments from the sinkholes of ʻEwa, and in pollen grains from the silty depths of ʻUkoʻa and Kawainui.

SEVEN · Voyaging Chiefs from Kahiki

Perched a thousand feet above sea level on the leeward flank of Haleakalā volcano, in the windswept district of Kahikinui, Maui, stands the ruin of St. Ynez, a small Catholic church built about 1836. The volcanic stones, held together with lime mortar painstakingly made from coral, define a small rectangular sanctuary. One can still discern the impressions in the plaster where narrow wooden windows and an entryway were once set. St. Ynez sits upon a small rise, Puʻu Ānuenue (Hill of the Rainbow). From this vantage point I have many times admired the late afternoon rainbows that adorn the skies eastward over Kaupō. Local tradition has it that the little church was built upon the foundation of an ancient temple; temples were often constructed upon such promontories with commanding views.

In the early 1990s, a group of Native Hawaiians calling themselves Ka ʻOhana o Kahikinui (the family of Kahikinui) cut away the thick *koa haole* brush engulfing the ruins of St. Ynez. They put up a simple plywood roof to make the sanctuary a usable space for people to gather once again. Mo Moler, Aimoku and Lehua Pali, Donna and Walter Simpson, Kawaipiʻilani, and others of that grassroots group had come to Kahikinui to make a claim on their ancestral lands. Kahikinui is part of the Hawaiian Homelands established by the U.S. Congress at the urging of Prince Jonah Kūhiʻō Kalanianaʻole in 1920. Unfortunately, many of the Hawaiian Homelands, including Kahikinui, had never been turned over to Hawaiians. Instead, Kahikinui had for decades been leased to Haole cattle ranchers. The members of Ka ʻOhana o Kahikinui were determined to end this practice. In taking over the ruin of St. Ynez as a meeting place, they firmly established their presence on the ʻāina.

On a chilly morning in early February 1997, I pulled my jeep over at the St. Ynez church site, by now outfitted with its new roof and converted into a cozy shelter from the incessant winds. Two years earlier I had started a long-term archaeological study of Kahikinui. This morning, Donna Simpson of Ka ʻOhana had invited me to join some of them in a ceremony rededicating the stone-walled ruin of the priest's house adjacent to the St. Ynez church. For this she had asked the assistance of the respected Reverend Kawika Kaʻalakea, a pastor of the Hawaiian Congregational Church and a *kupuna* (elder). Reverend Kaʻalakea had been raised as a *punahele* (favored grandchild) of his *tūtū wahine* (grandmother), who was herself a *kamaʻāina*, native of the neighboring district of Kaupō. About 1890 the last inhabitants of Kahikinui left the area and moved to Kaupō, abandoning little St. Ynez and joining the congregation of St. Joseph's Church. Hence the people of Kaupō had become the custodians of traditional lore for this vast leeward side of southeastern Maui.

Donna and I greeted each other with the traditional Hawaiian *honi* (pressing of noses and sharing of breath); she then introduced me to Reverend Kaʻalakea. We were a bit early for the ceremony; the pastor and I started to pass the time chatting. He spoke a thick pidgin English, heavily laced with Hawaiian words, the kind of speech I remember well from my childhood growing up in the islands but is rarely heard anymore. Many of his phrases were more Hawaiian than pidgin; naturally, he was fluent in the language of his ancestors. He politely asked what I was doing out here in this rarely visited landscape. I explained the nature of our research project, and told him about some of the ancient sites we had been mapping. "*Maikaʻi,*" he said, "*maikaʻi.*" Good that you are working to save the knowledge of those ancient times. Looking intently into my eyes, he cautioned me that Kahikinui could be a dangerous place, a place of wandering spirits, *ʻuhane*, who might "whistle at us," he said, and throw us off our work. Ignore the *ʻuhane*, he advised, they are just curious about you Haole folks. Do your work, leave the land each day as you found it, and no harm will come to you. I appreciated his words.

As we chatted, standing on the edge of the hill with its magnificent views over the coastline, Reverend Kaʻalakea gestured off in the distance toward a grassy flat that we could barely discern between patches of dark lava rock. It was about three miles away as the crow flies. I had several times negotiated the tortuous four-wheel track that passed for a road and provided access to that area; it is just about the worst jeep road that I know of in the islands. But the horrendous drive is worth

FIGURE 5.
The *pānānā* (sighting wall) at Hanamauloa, Kahikinui, Maui Island. Photo by P. V. Kirch.

it, for when one arrives at the coast there are many well-preserved archaeological sites, including fishing shrines, small *heiau*, and clusters of house foundations. One site is distinctive, standing out by itself on the grassy plain. It consists of a particularly well constructed lava rock wall, the stones set without mortar, about twenty-six feet long and six feet high. I have often marveled at the craftsmanship of the stonemason who directed the work of selecting and setting the rough ʻaʻā stones so that they locked together perfectly. In the exact middle of the wall is a notch, also expertly constructed, so that from a distance the wall gives one the impression of a gun sight (see figure 5). A curious structure, it is clearly not a habitation site or a windbreak, for its long axis is east-west, in line with the incessant prevailing winds. Nor is it obviously a *heiau*, or a canoe shed. The site is one of a kind, unique.

Reverend Kaʻalakea gestured again toward the distant grassy flat, calling the place by its proper name, Hanamauloa. He asked whether I had noticed this notched wall, his weather-beaten eyebrows rising subtly in a classic Polynesian gesture. This was the first time I had heard anyone give a name to this forsaken place, which in itself was new and valuable information. His demeanor told me that he was about to share something important, some knowledge that he had, no doubt, learned from his grandmother in Kaupō many long years ago. Why he suddenly decided to share this with me, a Haole archaeologist he had only just met, I do not know. Maybe he trusted me.

Straightening his back and gesturing toward the soaring ridgeline of Haleakalā, the majestic "House of the Sun" that looms over Kahikinui, Reverend Kaʻalakea recounted the tradition that here the first people had arrived from Kahiki, the ancestral lands to the south, beyond the horizon. The place had been nameless before voyagers "awakened the land" *(e ala ka ʻāina)*, calling it Kahikinui. Literally, this translates as "Great Kahiki" or "Great Tahiti." Coincidentally, Maui bears the same shape as the island of Tahiti, also composed of two separate volcanoes linked by an isthmus. The larger part of Tahiti is called Tahiti Nui. One can well imagine that when voyagers from Tahiti arrived off Maui and saw that it was identical to the great island they had left, they appropriately bestowed this name on eastern Maui, capped by its great mountain Haleakalā.

Pointing again toward the flat of Hanamauloa, where we could barely make out the notched wall as a dark blotch in the light brown grass, Reverend Kaʻalakea told me a startling fact: that the wall pointed out the sea path to Kahiki, the ancestral homeland. He called the wall a *pānānā*, repeating it several times so as to emphasize this point: "You *nānā*, see, go Kahiki." As he said these words his hand flowed from his eyes out toward the horizon, in the gesture of a *hula* dancer. I was startled by his words. I can still sharply recall the mental image of that wise *kupuna*, gesturing and repeating *"nānā, ala i Kahiki."*

I was familiar with the Hawaiian word *pā*, this being the general term for "wall" or "enclosure," a word that can be traced back to Proto-Polynesian language. But Reverend Kaʻalakea made it clear, through his evocative hand gesture, that *pānānā* must mean, in one sense at least, to "look, sight, or take a bearing from the wall." "You *nānā*, see, go Kahiki." *Pānānā*. A wall for sighting or seeing, I surmised. In all my years of doing archaeology in Hawaiʻi and elsewhere in Polynesia I had never heard of such a thing. But I had no reason to doubt what Reverend Kaʻalakea was sharing with me.

At this point our conversation was interrupted by the arrival of the remaining guests. The reverend slipped a simple white robe over his *palaka* shirt and blue jeans, instantly transforming himself from an aged *kupuna* to a stately minister of the church. From his car he procured his Hawaiian Bible, wooden calabash with water, and several leaves of the *kī* plant. After the prayers and a couple of hymns sung in Hawaiian, Reverend Kaʻalakea walked around St. Ynez dipping his *kī* leaves into the calabash, shaking the sacred water indiscriminately onto stone walls and assembled company. I was delighted that my Berkeley students who had joined me could share this experience. Later we talked about the fascinating incorporation of the ancient Polynesian *kī* plant, often found on *heiau* sites, into the Hawaiian Congregational religious practice.

When we got back to our field house later that day, I immediately delved into my copy of the definitive Hawaiian dictionary compiled by Mary Kawena Pukui with the assistance of Polynesian linguist Sam Elbert. I was astounded to find an entry for the compound word *pānānā,* with the definition "compass; pilot; to row here and there irregularly." Later, when I was able to consult the rare 1865 Andrews's dictionary, the first published dictionary of the Hawaiian language, I again found the term, with this entry: "PA-NA-NA, s. [substantive noun]. *Pa* and *nana,* to look. A compass, especially a mariner's compass. 2. A pilot; one who directs the sailing of a vessel; he mea kuhikuhi holomoku."

Now, this was a curious thing indeed: why would Native Hawaiians, upon first being introduced to Western seafaring and voyaging concepts in the late eighteenth and early nineteenth centuries, call the mariner's magnetic compass by a combination of words that meant "sighting wall"? Unless—unless indeed—they already had an indigenous tradition of walls constructed expressly as sighting devices to provide orientation for cardinal, or sailing, directions. If they had such a tradition of building "sighting walls" (even though I had never read of this in the literature on Hawaiian anthropology and archaeology), it would of course make sense for Hawaiians to have called a Western magnetic compass by their word for such navigational walls. This hypothesis gained credence when I consulted nineteenth-century Hawaiian texts on astronomical knowledge. There again I found the term *pānānā* used, for example in the nineteenth-century sage Kepelino's notes about the sun, in which he refers to *ke panana kahiko a Hawaii,* "the old Hawaiian compass."

I have returned to study the *pānānā* wall at Hanamauloa many times since that fateful encounter with Reverend Kaʻalakea, including a field trip in the company of Professor Clive Ruggles, one of the world's leading archaeo-astronomers, who specializes in the ways in which ancient peoples observed the heavens. Using his theodolite, Clive and I determined that the wall itself is oriented along cardinal east-west directions. About 160 feet seaward of the wall there is a stone cairn, what the Hawaiians call an *ahu*, constructed in the same precise masonry style as the wall itself. When one stands slightly inland of the wall and moves so that the *ahu* is visible through the notch in the wall (as both structures seemed designed to enable), the orientation is 189 degrees, or just slightly to the west of true south. About A.D. 1400, this would have been the precise orientation of the Southern Cross when the main axis of the Cross was at its vertical position, just after sunset. The Hawaiians called the Southern Cross Hōkū-keʻa and knew that it pointed the way back to Kahiki. Thus, whoever built the *pānānā* at Hanamauloa designed it so that the alignment between the sighting notch and the *ahu* gave the precise orientation of Hōkū-keʻa when the star cluster hung vertically in the sky.

Walking along the rugged cliff-bound Hanamauloa coastline with Clive back toward our campsite at Ninialiʻi, I commented that here in Kahikinui we were surrounded by place-names that hark back to the ancient homeland of Hawaiki. Off to the southeast loomed the distant peaks of Mauna Kea and Mauna Loa on Hawaiʻi Island, named in honor of the ancient homeland itself. The point of Hawaiʻi closest to us across the ʻAlenuihāhā Channel is called ʻUpolu, in memory of the large island of ʻUpolu in Samoa. If we were to walk eastward along the Maui coastline for a day or two, I told Clive, we would arrive at the bay of Hāmoa, the Hawaiian equivalent of Samoa. Or if we were to walk a shorter distance to the west, we would cross over into the district of Honuaʻula, the Hawaiian version of a truly ancient place name, Fenuakula, a fabled island whose birds provided the precious red feathers that adorn chiefly mats and headdresses. And looking directly out to sea toward the southwest, there in front of us stood the waterless island of Kahoʻolawe, whose southern point is called Lae o Ke Ala i Kahiki (the Road to Tahiti). Tradition has it that from this cape were launched voyages to the ancestral homeland of Kahiki. Thus our *pānānā*, here on the shores of Kahikinui, Great Tahiti, was surrounded on all sides by places that had been named in honor of Polynesian ancestral homelands.

No hieroglyphic inscription, no memorial plaque exists to tell us who built the sighting wall of Hanamauloa. Nor can we directly date the era of its construction. But I have no reason to doubt that it was constructed by someone who had undertaken a voyage from or to Kahiki. It seems to be a visual memory or marker of such a voyage, retaining in stone the knowledge that beneath the shaft of the Southern Cross, Hōkū-keʻa, lay the path to Kahiki, *ke ala i Kahiki*.

Hawaiian traditions do contain a clue to who may have built, or been memorialized by, the *pānānā* of Hanamauloa. Buried in the yellowing pages of *Fornander's*

Collection of Hawaiian Antiquities, published in 1916 by the Bernice Pauahi Bishop Museum, one finds the *moʻolelo* of Moʻikeha. Reading through this treasure of Hawaiian lore, some years after my encounter with Reverend Kaʻalakea, I came across this passage: "Laamaikahiki lived in Kauai for a time, when he moved over to Kahikinui in Maui. This place was named in honor of Laamaikahiki. As the place was too windy, Laamaikahiki left it and sailed for the west coast of the island of Kahoolawe, where he lived until he finally left for Tahiti. It is said that because Laamaikahiki lived on Kahoolawe, and set sail from that island, was the reason why the ocean to the west of Kahoolawe is called 'the road to Tahiti.'"

Laʻamaikahiki—one of the great voyaging chiefs of Hawaiian tradition. Was it he who directed the careful placement of the stones in the *pānānā* before he left Kahikinui for Kahoʻolawe, thence to return to his homeland in Great Tahiti? Or was the *pānānā* built later, years after his departure, after the great interarchipelago voyages had ceased, by someone who felt fervently that the knowledge of the star-path to Kahiki should not be lost? We shall never know the answer to that question. Nonetheless, the link between Kahikinui and Laʻamaikahiki was preserved in tradition, passed down from generation to generation. Laʻamaikahiki's story remains one of the greatest sagas of Polynesian seafaring.

· · ·

Laʻamaikahiki's voyages come at the close of a great saga relating to a line of seafaring chiefs, a saga that spans several generations. Applying the relative chronology of the genealogies, these chiefs can be traced to roughly the fourteenth century. The tale begins with Māweke, a famous ruling chief of Oʻahu. Māweke was the lineal descendant of the fabled Kiʻi (Tiki), through Kiʻi's son Nānāʻulu, to whom many of the great Hawaiian chiefs traced their ancestry. The traditions conflict regarding whether Māweke was born in Kahiki and voyaged to Hawaiʻi to lay claim to Oʻahu, or his ancestors had arrived earlier to settle that fertile land. But about the beginning of the fourteenth century, Māweke was the acknowledged ranking chief of Oʻahu. He had three sons, Muliʻelealiʻi, the firstborn, Keaunui, and Kalehenui, among whom he divided Oʻahu's rich resources. Muliʻelealiʻi ruled over the Kona district, including the famed lands of Waikīkī; Keaunui controlled the districts of ʻEwa, Waiʻanae, and Waialua; and Kalehenui received the windward Koʻolau district with its reefs, fishponds, and irrigation lands.

Muliʻelealiʻi, in turn, also had three sons: in birth order they were Kumuhonua, ʻOlopana, and Moʻikeha. Following Polynesian rules of patrilineal succession, the oldest son, Kumuhonua, took over control of the Kona estates after the death

FIGURE 6.
View of Waipi'o Valley, ancient seat of the Hawai'i Island kings, where the O'ahu chiefs 'Olopana and Mo'ikeha settled after being expelled by Kumuhonua. Photo by P. V. Kirch.

of his father. But in a classic pattern that would be repeated many times over, the younger, junior-ranked siblings were not happy with this arrangement (presumably because they inherited only minor estates within the Kona district governed by their older brother). The younger brothers plotted against Kumuhonua and organized a sea attack of war canoes against him. The would-be usurpers were defeated and captured.

'Olopana and Mo'ikeha departed their birth island of O'ahu and sailed southeast down the archipelago to Hawai'i, where they settled in Waipi'o, the largest valley on that island's windward coast (see figure 6). The newcomers from O'ahu likely chose this location because of its great potential for irrigation; indeed, Waipi'o with its extensive alluvial floodplain has more irrigated taro lands than any other place on Hawai'i Island. In later centuries, Waipi'o became the seat of the ruling house of Hawai'i, especially under the famed king Līloa.

In these early years, vast Hawai'i was still but sparsely populated, though it was certainly occupied. The traditions tell us that the chief of Kohala district, comprising the northern part of the island, was Hikapoloa, who with his wife, Mailelauli'i, came from a line of Kona or leeward chiefs. The latter two were, like 'Olopana and Mo'ikeha, descended from Nānā'ulu in the ancestral Kahiki homeland. It may have been this kinship tie that induced Hikapoloa to welcome the two outcast brothers from O'ahu. He allowed them to become the *haku* (lords) of Waipi'o.

Hikapoloa, the Kohala ruler, had a granddaughter named Lu'ukia, who is reputed to have been a great beauty. 'Olopana took her for his wife. The chiefly pair ruled over Waipi'o, along with 'Olopana's brother Mo'ikeha, for some years. The brothers applied the knowledge of taro irrigation that they had brought from O'ahu to develop the fertile alluvial floodplains of Waipi'o into a highly productive landscape. But Waipi'o is not at all like O'ahu's gentle valleys. Being much younger geologically, Waipi'o has steep valley walls that can give way in massive landslides when heavy rainstorms hit the islands, or when earthquakes shake the still-active volcanic island. The hydrology of the steep and short streams leading down the waterfalls into Waipi'o is conducive to massive flooding during storms. In historic times, such events have created major floods in Waipi'o. Thus it is not surprising that the traditions of 'Olopana and Lu'ukia recount that after some years of happy residence in Waipi'o, a series of storms and flooding wreaked havoc across the valley floor.

'Olopana and Lu'ukia, accompanied by Mo'ikeha, decided to quit Hawai'i, to voyage south to their ancestral lands in Kahiki. The place where they ultimately settled is celebrated in Hawaiian tradition as Moa-ula-nui-ākea. Scholars have debated over the years about the location of this place. Edward S. C. Handy, who conducted fieldwork in Tahiti, the Marquesas, and throughout the Hawaiian Islands, thought that Moa-ula-nui-ākea was part of the fertile Puna'auia district on Tahiti-Nui, the larger section of Tahiti Island. Until modern times this district has been the seat of the 'Oropa'a lineage, which is the Tahitian spelling of 'Olopana. At the heart of the Puna'auia district lies the deep valley of Punaru'u, another name shared with 'Olopana's O'ahu roots; this is the same place-name as Punalu'u on the windward side of O'ahu. Handy gives another link in the chain connecting 'Olopana to this part of Tahiti: the presence of a large temple (Tahitian *marae*) close to the shoreline called Taputapuātea, named after the most famous of all Society Islands temples at 'Opoa, on the island of Ra'iatea. On O'ahu, as well, there was a most sacred *heiau* by the name of Kapukapuākea, the Hawaiian version of Taputapuātea, situated at Waialua on that island's northern shore. 'Olopana may well have participated in rites at both of these temples. In short, the evidence, though circumstantial, is sufficient to suggest that 'Olopana, Lu'ukia, and Mo'ikeha established themselves in the Puna'auia district of Tahiti Nui.

At this point love—and lust—enter the saga. Luʻukia, married to ʻOlopana, had also become the lover of her brother-in-law, Moʻikeha. While this might appear immoral in Western eyes, as Polynesian anthropologist Te Rangi Hiroa pointed out, Luʻukia may simply have been following the accepted Hawaiian practice of *punalua*, the sharing of a wife by two friends or brothers by mutual consent among all parties. After all, the three partners had lived together for many years in Waipiʻo and now Tahiti. The contented *ménage à trois* was disrupted, however, by the actions of a local Tahitian chief named Mua, who lusted after the beautiful Luʻukia but had been rebuffed in his advances. Mua led Luʻukia to believe that Moʻikeha had publicly defamed her, and she turned against Moʻikeha. To assure that Moʻikeha would no longer have sexual access to her, Luʻukia had her female attendants weave a kind of chastity belt made of sennit cord around her loins, from her waist to the middle of her thighs. This became known in tradition as the *pāʻū* of Luʻukia (skirt of Luʻukia), a name later given to the intricate woven network of cordage that was skillfully wrapped around Hawaiian water gourds, with no evident beginning or end. (Te Rangi Hiroa says that a Hawaiian fisherman, showing him a "very neat lashing on a pearl-shell hook," also called the lashing pattern *pāʻū-o-Luʻukia*.)

Not knowing that Mua had poisoned the mind of his lover against him, Moʻikeha was confounded by Luʻukia's actions. He pleaded with Luʻukia to tell him why she had had herself so tightly wrapped in her protective *pāʻū*, but she would not speak to him. After several days and nights of agony, Moʻikeha decided that he would have to depart Moa-ula-nui-ākea or go mad with unrequited desire. Calling together his immediate followers, he ordered that his double-hulled voyaging canoe be readied for the sea. "Let us sail for Hawaiʻi, because I am so agonized for love of this woman. When the ridge-pole of my house, Lanikeha, disappears below the horizon, then I shall cease to think of Tahiti." Kamahualele, who was Moʻikeha's trusted *kilo*, an expert in stars, astronomy, and reading the signs of nature, prepared the great canoe. When all was ready, they set sail at dawn, just as the star Hōkū Hoʻokelewaʻa (Sirius) was rising, leaving Tahiti and the ridgepole of Moʻikeha's house Lanikeha in their wake.

The long voyage passed uneventfully until Moʻikeha and his party caught sight of the looming peaks of Hawaiʻi. The steersman brought the canoe in to the deep bay of Hilo, on the southeastern side of the island. As they approached the land Kamahualele strode to the bow of one the vessel's twin hulls. In the high wavering voice used by Polynesian chanters, he voiced the lines of a *mele* that has come down through the centuries as one of the most famous in all Hawaiian oral literature:

Eia Hawaiʻi, he moku, he kanaka,
He kanaka Hawaiʻi—e.
He kanaka Hawaiʻi,
He kama na Kahiki,
He pua aliʻi mai Kapaahu,
Mai Moaulanuiākea Kanaloa,
He moʻopuna na Kahiko laua o Kapulanakehau . . .

Here is Hawaiʻi, an island, a man,
Here is Hawaiʻi—indeed.
Hawaiʻi is a man,
A child of Tahiti,
Royal offspring from Kapaahu,
From Moaulanuiākea of Kanaloa,
Grandchild of Kahiko and Kapulanakehau . . .

The chant continues, proclaiming, "My chief dwells in Hawaiʻi, / He lives! He lives!" and ends with the lines, "Kauaʻi is the island, / Moʻikeha is the chief!"

The canoe continued its voyage along the coast until they reached Hāmākua, the district where Moʻikeha had long before farmed in the great valley of Waipiʻo with ʻOlopana and Luʻukia. Moʻikeha's priest, Moʻokini, expressed a desire to remain in Kohala; Moʻikeha accordingly landed him there. The great war temple on the northern point of Kohala, facing Maui, still bears the name of Moʻikeha's priest, Moʻokini. Onward up the archipelago the remaining voyagers sailed, past fair Molokaʻi, to Moʻikeha's homeland of Oʻahu. Here Moʻikeha's sisters wished to stay, and Moʻikeha bade them farewell.

Now Moʻikeha left Oʻahu astern and guided his double-hulled canoe to Kauaʻi, arriving off the central district of Wailua, a fertile land of streams, rich in irrigated taro fields, its beaches noted for their fine rolling surf. The ruling *aliʻi* of Kauaʻi at this time was Puna, descendant of the original founding settlers of this island. Puna received Moʻikeha, who came ashore with all the trappings of a high chief. The people of Wailua had gathered that day to go surf-riding. Moʻikeha— according to the traditions a tall, commanding man with dark reddish hair— went along to join them. Also going surfing that day were Puna's two daughters, Hoʻoipoikamalanai and her younger sister, Hinauʻu. They were smitten with this handsome voyager chief from Kahiki; the attraction was mutual. With Puna's blessing Moʻikeha took the two young chiefesses to be his co-wives, and between them they bore him five children.

After Puna died, Moʻikeha became the *aliʻi nui* (paramount chief) of Kauaʻi. He dwelled contentedly with his two wives and five sons, forgetting his troubles with Luʻukia in distant Kahiki. As Moʻikeha grew older, however, he began to think about an infant son he had left behind in Kahiki, a son born to Kapo, another of his alliances. Moʻikeha became obsessed with the thought of being able to see this son, whom he had named Laʻa. When he had departed Kahiki, Moʻikeha had left Laʻa with his older brother ʻOlopana, who had adopted and raised Laʻa in the land of Moaulanuiākea. Kila, the youngest son of Moʻikeha with Hoʻoipoikamalanai, was chosen to lead the voyage back to Kahiki to fetch Laʻa so that he could be reunited with his father.

The double-hulled canoe was hauled out from the *halau* and made ready. Kamahualele, Moʻikeha's trusted astronomer and navigator, who had sailed with him from Kahiki and chanted the famous *mele* off Hilo, would return with Kila to guide him on the long and dangerous voyage. The canoe departed Wailua and sailed for Oʻahu, the travelers stopping to pay their respects to Kila's aunts. Kamahualele safely guided Kila and his party to Kahiki, where they landed at Moaulanuiākea. They saw the house site Lanikeha, which Moʻikeha had left behind so many years earlier. Kila went in search of Luʻukia and, finding her, called out his greeting. Luʻukia, not recognizing the youth, asked who he was. Kila replied, "I am Kila of the uplands, Kila of the lowlands, Kila-pa-Wahineikamalani. I am the offspring of Moʻikeha."

One can imagine Luʻukia's shock at these words, as she replied, "Is Moʻikeha then still living?"

"He is still living," answered Kila. "He is indulging in ease in Kauaʻi where the sun rises and sets; where the surf of Makaiwa curves and bends; where the *kukui* blossoms of Puna change; where the waters of Wailua stretch out. He will live and die in Kauaʻi."

Kila inquired of Luʻukia where he might find Moʻikeha's son Laʻa. Luʻukia replied that she had not seen him for a long time, that he was hidden in the mountains. With the aid of an aged sorceress named Kuhelepolani, Kila and Kamahualele found Laʻa. They relayed to him the message that his aged father, Moʻikeha, still dwelling in Kauaʻi, wished Laʻa to voyage there to see him once again.

Laʻa consented to make the long sea journey to Kauaʻi. As he came from Kahiki, Laʻa became known thereafter as Laʻamaikahiki (Laʻa-from-Tahiti). Laʻamaikahiki had adopted the new Tahitian temple rituals involving the use of a large cylindrical drum, of a type known as *pahu*. When Laʻamaikahiki sailed with Kila to visit

Moʻikeha, he was accompanied by his priests, the image of his god Lonoikaoualiʻi, and his *pahu* drum and drummer. The Hawaiian traditions recount that this is how the *pahu* drum was introduced into Hawaiian temple ritual. It is indeed the case that the Tahitian and Hawaiian forms of these drums are very similar in shape, high cylinders covered with shark's skin.

Laʻamaikahiki arrived on Kauaʻi, to Moʻikeha's great joy. He was taken by the high priest of Kauaʻi, Poloahilani, into the sacred temple at Wailua, where Laʻa's god Lonoikaoualiʻi was set up (see figure 7). The deep, sonorous beat of the *pahu* drum was heard by the people of Kauaʻi for the first time. Laʻamaikahiki stayed awhile in Kauaʻi. He then voyaged to Oʻahu, where he stayed with relatives descended from Māweke, principally at the chiefly residence of Kualoa on the windward side of the island. While at Kualoa Laʻamaikahiki took three high-ranking and beautiful young chiefesses as wives, marrying all three on the same day, in an event famous in Hawaiian lore. Nine months later each of the three women bore him a son, again on the same day. These offspring, named Ahukini-a-Laʻa, Kukona-a-Laʻa, and Lauli-a-Laʻa, became the progenitors of important chiefly families of Oʻahu and Kauaʻi; Queen Kapiʻolani, wife of King Kalākaua, was a descendant of Ahukini-a-Laʻa.

Laʻamaikahiki did not remain in Hawaiʻi. After staying to see his sons born on Oʻahu, he decided to return to Kahiki. As we have already seen, the *moʻolelo* recount his visit to Kahikinui. Perhaps hearing the name of that place, which commemorated his own natal land of Tahiti-Nui, he was curious to see it. But Laʻamaikahiki found Kahikinui "too windy." Anyone familiar with southeastern Maui can appreciate Laʻamaikahiki's complaint, for the winds that wrap around Haleakalā and sweep down the ʻAlenuihāhā Channel blow incessantly. So Laʻamaikahiki did not tarry in Kahikinui. He may have constructed the *pānānā* notched wall as a guide to following generations who might wish to follow his course back to Kahiki. Or perhaps it was built later, in memory of Laʻamaikahiki. In any event, Laʻamaikahiki left the Hawaiian Islands, departing from Kahoʻolawe, along the route called Ke Ala i Kahiki.

Laʻamaikahiki was among the last chiefs to make the long, arduous voyage between Kahiki and Hawaiʻi. It is said that a grandson of Moʻikeha's, named Kahaʻi-a-Hoʻokamaliʻi, carried on the family tradition by sailing to Kahiki, departing from Kalaeloa, on Oʻahu. Kahaʻi returned with the breadfruit tree, which formerly had not been present in the islands. This was an important addition to the agricultural subsistence base. But after Laʻamaikahiki and Kahaʻi, the era of long-distance voyages linking Hawaiʻi with the Polynesian populations of

FIGURE 7.
The ruins of the ancient temple of Hikinaakalā (Rising of the Sun) may still be seen on a point of land next to the Wailua River mouth on Kauaʻi Island. It may have been at Hikinaakalā that the *pahu* drums brought by Laʻamaikahiki from Tahiti were first beaten. Photo by Thérèse Babineau.

central Eastern Polynesia, especially the Society Islands, came to a halt. Kahiki became a place of memory and myth, recalled in *mele* and *moʻolelo* but no longer visited. It was the beginning of a long period of isolation, when Hawaiʻi was cut off from external communication. Whereas the voyages between Kahiki and Hawaiʻi had continued to bring newcomers and new ideas to the islands—such as the *pahu* drum of Laʻamaikahiki and its associated temple ritual, as well as the breadfruit of Kahaʻi—in the future all innovations would occur internally on the island chain.

The dynamics of historical change would now be driven solely by circumstances and events indigenous to the archipelago. The last long-distance voyages probably ended about 1400. It would be nearly four centuries before Cook burst upon the scene, once again opening up the seaways to Kahiki.

. . .

The *moʻolelo* of ʻOlopana, Moʻikeha, Kila, Laʻamaikahiki, and Kahaʻi-a-Hoʻokamaliʻi are rich in their narrative details, linked with unbroken genealogies that descended to the time of the ruling chiefs who met Captain Cook. These genealogies, going back between twenty-three and eighteen generations before Cook's arrival, allow us to estimate that the events described in the great saga of Māweke's descendants took place during the fourteenth century. But did these epic voyages really take place? Or were they invented by later Hawaiian poets and storytellers, as some have suggested? Similar questions, it might be noted, have at times been raised about the great Norse sagas of Eric the Red and Leif Erikson, who sailed in open wooden boats across the North Atlantic to settle Iceland and Greenland, and to touch upon the shores of the New World. Although these similar traditions of great transocean voyaging were for many years dismissed as mere myth, archaeological research has recently proved their veracity.

Nineteenth-century scholars such as Abraham Fornander had no doubts about the historical basis of the Hawaiian traditions. Early twentieth-century anthropologists, including Edward Handy and Te Rangi Hiroa, also accepted the historical core of the voyaging traditions, even if they admitted the possibility of some embellishment to add luster to the stories. Later, however, others began to look skeptically on these sagas of long-distance voyaging between the far-flung archipelagoes of Eastern Polynesia. Norwegian adventurer Thor Heyerdahl, who in 1947 drifted from Peru to the Tuamotu Islands on his balsa raft, *Kon Tiki*, was one skeptic. Andrew Sharp maintained that the Polynesians had simply drifted to their islands by accident. This skeptical attitude was tempered after the first dramatic voyage of the replicated double-hulled canoe *Hōkūleʻa*, as recounted in chapter 3. Subsequent experimental voyages of the *Hōkūleʻa*, under the noninstrumental navigational guidance of Nainoa Thompson, established the technical ability of Polynesian double-hulled canoes to sail between any of the archipelagoes of the eastern Pacific, even as far as remote Rapa Nui and beyond, to the Americas.

Recently, a striking piece of new evidence emerged, evidence of a voyage that started in Hawaiʻi and ended in the coral atolls of the Tuamotu Archipelago, 2,500 miles to the south. It consists of a stone adz, the indispensable woodworking tool of the ancient Polynesians, made from rock found uniquely, of all places, on Kahoʻolawe Island, the very place where Laʻamaikahiki began his return voyage to Kahiki, following Ke Ala i Kahiki, the sea path to Tahiti. This humble stone adz provides incontrovertible evidence for at least one return voyage from Hawaiʻi back to central Polynesia.

Archaeologists have for a number of years been intrigued by the possibility of tracing stone tools to their points of geological origin, a method we refer to as sourcing. Particular sources of useful stone, such as fine-grained basalt, which can hold a sharp cutting edge, have distinctive mineralogical and chemical characteristics, or signatures. The basalt that was erupted from, let us say, the volcano of Mauna Kea on Hawaiʻi Island differs from that erupted from Haleakalā volcano on Maui, or indeed from ʻEiao Island in the northern Marquesas. But the differences are minute, requiring the technical ability to discriminate quantities of key elements, such as strontium or rubidium, down to measurements in parts per million (ppm).

In the 1980s, a new method for geochemically analyzing rocks began to be developed and refined, known as X-ray fluorescence, XRF for short. The method is based on the principle that when an object is exposed to X-rays, or gamma rays, the component atoms of which that object (such as a basalt adz) is made up become ionized, meaning that the atoms shed one or more of their electrons. In doing so, energy is released in the form of photons. The material that has been bombarded with X-rays emits radiation back (the fluorescence referred to in the name XRF). This back radiation has energy levels characteristic of the particular atoms present. The re-emitted energy can be captured and the elemental composition of the material calculated. This is done in a laboratory machine called an X-ray fluorescence spectrometer.

In late 1989, when I had first established my Oceanic Archaeology Laboratory at Berkeley, one of my graduate students, Marshall Weisler, became interested in applying the XRF method to the problem of sourcing Polynesian basalt adzes and other tools. The Berkeley Geology Department conveniently had an XRF facility, which Weisler learned how to operate. He began to process samples from both archaeologically excavated stone adzes and known stone quarries. At first he focused on specimens from the Hawaiian Islands. Soon he was able to discriminate among adzes that had been manufactured from stone at different quarries, on Molokaʻi, Lānaʻi, and the big island of Hawaiʻi. A major test of the method came when Weisler applied the XRF technique to adz materials that I had dug up in 1991 from a rock-shelter site on the island of Mangaia, in the southern Cook Islands.

As we had anticipated, most of the basalt flakes and polished adz chips (flakes that come off of the blade of an adz during its use) in the rock-shelter deposits could be traced to a local quarry source on Mangaia itself. But four of the adz chips were of an entirely different kind of basalt, with a distinctive chemical signature. Weisler demonstrated that these four chips, from the deepest levels of the site, came from stone adzes that had been manufactured at a large basalt quarry on the Samoan island of Tutuila (the Tatanga-matau quarry), one thousand miles to the west of Mangaia. This was very exciting, providing some of the first hard scientific evidence for the long-distance, or interarchipelago, transport of stone tools in Polynesia.

Marshall Weisler has gone on to become a senior faculty member at the University of Queensland, where he continues to pursue the application of geochemical analysis to prehistoric stone tools, becoming a leader in this field. Subjecting stone adzes from various archaeological sites throughout eastern Polynesia to the X-rays of his spectrometer at Queensland, Weisler has now shown that there was a remarkable amount of movement of these prosaic artifacts among the island groups. Adzes made at a large quarry on ʻEiao Island in the Marquesas have turned up in the Society Islands, and as far away as Mangareva. Adzes made on tiny Pitcairn Island (where there was a particularly good source of basalt, at a place called Tautama) were also distributed westward to the major islands. These results, added to the experimental voyages of the *Hōkūleʻa*, have helped to revise our thinking about just how readily and frequently the Polynesians moved about among their far-flung islands.

Until recently, none of these findings had yet linked Hawaiʻi to the southern lands of Kahiki. Then Weisler cleverly decided to study a collection of adzes that had been obtained by the pioneering Bishop Museum archaeologist Kenneth Emory in the Tuamotu archipelago in the 1920s and 1930s. Because the Tuamotu islands are coral atolls, lacking basalt rocks of any kind, Weisler knew that the adzes found there by Emory had to have been imported from other Polynesian locations. Working with Weisler in this latest project was Queensland geochemist Kenneth Collerson, who brought to bear an even more powerful and accurate technique, called inductively coupled plasma mass spectrometry. Collerson crosschecked his results by also analyzing radiogenic isotope ratios. Together, they analyzed nineteen adzes from nine different coral atolls, comparing the results with the chemical compositions of potential source rocks from twenty-eight different volcanic sources around Polynesia. The Tuamotu adzes had come from islands scattered across southeastern Polynesia, including the ancient quarries of ʻEiao, Pitcairn, Rapa, and Rurutu islands in the Austral group, and the Society Islands.

But with one particular adz, bearing Bishop Museum catalog number C7727, Weisler and Collerson hit archaeological pay dirt.

The little adz numbered C7727 had been obtained by Emory on the atoll of Napuka, at the remote northeastern fringe of the Tuamotus, one of the least visited of that archipelago. Emory spent several months there in 1934, obtaining valuable information on the ancient ceremonies, ritual practices, and traditions of the Napuka people. Adz C7727 is among the smallest in the Tuamotu collection, only ninety millimeters in length. It has a distinctive shape, with the butt end (opposite the bevel) being reduced to a tang in order to facilitate hafting onto a wooden handle. Emory described the cross-section of the tool as "inverted-triangular," meaning that the apex of the triangle pointed downward when the adz was hafted to its handle. This is a significant detail, for the inverted-triangular cross-sectioned type of adz is almost never found in Hawai'i (although one was found at the early Waimānalo dune site) but is characteristic of adzes made and used in the Society Islands, including Tahiti.

When the chemical signature of adz C7727 was compared with all possible known sources in Polynesia, it immediately fell into a group of Hawaiian rocks that are distinctively different from anything else in Polynesia. These include rocks on the islands of Kaho'olawe and Lāna'i. Collerson and Weisler were able to discriminate further between these two possible sources by looking in detail at various elements and their ratios. They determined that adz C7727 "is consistent with derivation from a hawaiite [a specific kind of rock] source on Kaho'olawe." Moreover, this hawaiite occurs on Kaho'olawe only in a few places, one of which is close to Lae o Ke Ala i Kahiki, the headland overlooking La'amaikahiki's famed sea road to Tahiti.

Because adz C7727 is not made in a style known to Hawaiian craftsmen, but rather in a typical Tahitian style, its maker was probably someone from the Society Islands, perhaps Tahiti itself. Indeed, the adz could have been made by a member of La'amaikahiki's crew on Kaho'olawe before they departed on the voyage south. Or as Collerson and Weisler hypothesize, "rock from Kaho'olawe may thus have been taken as a gift or memento (as is done today by modern traditional voyagers) or used as ballast, and fashioned into adzes in the Tuamotus." Either way, that piece of rock, quarried by a Polynesian craftsman of unknown name near Lae o Ke Ala i Kahiki, was without doubt carried on a voyaging canoe from its point of origin over a distance of at least 2,500 miles to the south, where it ultimately ended up on tiny Napuka. It is the "smoking gun" that archaeologists had long searched for, the incontrovertible evidence of two-way voyaging between Hawai'i and Kahiki.

Was the little adz of Kahoʻolawe rock actually carried on Laʻamaikahiki's double-hulled canoe, on his return voyage? And did Laʻamaikahiki construct the *pānānā* that stands to this day sentinel along the coast of Kahikinui, pointing out the seaways to Kahiki? I would like to think that the answers to both questions are affirmative. But science cannot prove either one. Ultimately, however, it matters not whether Laʻamaikahiki, or some other Polynesian voyager, was responsible for those artifacts. The *pānānā* of Kahikinui and the little basalt adz from Napuka are priceless treasures, material witness to a time when fearless and learned *kilo* plied the seaways between Kahiki and Hawaiʻi. That time ended at the close of the fourteenth century. Kahiki became a distant memory. Thereafter, Hawaiian civilization developed in complete isolation.

EIGHT · Māʻilikūkahi,
O'ahu's Sacred King

By the opening years of the fifteenth century, the days of heroic voyages between Hawai'i and Kahiki had faded into memory. Five generations of ruling chiefs had come and gone since the time of Kahaʻi-a-Hoʻokamaliʻi, Moʻikeha's grandson, who made the final round-trip voyage to Moa-ula-nui-ākea, returning with breadfruit seedlings. Oʻahu and Kauaʻi, favored lands of the Kahiki chiefs, continued to hold primacy in the archipelago as the centers of population and chiefly high culture—O'ahu probably more than Kaua'i, considering the latter's isolation at the western end of the island chain. In fact, the fifteenth century would become Oʻahu's golden age, a time of peace and great prosperity, when chiefs from all the other islands would visit and return to their own lands filled with desire to emulate what O'ahu's rulers had accomplished.

The story of O'ahu in these times is inextricably bound up with a remarkable *aliʻi kapu* (sacred chief) by the name of Māʻilikūkahi, about whom the *moʻolelo* and *mele* sing high praises. He is the first in the line of O'ahu rulers about whom the traditions give more than passing mention. His reign was still being valorized when Cook arrived three centuries later. Māʻilikūkahi's story begins in the land of Wahiawā, on the rolling central plateau of O'ahu, between the long cloud-capped summit of the Koʻolau Mountains to the east, and the majestic peak of Kaʻala to the west (see map 2). Here, in the cool uplands, only a few hours' walk from the fertile lowlands of Waialua with their rich taro fields, the sacred chiefs descended from Nānāʻulu held court, at a fabled place called Līhuʻe. The exact location of Līhuʻe is

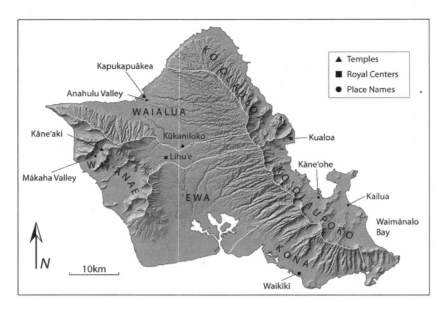

MAP 2.
Oʻahu Island with important places mentioned in the text.

lost to us now, plowed under by the pineapple plantations of Dole and Del Monte early in the twentieth century. But thanks to the foresight of one of the early managers of Waialua Plantation, a small part of the chiefly center of Līhuʻe—one of the most sacred places of ancient Hawaiʻi—has been preserved to this day.

Kūkaniloko was the birthing place of the *kapu* chiefs. Here Māʻilikūkahi was born to a chiefess named Kokalola or perhaps Nononui (the sources disagree about his parents' names), about the year 1425. I have seen Kūkaniloko many times, but one of my most memorable visits was in 1993, when I took my wife, Thérèse, there so she could photograph it for our book on ancient Hawaiian sites. While Thérèse set up her Hasselblad camera, I wandered respectfully among the weathered stones, trying to visualize what it might have looked like in Kokalola's day. There are no constructed stone walls or platforms at Kūkaniloko, only a cluster of deeply weathered boulders half-buried in the red, iron-rich earth of the Wahiawā plain, sculpted by nature into sensuous curved forms and pockmarked here and there by dimples and depressions (see figure 8). Why only here, in this spot, the ancients must have pondered, does this family of *pōhaku* (stones) appear to be timelessly trapped in the process of emerging out of the earth, of being born? The nat-

FIGURE 8.
The sacred birthing stones at Kūkaniloko, on the central plateau of Oʻahu Island. Photo by Thérèse Babineau.

ural symbolism is indeed poignant. Tradition tells us that Nanakaōko first chose this spot for his wife to give birth to their son Kapawa. By the time Māʻilikūkahi's mother was carrying his fetus, Kūkaniloko had long been established as the birthing place of Oʻahu's sacred *aliʻi*.

From the scraps of tradition that have come down to us, thanks to Samuel Kamakau and others, we can imagine some of what transpired during the birth of a sacred chief such as Māʻilikūkahi. When the midwives determined that Nononui's time was near, a small house was erected to cover the particular rock, Kūkaniloko proper, against which the chiefess would rest her back while in labor. Only the finest *loulu* palm leaves were selected for the roof thatching, while the floor was softened with layers of *hala* mats, woven by Nononui's female relatives. Word spread across the district that Nononui had gone to Kūkaniloko. People dropped their digging sticks in their gardens and left the fishponds at Waialua, trekking up the slippery paths to assemble on the eastern side of the stream, opposite Kūkaniloko. The birth of a sacred chief must be witnessed by the people, for in the young child, especially should he prove to be a male, lay their own future.

When the head midwife emerged from the little hut, she gave word to the waiting messenger to run and inform the anxious priests waiting at nearby Hoʻolonopahu temple. A son was born to Nononui. Hearing the wonderful news, the *kahuna* signaled his drummer: beat out the deep, sonorous rhythm on the *pahu* drums named ʻŌpuku and Hāwea, announcing to the assembled crowd that indeed, a male heir had been born. Laʻamaikahiki had brought the first *pahu* temple drums from the ancestral lands of Moa-ula-nui-ākea. Their sounding now told of the birth of a new *aliʻi kapu*, to carry on the lineage that traced back to Nānāʻulu. The infant was carried to Hoʻolonopahu temple, where his umbilical cord was severed with a bamboo knife.

The traditions are silent regarding Māʻilikūkahi's youth. But they do speak of the troubled times in which he grew up. Oʻahu in this early time was not yet united under a single ruler, but divided into three districts *(moku)*, each under the control of a chief who traced his descent back to Māweke. Huapouleilei was the chief of Waialua and Wahiawā, along with ʻEwa, the *moku* where Māʻilikūkahi was born. Kona on the leeward side was ruled over by Kahuʻoi. The lush windward coast from Kahuku to Makapuʻu was the land of Moku-a-Loʻe. There was a loosely organized council of chiefs, the *ʻaha aliʻi,* who met from time to time to consider affairs of mutual interest, and to concern themselves with keeping the sacred genealogies. But lacking a unified, central leadership, Oʻahu was vulnerable to potential enemies from beyond its shores. Such an enemy would soon darken Oʻahu's famed beaches. Kalaunuiohua, paramount chief of Hawaiʻi Island far to the east, was plotting war against Oʻahu.

In the nineteenth century, David Malo recounted the story of how Kalaunuiohua acquired his warlike propensities. Kalaunuiohua had many times tried to put to death a *kāula* (prophetess) named Waʻahia. She had repeatedly survived beatings, being cast into the sea, and being thrown down into deep gulches. Finally Waʻahia told the chief that if he truly wished for her death he must put her in his *heiau* and set fire to the temple house, immolating her. But, she warned him, the smoke from the burning *heiau* would rise up into the sky, where her god Kāne-ope-nui-o-alakai would transform the billows into wondrous shapes. During this time, Kalaunuiohua must remain closeted in his house. No matter how much the populace outside might cry at the spectacles in the heavens, he dare not emerge to look himself upon the phenomenon until the day was ended.

Kalaunuiohua commanded that Waʻahia be burned in the temple house. True to her prophecy the smoke rose in the sky to take the shape of two giant fighting cocks. The people were wild with excitement, but Kalaunuiohua remained in his house, cautioned by his frightened retainers. Other marvelous forms emerged in the heavens: a giant pig dashing back and forth, and wondrous colors and lights. Finally, two clouds resembling ʻalae birds (mudhens), black with red crests, flew down from the heavens and alighted near the chief's house. Unable to restrain himself, Kalaunuiohua thrust his hand through the wall and tore away the thatch to gaze upon the ʻalae. At that instant Waʻahia's spirit took possession of Kalaunuiohua. Thereafter, as Malo put it, "he had only to point with his hand and direct war against another country and that country would be at his mercy."

Perhaps the account of Waʻahia's possession of Kalaunuiohua was a clever invention spread by word of mouth up the island chain to instill fear in the hearts of his intended enemies. Certainly, the Hawaiians greatly feared the powers of *kāula*, who were not like regular temple priests *(kāhuna)*, but rather loners, hermits often, who became possessed with powerful spirits and spoke their oracles. Receiving the news that Kalaunuiohua was on the warpath, Huapouleilei passed many sleepless nights in his ancient home of Līhuʻe on the high plateau of Oʻahu. Before long the news spread up the archipelago that Kalaunuiohua had moved against the *aliʻi* of Maui, Kamaluohua, defeating his warriors and making Maui subservient to Hawaiʻi. Molokaʻi fell next. Kalaunuiohua now led his war canoes across the Molokaʻi channel to Waikīkī. Huapouleilei and his loose Oʻahu coalition were no match for the Hawaiʻi warriors; they quickly surrendered. Kalaunuiohua's policy was not to kill his vanquished foe, but to co-opt them to join him in an ever-larger force. This he now directed to attack the remaining stronghold of Kauaʻi. Thus Kamaluohua of Maui, Kahakuohua of Molokaʻi, and Huapouleilei of Oʻahu all accompanied the fearless Kalaunuiohua in his fleet across the wide channel between Oʻahu and Kauaʻi, bringing war to that island's chief, Kūkona. But there, in the sacred lands of Wailua on Kauaʻi's eastern shore, Kalaunuiohua's dark power deserted him. Kūkona and the Kauaʻi forces were victorious; Kalaunuiohua and his subjugated chiefs were taken captive. They remained prisoners on Kauaʻi for many years before Kūkona released them to return to their home islands.

· · ·

With Huapouleilei captive on Kaua'i, the O'ahu council of chiefs chose Haka to be their nominal leader, the *ali'i nui*. Haka was in the direct line of patrilineal descent from Muli'eleali'i and Kumuhonua, the son and grandson of Māweke, founder of the O'ahu dynasty of chiefs, and thus a rightful heir to this fertile land. But Haka soon proved to be a bad choice, his personality not living up to his noble pedigree. The *mo'olelo* describe him as being "stingy," an oppressor of the people. Polynesian nobility receive their *mana* (spiritual power) by virtue of their descent from high-ranking ancestors, traceable ultimately back to the gods of creation. But sacred chiefs must not abuse that power. It is incumbent on them to act in ways that manifest *mana* benevolently, in the interests of the people. Haka, as with other chiefs infamous in certain traditions, abused his *mana*. He aroused the anger and resentment of the people and their chiefs. The *'aha ali'i* began secretly to plot against him.

The rebellious chiefs led the attack against Haka at Līhu'e, in the Wahiawā uplands. With the few warriors still loyal to him, Haka retreated to the high mountain fortress of Kawiwi, along the crest of the Wai'anae mountain range. Many years ago, I climbed up to one such mountain fortress high on the Wai'anae ridge, with Peter Chapman and Bill Kikuchi of the Bishop Museum. Following reports by local hunters, we found the deep trenches that had been painstakingly excavated with wooden digging sticks, cutting across the sharp, narrow ridges that lead up to the central peak. As I traced one knife-edge ridge down through thick *uluhe* fern up to my neck, the ground suddenly disappeared beneath my feet and I slid on my backside into the bottom of another defensive trench, some fifteen feet down. Whether this fortress was the same Kawiwi where Haka holed up for his last stand I cannot say. But such fortified peaks would have been difficult to take by sheer force. The only access is by means of two or three knife-edge ridges. When these were cut with moats and palisaded, as they doubtless were, anyone trying to scale them from below was at the mercy of slingstones and spears from the defenders towering over them.

Haka, his *mana* having abandoned him, was betrayed from within his own mountain refuge. The young *koa* (warrior) who had been left to keep watch during the night while Haka slept was disaffected, having been slighted by the stingy chief. He called out reassuringly to Haka and the other warriors, "O Haka, O Haka, sleep, sleep." Then the youth whispered to the rebel mob below to steal up through the defenses, one at a time. They crowded into the mountain stronghold. Haka—he who had offended his birthright—was put to death. All the others were spared.

Now the ʻaha aliʻi again faced the decision of who should be chosen as their leader. Māʻilikūkahi, then in the prime of young adulthood, must have seemed the obvious candidate. Although his descent line was slightly junior to that of the defeated Haka, Māʻilikūkahi was nonetheless an aliʻi kapu, who had been born at Kūkaniloko. The pahu drums ʻŌpuku and Hāwea had heralded his birth. His lineage traced back to Māweke, through the illustrious Moʻikeha, and before them to Nānāʻulu. The chiefs and their kāhuna approached the young aliʻi, who no doubt had been among those fearless enough to scale the heights of Kawiwi fortress and take the kingdom back from the evil Haka. Māʻilikūkahi consented to the will of the ʻaha aliʻi.

Māʻilikūkahi was led by the priests from Līhuʻe high in Oʻahu's interior to the shores of Waialua. There the most sacred temple of all, Kapukapuākea, stood in the land of Paʻalaʻa-kai, on the eastern side of Kaiaka Bay. Kapukapuākea was to the Nānāʻulu line of Oʻahu aliʻi what Westminster Abbey is to the kings of England, the site of installation and ritual acknowledgment of their divine right to rule. It was here that the sacred chiefs of Oʻahu had been installed with special rites from remote times, mai ke pō mai as the traditions say. The temple was named after Taputapuātea in Kahiki, in the land of Moa-ula-nui-ākea. The priests led Māʻilikūkahi into Kapukapuākea to have his navel cord ceremonially cut. Of course, this had been done at his birth at Kūkaniloko, and hence was now only a reenactment of the rite, necessary to "cleanse and purify him," as the moʻolelo says. The act of circumcision was likewise reenacted by the priests, chanting the ʻUlunokū prayer, after which Māʻilikūkahi was installed as ke aliʻi o ka moku, the ruler of all of Oʻahu. Oʻahu's golden age had commenced.

One of Māʻilikūkahi's first acts was to move the chiefly seat from Līhuʻe in the uplands of Wahiawā to Waikīkī in Kona. The long rolling breakers of Waikīkī, under the shadow of Lēʻahi (known to tourists as Diamond Head), were already favored by the island's young aliʻi who competed on their wooden surfboards. Stately coconut trees graced the sandy ridge behind the curving beach, offering not only shade but providing nuts whose rich, oily meat went to make the kūlolo and haupia puddings that fattened chiefly bodies. More important, inland from Waikīkī stretched a great low-lying swampy plain, watered by the confluence of streams from Mānoa and Pālolo valleys. Drawing their waters from Kōnāhuanui and the highest peaks of the Koʻolau Mountains, almost constantly drenched in rain, these permanent streams endowed the plains of Waikīkī with the life-giving water to sustain hundreds of acres of taro fields. Between the sandy beach where Māʻilikūkahi established the chiefly compound and the taro fields inland were also freshwater ponds stocked with mullet and milkfish. And that was not all, for the higher slopes of Kaimukī and the rounded dome of Puʻuʻualakaʻa (today known as Round Top) overlooking all of this also boasted some of the most fertile soil for growing sweet potatoes and yams. Waikīkī truly was a blessed land, at the center of all that was necessary for a bountiful life in the Polynesian style. From here Māʻilikūkahi ruled his island wisely.

· · ·

In the early fifteenth century, Hawaiian society was still configured in the mold of ancient Polynesia, following norms and customs that had developed in the Hawaiki homeland during the first millennium B.C. To be sure, there had been some changes after the first settlers arrived in Hawaiʻi, for no human society is ever truly static. Some adaptations to local conditions had been necessary. But Hawaiian society at this time still more closely resembled its ancestral form in Hawaiki than it would the society that greeted Cook just three centuries later. By the time of Māʻilikūkahi's reign, the pace of social and economic change was beginning to heat up. Māʻilikūkahi would hasten the transformation to a new order. Before his installation at Kapukapuākea, an *aliʻi nui* (ruling chief) was paramount in name only, with little direct power over the chiefs of the various districts. Māʻilikūkahi asserted his leadership in ways that no predecessors had dared to. In particular, he implemented a sweeping reform of the system of land tenure.

In the ancient homeland of Hawaiki, the right to reside at a particular house site and to farm the lands attached to that house (the house and its residents being called a *kāinga*) was validated by invoking one's genealogy, from either one's father's line or one's mother's, up from the founding ancestor of that house. (Whereas Europeans tend to talk about being "descended" from ancestors, Polynesians use a botanical metaphor of growth. Ancestors form the base and trunk, and later generations the branch tips. Thus Polynesians "ascend" from their ancestors.) In addition, the different *kāinga* of a place collectively made up a more inclusive social group called the *mata-kāinanga*. (In ancient Hawaiki, the word for this collective social group was *kāinanga*. After the movement of people from Hawaiki into Eastern Polynesia, the word *mata* was prefixed to *kāinanga* to form *mata-kāinanga*.) Belonging to particular *kāinga* and *mata-kāinanga*, each with its own proper name, defined who you were, what lands you could farm, from which areas of the reef you had rights to fish and gather, and to which ancestors you made offerings.

Just why this ancient system, which had been transferred from Hawaiki to

Hawai'i by the islands' first settlers, began to break down we cannot be sure. Conflicts between local groups may have resulted in some people being displaced, some ancestral land rights being abrogated. Some chiefs may have begun to wield their status and power over the commoners, trying to take the best lands for themselves or their close followers. Whatever the initial causes, the traditions tell us that by Māʻilikūkahi's time the old system of land rights was "in a state of confusion."

Māʻilikūkahi made the momentous decision to impose a new hierarchical order over the entire island. Rather than let the people work out their territorial rights according to ancestral claims based on membership in their *mata-kāinanga* and *kāinga* groups, he would assign the land to the chiefs, lesser chiefs, and through them down to the common farmers and fishermen. This was a radical departure from the old Polynesian system, in which rights to land were tied to kinship. From this point on, the land tenure system became one of chiefly territories, in which the common people's rights to land depended on their relationship with their chief. This new system would become fundamental to the economic and political order of the islands, from the time of Māʻilikūkahi until it was abolished four centuries later during the fateful enactment of the Great Mahele by King Kamehameha III in 1848.

The system that Māʻilikūkahi imposed on the lands of Oʻahu, from Waikīkī to Waialua, from Waiʻanae to Waimānalo, envisioned the island as a great circle subdivided into nested sets of radial units, like ever-smaller slices out of a giant pie. The entire island was for the first time an integrated political unit, a *moku*, using the ancient Polynesian word for island. The three great districts into which it had for a long time already been divided (Kona in the south, ʻEwa-Waialua in the north, and Koʻolau to the east) were also themselves referred to as *moku*, but instead of being independent polities, each was now assigned to a high chief under Māʻilikūkahi's rule. These great chiefs were henceforth to be called *ali'i 'ai moku*, the word *'ai* meaning "to eat," thus signifying he who eats from, or metaphorically "consumes," the district.

The most fundamental building blocks in Māʻilikūkahi's land reform, however, were territories called *ahupua'a*. This word did not exist in the older Polynesian language of Hawaiki. Quite possibly the word was invented by Māʻilikūkahi or his contemporaries, and it was at the heart of the new land system. *Ahupua'a* is a compound word, formed from the separate words for a stone altar, *ahu*, and for pig, *pua'a*—hence "pig altar," probably referring originally to shrines or altars erected at the boundaries of each *ahupua'a* to receive pigs and other offerings, *ho'okupu*, at the annual Makahiki festival. The Makahiki marked the new year, initiated by the first rising of the star cluster Pleiades (Makali'i).

In this new land system, each *moku* or great district was now subdivided into numerous *ahupuaʻa*, each of which corresponded to an individual valley, or equivalent section of an island. Like the larger *moku* within which they were nested, these *ahupuaʻa* were each radial segments, dividing an island into many small units. Thus the lands of an *ahupuaʻa* included part of the coast, a valley's alluvial lowlands where irrigated taro was grown, the steeper slopes where dryland crops could be planted, and parts of the hills and mountains where forest resources such as timber and birds could be procured. Ideally, each *ahupuaʻa* constituted a self-sufficient economic unit, cross-cutting the concentric ecological zones of an island. The inhabitants of an *ahupuaʻa* thus had access to all essential resources. It was a brilliant way to organize land rights, an administrative stroke of genius. Instead of letting individual families hold the rights to *ahupuaʻa*, Māʻilikūkahi assigned to each territory a chief, who was responsible for the well-being of the *ahupuaʻa* and its people. These chiefs came to be known as *aliʻi ʻai ahupuaʻa*, the chief who "ate" the *ahupuaʻa*. If they were loyal to Māʻilikūkahi, made sure that the land was well managed, and provided annual gifts of produce and other products of the land, their *ahupuaʻa* territories were secure. But there was a catch in this new system, for the paramount chief, Māʻilikūkahi and his successors, could dispossess an *aliʻi ʻai ahupuaʻa* who was rebellious, did not provide the expected tribute, or in some other way displeased the high chief.

Māʻilikūkahi's hierarchical new land system did not end with the *ahupuaʻa* territories of the mid-ranked chiefs, but continued on down with more slices out of the *ahupuaʻa* pie. The next level down was the *ʻili*, a named segment within the *ahupuaʻa*, of which a large valley might have ten or even twenty. These could be thought of as extended "neighborhoods" with groups of residents who organized various agricultural and other economic labors together, such as maintaining irrigation canals. Each *ʻili* had its own agricultural temple, *heiau hoʻoʻuluʻai*, or temple "to increase the food," dedicated to the gods of farming, Kāne and Lono. Here the people brought firstfruits offerings, and the priests prayed for rain and bountiful harvests. These *ʻili* were subdivided yet again into even smaller parcels, called *moʻo*. This low level in the hierarchy corresponded to an extended household group and its immediate farmlands. Individual garden plots within a *moʻo* were called *kīhāpai*.

Māʻilikūkahi's new way of ordering and organizing the lands of Oʻahu precipitated other social changes that affected the ways in which these lands were administered. The traditions say that Māʻilikūkahi kept the chiefs close to him. He particularly insisted on raising the firstborn sons of the chiefs in his extended household. In this way he accomplished several goals at once, keeping a watchful eye over *aliʻi* who might be tempted to become independent and rebellious were they left to their own devices on their estates. He made sure that the noble sons were properly trained in the arts of war, should Māʻilikūkahi need to defend his island against external attacks. The aggression of Kalaunuiohua of Hawaiʻi against Oʻahu was still fresh in the people's minds. In effect, under the shading palms of Waikīkī, Māʻilikūkahi created a royal court, gathering about him the various *aliʻi* and their sons and daughters. He was a Polynesian Louis XIV, the Sun King whose palace at Versailles may have been grander, but who followed the same essential social and political strategy as Māʻilikūkahi, gathering the French nobility about him, keeping them under constant surveillance and control.

With the chiefs mostly residing at Waikīkī and only periodically visiting their dispersed *ahupuaʻa*, these territories needed to have individuals who could manage them on a daily basis. This gave rise to another layer in the evolving hierarchy of Hawaiian society. Such individuals, typically drawn from junior siblings or collateral relatives of the *ahupuaʻa* chiefs, were called *luna* (overseers), derived from the old Polynesian word *runga*, meaning "above." In later times they came to be called *konohiki*, another innovation in the Hawaiian language. These *luna* and *konohiki* were responsible for seeing that the people maintained the irrigation ditches, kept the taro fields planted and the embankments weeded, and cultivated smaller plots whose harvests were reserved for their chiefs. The *luna* saw to it that the pigs and other tribute appeared on the *ahupuaʻa* altars at the territorial boundaries when the "Little Eyes" of Pleiades first twinkled in the eastern sky.

But the new, territorial land system provoked even more fundamental changes in the nature of Hawaiian society. With Māʻilikūkahi's formalized, hierarchical land system in place, two of the most important social groups in the old kinship structure of Hawaiki—*kāinga* and *mata-kāinanga*—now underwent radical transformations. *Kāinga* had originally been the local residential group that occupied a house site and whose members farmed its adjacent lands. In the new Hawaiian system, this ancient word now came to refer not to a group of people residing in a local area, but to "land" in the most general sense. Because of subtle changes in the Hawaiian language as well, the word was now pronounced as *ʻāina* (Hawaiian speech replaced the old Polynesian *k* with a glottal stop ʻ, and merged the original *ng* sound with *n*). *Mata-kāinanga*, similarly transformed in pronunciation to *makaʻāinana* (old Polynesian *t* changed to *k*), no longer referred to an "ascent group" tracing its pedigree up from a trunk of ancestors, but came to refer to the common people at large. The collective *maka'āinana* (commoners) were now defined in opposition to the collective chiefs, the *papa ali'i* or *nā li'i*.

O'ahu prospered greatly under Mā'ilikūkahi and his efficient system of land management. Although the commoners lost some of their ancient rights under this new regime, the traditions tell us that Mā'ilikūkahi was a benevolent ruler, not like his predecessor, Haka. The people loved Mā'ilikūkahi. They willingly gave him their *ho'okupu* (tribute), offering up these "gifts," it is said, with "joyous hearts." Although he was a religious chief and built many temples, Mā'ilikūkahi did not sacrifice men as the kings of Hawai'i did, following the cult of Pā'ao. He maintained the old rituals dedicated to Kāne, as well as a cult of Kū that emphasized his creator virtues, rather than war. Word spread down the archipelago to Maui and Hawai'i of Mā'ilikūkahi's accomplishments.

Ka'ulahea, paramount chief of western Maui, made a visit to O'ahu as a guest of Mā'ilikūkahi at his court under the palms of Waikīkī. Impressed with what he saw, Ka'ulahea returned to Maui with tales of a highly organized landscape, well administered under the new system of chiefs and subchiefs and their *luna*. Ka'ulahea's sons Kaka'e and Kaka'alaneo soon instituted a parallel land reform on Maui, dividing that great landscape into *moku*, *ahupua'a*, and *'ili*.

Meanwhile, the chiefs of Hawai'i and Maui, hearing of O'ahu's great prosperity, became jealous. They decided to wage war against Mā'ilikūkahi, hoping to seize his rich kingdom for themselves. The plot was hatched by Hilo and Punalu'u, chiefs of Hawai'i Island, who convinced the Maui chief, Luako'a, to join them. Their war canoes landed first at Waikīkī, and then at Kapua'ikāula in 'Ewa. But Mā'ilikūkahi was ready for them. His strategy of training the sons of O'ahu *ali'i* in his court, making sure they were well practiced in the arts of war, paid off. The invading forces were cut off from the rear. In a decisive battle the invaders were slaughtered in great numbers. Dead bodies "paved" the floor of a narrow gulch to this day called Kīpapa, the "pavement." O'ahu had been saved, and Mā'ilikūkahi's power and prestige enhanced even more. Under the innovative Mā'ilikūkahi, the ancient system of Polynesian chiefship, tracing back to Hawaiki, had taken its first bold steps toward kingship.

PLATE 1.
The twin-hulled voyaging canoe *Hōkūleʻa* at dawn. The volcanic crater of Diamond Head, Oʻahu, is visible behind the twin bows. Photo courtesy of Sam Low.

PLATE 2.
The windward coastline of Molokaʻi Island exemplifies the breathtaking landscapes of the Hawaiian archipelago. Here some of the highest sea cliffs in the world, created by massive landslides, are punctuated by deep, amphitheater-shaped valleys. Photo by P. V. Kirch.

PLATE 3.
A newly planted irrigated pondfield *(loʻi)* on the floor of Waipiʻo Valley, Hawaiʻi Island. Taro *(kalo)* has been grown in this manner in the valley since the time of ʻOlopana and Moʻikeha. Photo by P. V. Kirch.

PLATE 4.
A Hawaiian chief in feathered cloak and helmet, as drawn by French artist Jacques Arago in 1819. His left leg and chest are extensively tattooed. From Louis de Freycinet, *Voyage autour du monde* (Paris: Imprimerie Royale, 1824, Atlas). Courtesy of the Bancroft Library, University of California, Berkeley.

PLATE 5.
The Kohala field system as seen from the summit of Puʻu Kehena, Hawaiʻi Island. Note the parallel rows of ancient agricultural field embankments highlighted by the late afternoon sunlight. Photo by P. V. Kirch.

PLATE 6.
A feathered *mahiole* (helmet) in the collection of the Bishop Museum, Honolulu. This particular *mahiole* is reputed to have been given by Kamehameha I to Kaumualiʻi, the last king of Kauaʻi Island. Photo by Seth Joel, courtesy of the Bishop Museum Library and Archives.

PLATE 7.

Three war canoes of King Kalaniʻōpuʻu arriving at Kealakekua Bay, Hawaiʻi Island, to greet Captain Cook in 1779, as drawn by the expedition's artist, John Webber. From James Cook, *A Voyage to the Pacific Ocean Undertaken by the Command of His Majesty for Making Discoveries in the Northern Hemisphere* (London: G. Nicol, 1784, Atlas). Collection of P. V. Kirch.

PLATE 8.
A feathered image of the war god Kūkāʻilimoku, in the collection of the Bishop Museum, Honolulu. The image's eyes are of pearl shell and the mouth is formed by a ring of ninety-four dog teeth. Courtesy of the Bishop Museum Library and Archives.

NINE · The Waters of Kāne

Anyone who has ever participated in a *lū'au*, a traditional Hawaiian feast, has probably sampled *poi*, the grayish, glutinous paste made from cooked and pounded taro mixed with water. As bread is to the French, pasta to the Italians, or rice to the Chinese, *poi* is the essential food at the core of a Hawaiian meal. In the words of Edward Handy, who studied the waning world of the Hawaiian planters in the 1930s, taro was "the select food, the food of the elect, and . . . the staple of the *ali'i*." Though not always available owing to drought, or to a local scarcity of adequate irrigation water, taro and *poi* were the preferred foods of Native Hawaiians.

Taro, or *kalo* as the Hawaiians call it, a plant in the aroid family, was domesticated thousands of years ago in island Southeast Asia. Its rootstock was later carried by Lapita voyagers and their Polynesian descendants throughout the Pacific islands. The Hawaiians had as many as sixty-seven different named varieties of taro. The large heart-shaped leaves are edible after cooking (resembling spinach), and an excellent source of vitamins. But the most important part of the plant is its tuber, botanically speaking a "corm," that forms at the base of the petiole (leaf stalk). The taro corm is a source of high-quality carbohydrates, rich in potassium and in vegetable protein. After cooking (usually in an earth oven), taro can be eaten without further preparation, but more often a Polynesian cook transforms the corm's glutinous flesh by pounding or mashing. In Tahiti or the Marquesas Islands, pounded taro is often mixed with coconut cream to make a dish called *popoi*. With coconut trees relatively scarce in Hawai'i, the Hawaiians substituted water, resulting in the classic *poi*. For special occasions, however, they made a sweet pudding called *kūlolo* by mixing grated *kalo* with coconut cream and baking this in the earth oven.

Traditionally, the preparation of *poi* was an art, passed on from generation to generation. The peeled and cooked *kalo* corms were expertly prepared using stone pounders carved from lava rock. The pounders of Oʻahu and the southeastern islands were of conical shape with a flaring base, while those of Kauaʻi were uniquely of ring and "stirrup" forms. The cook placed a corm on a special wooden board before him, and began to pound the starchy mass with the heavy stone implement. At the same time, fresh water from a wooden calabash was splashed onto the taro, becoming incorporated into the *poi*. Knowing just how much water to add so as to obtain the proper consistency was part of the skill of an expert *poi* maker. *Poi* might be eaten fresh or left to stand for several days, allowing it to ferment mildly and yielding a slightly sour taste. True connoisseurs prefer sour *poi*, a taste I admit to having acquired over the years.

Food and religion are intimately intertwined in most of the world's cultures. The Christian practice of breaking bread and drinking wine to symbolize the body and blood of Christ has ancient roots in the wheat- and grape-based agricultural economy of the Near East and Mediterranean. The Chinese offer rice wine to their deities and ancestors. Not surprisingly, to the Hawaiians taro was a sacred plant, its cultivation and consumption accompanied by appropriate rituals. Hawaiian myth traces the origin of taro to the prematurely born son of Wākea and Hoʻohokukalani (in one version Hoʻohokukalani gave birth to a root, rather than to a baby boy). The premature infant, named Haloa, quickly expired. Wākea buried the little body in the eastern corner of their house. East, notably, is sacred to the creator god Kāne, who follows the path of the sun daily across the heavens. Not long after Haloa's body was interred, a plant emerged from the spot. Wākea named the plant Ha-loa, "long rootstalk." This was the first *kalo*.

A peculiar property of the taro plant is that while it can be grown in "dry" soil conditions (as long as there is sufficient rainfall, usually at least forty inches annually), it responds especially well to irrigation. Much like rice, taro can be grown in flooded paddies or "pondfields" (*loʻi* in Hawaiian) to which water is diverted from springs or streams (see color plate 3). As long as the water flows freely through these pondfields and does not become stagnant (there must be a continual pass-through of cool water), the taro grows rapidly and produces excellent corms.

A *loʻi* pondfield (or "taro patch" as these are often referred to by local Hawaiʻi residents) is a marvelous microecosystem. Cool water, diverted from a mountain stream by a simple dam of boulders, runs through a canal *('auwai)* following the gentle contour of a slope, until it enters the first *lo'i*. From the uppermost *lo'i*, the water passes from pondfield to pondfield in a stair-step series until it reenters the stream. This stream water, which originates in the upland mountains, brings with it dissolved nutrients, fertilizing the *lo'i* and nurturing the growing taro plants. In addition, nitrogen-fixing algae bloom in the pondfield water, further adding to the rich growing medium.

With these ideal nutrient inputs, the yields of irrigated taro can be as great as ten tons per acre. Given the crucial importance of water supply to taro fields, taro and water, *kalo* and *wai*, are closely linked in Hawaiian culture. As we might expect from the story that Haloa's body produced the first *kalo* in the eastern part of Wākea's house, life-giving water is also sacred to the creator god, Kāne. In the late nineteenth century, Nathaniel Emerson collected a sacred chant *(mele)* from Kaua'i Island. The first and last stanzas perfectly capture the deeply rooted associations among Kāne, the sun, water, and life:

He u-i, he ninau:
E u-i aku ana au ia oe,
Aia i-hea ka wai a Kane?
Aia i ka hikina a ka La,
Puka i Hae-hae;
Aia i-laila ka Wai a Kane.

A query, a question,
I put to you:
Where is the water of Kāne?
At the Eastern Gate of the Sun,
At Hae-hae;
There is the Water of Kāne.

E u-i aku ana au ia oe,
Aia i-hea ka Wai a Kane?
Aia i-lalo, i ka honua, i ka Wai hu,
I ka wai kau a Kane me Kanaloa—
He wai-una, he wai e inu,
He wai e mana, he wai e ola.
E ola no, e-a!

> One question I ask of you:
> Where flows the water of Kāne?
> Deep in the ground, in the gushing spring,
> In the aqueducts of Kāne and Kanaloa,
> A wellspring of water, water to drink,
> Powerful water *[mana]*, life-giving water.
> Life, give us this life!

Kāne, god of sunlight and water, reigned supreme over the Hawaiian *loʻi* fields.

The practice of cultivating taro in irrigated pondfields is found throughout Polynesia, wherever local environmental conditions are suitable. But the Hawaiians developed taro irrigation to a greater extent than anywhere else in Polynesia. The highly organized civilization that arose under Māʻilikūkahi and his descendants was based on the high degree of productivity that such irrigation agriculture could confer. By the time of Captain Cook's arrival in 1778, taro irrigation fields covered all of the valley bottoms and lower slopes, wherever streams had sufficient water to supply the *loʻi*. Some irrigated field complexes have remained under cultivation throughout the nineteenth and twentieth centuries, as in the famous Hanalei Valley on Kauaʻi. Today, the renaissance in Hawaiian culture, along with a realization that *poi* is an excellent and nutritious food, has inspired a new generation of Hawaiian *kalo* farmers.

The vast taro fields of Hanalei on Kauaʻi, or those that spread across the floodplains of Koʻolaupoko on windward Oʻahu, were truly major feats of agricultural engineering. Some of these irrigation works covered dozens of acres, fed by stone-lined canals up to a mile or two in length, with as many as one thousand or more individual *loʻi* ponds. When and how were these impressive terrace systems constructed? The Hawaiian traditions offer only tantalizing hints. The story of Māʻilikūkahi suggests that the great king encouraged works of cultivation; Oʻahu was said to have been densely populated during his reign. Such high population numbers could have been supported only by an intensive agricultural economy. But the history of this remarkable economy, upon which so much of the rest of Hawaiian civilization depended, had to be painstakingly extracted from the ground by archaeologists. Teasing out the story of Hawaiʻi's ancient farmers and their feats of transforming valley after valley into managed landscapes of unparalleled productivity, from scraps of buried charcoal and the layered sequences of abandoned terraces, has taken forty years of hard research.

· · ·

Cold water seeped into my mud-caked boots as I squatted in the several-meter-long trench we had dug with pick and shovel through the sticky volcanic clay in the upper valley of Mākaha, deep in Oʻahu's Waiʻanae mountains. With the point of my trowel I was trying to tease small black specks of charcoal out of the reddish-brown soil in the trench face. These charcoal fragments were all that remained of forest trees that had once covered the adjacent hillslopes. The trees had been burned to clear the land to plant gardens. As a result, charcoal had washed down to be incorporated into the soil of an ancient *loʻi* taro terrace. If I could recover enough of these carbon specks, together they would provide a sufficient sample to send off to the radiocarbon-dating laboratory. Our goal was to be able to determine when the forests were cleared and the wood charcoal incorporated into the newly constructed taro terraces, thus providing a chronology for the development of irrigation in the Mākaha Valley. This kind of dating had never been attempted before. Although the muddy trench was not very glamorous, it was exciting to know that we were pursuing methods new to archaeological science.

The rain was now coming down in sheets. The tangle of alien guava trees overhead offered little protection. A few feet away Doug Yen huddled under a canvas tarp strung between several large guava trunks, heating a battered kettle perched precariously on a Primus stove, intent on brewing a pot of afternoon tea. Yen's New Zealand roots had instilled in him an unflinching belief in the restorative powers of a cup of strong tea. Soaked and beginning to shiver in my trench, I too was looking forward to placing a hot mug between my mud-smeared palms. The kettle whistled and Yen called out to Paul Rosendahl and me to put our charcoal hunt aside for the moment. We joined him under the tarp. I plopped down on a basalt boulder for a seat, and gratefully accepted an enameled mug of strong Kiwi tea.

It was the spring of 1970. Rosendahl and I, along with Tom Riley, had been enlisted by Yen to help develop archaeological methods to trace the history of agricultural development in the Mākaha Valley. Yen was a respected ethnobotanist on the staff of the Bishop Museum, where he had begun to collaborate with archaeologist Roger Green. Green was in the midst of directing a major archaeological survey of the Mākaha Valley, which was slated to become a model "planned community," with golf course, condominiums, and gated residences. Honolulu developer Chinn Ho had agreed to finance a study of the valley's rich archaeological sites prior to the planned land development. Green was applying the settlement pattern approach in Mākaha: mapping and recording not just the larger *heiau*, but every kind of surface feature, down to the smallest terraces and C-shaped shelters. The aim was to reconstruct ancient Hawaiian life through the local landscape of houses and fields. Of course, this meant that the remains of ancient agricultural terraces also needed to be investigated. Green asked Yen if he would join in and lend his expertise in Pacific ethnobotany (the study of human uses of plants) to aid in the Mākaha project.

We sipped our tea while the rain rolled off the tarp. Yen explained once again what he hoped we would accomplish. It was not enough, he asserted, just to find the surface remains of ancient Hawaiian taro fields and map them along with other kinds of archaeological sites such as houses and temples. Such surface information would tell us something about the extent of ancient agricultural practices, but nothing about their development over time. When did the Hawaiians first begin to construct stone-faced terraces and divert stream waters into them for taro irrigation? Did they bring knowledge of such practices with them from their homeland islands to the south, or did these sophisticated horticultural techniques develop later, perhaps in response to the need to feed a growing population? These were the sorts of questions that Yen thought we might be able to answer. But to do so, we had to dig.

Previously, no one had thought of digging into an ancient Hawaiian taro field. To Kenneth Emory and Yosi Sinoto, who had pioneered stratigraphic archaeology in the islands two decades earlier, this would have seemed pointless. Their aim had been to obtain artifacts, such as fishhooks and adzes, with which to trace Polynesian migrations and reconstruct the material culture of the islanders. Consequently, they targeted sites such as rock shelters and sand dunes, where people had lived and where artifacts of bone, shell, and stone were concentrated and well preserved. The idea of digging in an old taro field, where one could expect to find nothing but muddy soil, would have seemed absurd. But to Yen, it was exactly what we needed to do.

A reconnaissance team led by Rob Hommon had identified several sets of ancient agricultural terrace walls in the upper reaches of the Mākaha Valley. Situated on the drier, leeward side of O'ahu, Mākaha's stream carries flowing water only in the upper reaches of the valley, where rainfall is heaviest. The drier and more expansive lower valley had lots of house sites and many signs of intensive dryland planting, such as stone mounds, but no indications of permanent irrigation. Only in the upper valley, where stream flow was permanent year-round, had Mākaha's farmers been able to tap into Kāne's life-giving waters to irrigate their beloved taro plants.

Yen decided that our team would focus on a single set of stone-faced terraces, identified in Hommon's survey as "Area 17," on the western side of the small stream. Extending over a straight-line distance of about one hundred yards, some fifty terraces made up a single irrigation complex. Our first task was to map the complex. Since I had been trained in plane table surveying, Yen assigned this job to me. Over several days we traced the lines of stone-faced embankments through the tangle of guava and other exotic vegetation that partly obscured the ruins. When completed, my map gave us an accurate picture of the size and layout of the irrigation complex, with the highest terraces at the upstream end. It was evident that water had been fed into the system by means of these uppermost terraces, with field-to-field flow after that.

With the map completed, the hard work of digging began in earnest. We laid out trenches across the uppermost sets of terraces, usually at right angles to the stone-faced retaining walls. We needed to expose the stratigraphy, the sequence of sedimentary layers, within the terraces. Yen thought it quite likely that this irrigation system had developed over a considerable time span. He reasoned that it might have been rebuilt, perhaps more than once. If so, we should be able to see the evidence for such a sequence of construction phases in the form of a succession of buried terrace walls, or of fill deposits.

Yen's hunch proved to be correct, but the work of reading the subtle clues in the soil strata and buried foundations of previous terrace walls was difficult. No one had done this kind of archaeology before; we had no prior model to follow. To help us, Yen brought in two young geology graduate students from the University of Hawai'i. They found additional clues in geochemical traces left by ancient taro roots in the former cultivation layers. From their analyses we learned how the inundated soil conditions in a flooded pondfield create distinctive "reduction" and "oxidation" horizons in the soil. We argued long and hard over how to interpret particular sections of the trench walls. Yen is a firm believer in the intellectual value of argument and counterargument. Our afternoon tea sessions under the tarp became impromptu seminars in archaeological science! Gradually, consensus emerged among the members of the field team about what had transpired hundreds of years ago high up in the far reaches of the Mākaha Valley.

The seeking after bits of precious charcoal had paid off, giving us the first radiocarbon dates for an irrigated terrace site in the Pacific. Our oldest date, 560 plus or minus 110 years, came from the construction fill of a terrace wall. This was closely matched by two other dates from a soil layer that we identified as the first phase of irrigated taro cultivation at the site. The three dates indicated that the irrigation complex had been built sometime around the beginning of the fifteenth century. After that initial phase, disaster struck. The immediate cause was probably the clearing of forest on the steep slopes above the irrigation complex in order to plant dryland crops such as bananas (*mai'a*), or kava roots (*'awa*). With the forest removed, the steep slopes were exposed to erosion. The painstakingly built stone taro terraces were buried in mud and debris that came down following a particularly intense storm. But the taro fields were too valuable to abandon. Their owners rebuilt the system; the remnants of the older terrace foundations were now covered by a new set of retaining walls. A charcoal sample extracted from a cultivation layer of this later phase gave a radiocarbon date of 295 years, in the mid seventeenth century. Thus the wet cultivation of taro in the upper Mākaha, drawing on Kāne's life-giving waters, had spanned more than two hundred years. The *lo'i* system had first been built about the time of Mā'ilikūkahi, perhaps as part of his efforts to inspire agricultural development around the island of O'ahu. It had survived local disaster to continue to provide food and sustenance for several generations of farmers. Most likely, it was not abandoned until sometime in the nineteenth century, when many Hawaiian valleys were depopulated after the ravages of diseases introduced from the West.

· · ·

Our study of abandoned taro fields in the upper Mākaha Valley, guided by Doug Yen's inspiration, ushered in a new era in the archaeological study of ancient agriculture in the Pacific. Other archaeologists began to take up the question of how indigenous farmers had transformed the Pacific islands into highly productive agro-ecosystems. Our Mākaha study demonstrated that archaeological excavation in ancient terraced fields could yield key information on the history of irrigation. We showed that the terraces could be radiocarbon-dated. But while breaking ground in terms of methods, our study of the little Area 17 terrace system in the upper reaches of Mākaha Valley did not address the bigger question: when did truly large-scale irrigation develop and expand in Hawai'i? For this it would be necessary to tackle one or more irrigation systems in the real taro heartlands, such as in Ko'olaupoko district on windward O'ahu.

Ko'olaupoko, extending from Kualoa to Makapu'u, is truly one of the greatest zones of ancient taro irrigation in the archipelago, exceeded in total area of *lo'i* fields only by the vast taro lands of Puna district on central Kaua'i. Moisture-laden trade winds blow straight into the steep *pali* (cliff) of the Ko'olau Mountains, the spectacular backdrop to Ko'olaupoko. Incessant rains feed innumerable streams descending to rich alluvial plains, eventually to arrive at Kāne'ohe Bay. The amphitheater of valleys lying inland of Kāne'ohe Bay (Waiāhole, Ka'alaea,

Waiheʻe, Heʻeia, and Kāneʻohe proper) boasted thousands of irrigated pondfields. Before they entered the sea, the fresh waters also fed large fishponds along the shores of the bay, where milkfish and mullet were raised. Nathaniel Portlock, one of the first fur traders to arrive in the islands, visited the area in 1787, describing the "low land and valleys being in high state of cultivation, and crowded with plantations of taro, sweet potatoes, sugarcane, etc." In all, Portlock found "the prospect truly delightful." So did the high chiefs and kings of Oʻahu, such as Kualiʻi, who made his home at Kualoa.

The opportunity to investigate ancient irrigation in Kāneʻohe arrived in 1984, not through the generous donation of some benefactor of science, but because the state of Hawaiʻi's Department of Transportation was planning to build a huge new freeway across Oʻahu. "Interstate" highway H-3 (many have remarked on the oxymoron of an interstate highway that is confined to a single island) was the pet project of Hawaiʻi senator Dan Inouye. A World War II veteran who had risen to considerable power in Congress (and had become nationally known during the Watergate hearings), Inouye pressed hard for federal funds to build this multilane freeway. With impeccable Cold War logic, Inouye and his followers argued that H-3 was essential to link the Pearl Harbor navy base with the Kāneʻohe Marine Air Station on the opposite side of the island. The project became the center of a huge political and legal battle, with significant repercussions for Hawaiian archaeology along the way. In the end, Inouye and the powerful construction lobby won out. The H-3 cost $1.3 billion, one of the most expensive highways per linear foot ever built in the United States.

In 1984, however, H-3 was still on the drawing boards of the Department of Transportation's engineers. The necessity of providing an environmental impact statement led them to the archaeologists at the Bishop Museum. Jane Allen, a doctoral student at the University of Hawaiʻi who had worked with me on irrigation systems in the Anahulu Valley, was the logical choice to head up the field team. Allen was asked to identify and study any archaeological sites that might lie in the path of a huge highway interchange that the engineers had proposed for linking the new H-3 with the preexisting Likelike Highway. Allen, of course, could not just work anywhere within the Kāneʻohe area. The contract with the Department of Transportation dictated that her field team focus on the proposed route for the H-3 highway interchange, which ran along the base of the mountain. This was considerably inland from the largest taro irrigation systems that had once blanketed the lower alluvial plains. Nonetheless, Allen found abundant evidence of ancient terracing, especially in a section of Kāneʻohe called Luluku. Her team found rank after rank of beautifully preserved stone-faced terraces. Many of the terraces had faces with multiple courses of carefully laid basalt stones, up to four feet in height. Covered in brush and bananas, the terraces had not held water or been used for taro cultivation for many generations. But when one looked at these well-constructed retaining walls and at the perfectly leveled terraces they defined, it was not hard to imagine a landscape of high productivity. Generations of Hawaiian farmers had labored tirelessly to cut and fill the clay of these volcanic slopes, hauling boulders and cobbles and fitting them with great skill into the sturdy retaining walls. When had all this activity taken place? Jane Allen was poised to find out.

Allen instructed her crew to begin cutting trenches into several of the terraces, working back from the stone retaining walls into the deep soil forming the fill of the ancient pondfields. She used the same methods that we had pioneered in the upper Mākaha Valley. In the large *loʻi* complex of Luluku, however, the story was more complicated, historically richer. Where in Mākaha we had found only two phases of agricultural activity with a single rebuilding episode, Allen found a lengthy history of use and rebuilding. She discovered a complex sequence of soil layers, some dark gray and flecked with charcoal, others a reddish-yellow color. The darker layers indicated periods of cultivation, with rich organic material accumulating in the pondfield soil, while the reddish-yellow horizons had formed through oxidation beneath the waterlogged soil, where oxygen brought down by the taro roots resulted in a chemical reaction with iron in the volcanic soil. Carefully tracing the layers, Allen could make out at least nine different phases of soil formation, with four discrete phases of taro irrigation. As in Mākaha, Allen also found buried terrace walls representing earlier construction phases. Clearly, the Luluku terraces had been in use for a long period.

Radiocarbon dates from charcoal samples collected from the different cultivation layers in the Luluku terraces revealed when these irrigation works had been constructed. Dryland gardening had commenced as early as nine hundred years ago, about the eleventh century (about the same time that the little hamlet at Bellows Beach in Waimānalo was first occupied). The oldest firm dates for construction of irrigated terraces are from the thirteenth century, more than a hundred years before the time of Māʻilikūkahi. By the fifteenth century, when Māʻilikūkahi consolidated his rule over Oʻahu and established his innovative system of *ahupuaʻa* land divisions, the Luluku irrigation works were already in full production.

Oral traditions and the evidence from archaeology converge nicely to tell the story of Oʻahu's economic development. Three to four centuries after the island was first discovered and settled by Polynesians, Oʻahu had become densely populated throughout its lowlands. The palm forests—and the flightless ducks that had once roamed in their shade—were but a dim memory. In their place the industrious Hawaiian farmers had created a patchwork mosaic of gardens and house lots, joined by paths and crisscrossed by irrigation canals. Well-watered regions such as Koʻolaupoko had been converted to taro pondfields. And as the population continued to grow and the demand for taro and *poi* mounted, terraces were pushed back farther inland, right up to the base of the steep cliffs. Not surprising, then, that the traditions refer to Māʻilikūkahi's time as one of great prosperity, when the land was thick with people. The conditions were right, moreover, for this intelligent king to impose a new level of administrative control over the burgeoning apparatus of production. Organizing his people into a hierarchy of land units, the new *ahupuaʻa* system, Māʻilikūkahi laid the foundation for the later Hawaiian political state.

. . .

In a famous study of "oriental despotism," Karl Wittfogel teased out what he regarded as the three key elements of the "hydraulic economy," meaning an economic system based on irrigation: "Hydraulic agriculture involves a specific type of division of labor. It intensifies cultivation. And it necessitates cooperation on a large scale." Wittfogel regarded Hawaiian irrigation as a "rudimentary variant of simple patterns of hydraulic property and society," an early stage on the evolutionary path leading to the great irrigation-based states found in Mesopotamia and China. Timothy Earle, an archaeologist who has studied complex chiefdom societies in Hawaiʻi, Peru, and Denmark, likewise argues that irrigation was fundamental to the emergence of "an ideology of chiefly rule." The chiefs were the organizers of the social labor required to build and maintain the irrigation works; they were, in consequence, the "owners" of the system. As Earle writes, "From the commoners who lived on the pond fields' harvest derived a steady and predictable supply of labor that created the infrastructure and superstructure of the chiefs' ruling institutions."

No wonder that chiefs from Maui and Hawaiʻi came to marvel at Oʻahu's landscape, to take back to their own islands the lessons they learned. In the late fourteenth century, Maui and Hawaiʻi islands, in spite of their vast size, were still relatively undeveloped. They were economic backwaters compared with Oʻahu. Part of the reason for this lay in the relative lack of lands suitable for irrigation, especially on eastern Maui and Hawaiʻi. Because of their young geological age and consequent lack of erosion and weathering, the vast majority of Hawaiʻi and eastern Maui consist of lava flow slopes without permanent streams or other water sources to support irrigation. On Hawai'i, only the older Kohala peninsula in the northern part of the island, along with the narrow gulches of Hāmākua district, have the valley topography necessary to develop taro irrigation. Most of eastern Maui is also dominated by lava slopes suitable only for dryland cultivation (with small gulches along the windward side, and in Kīpahulu). Western Maui, which is geologically older, has valleys and streams capable of supporting large-scale irrigation.

The differences in the relative suitability of each island for development of irrigated taro agriculture are enormous. Using a geographic information systems (GIS) analysis, my research group has calculated the areas amenable to irrigated cultivation on each of the main islands. We did this by drawing on data sets for rainfall, stream flow, terrain slope, and other geographic factors that are essential for the construction and operation of irrigation systems. The statistics we obtained reveal the striking differences in the total areas in square miles of high to moderate potential for irrigation on each of the main islands:

Kaua'i	55.8
O'ahu	37.5
Moloka'i	1.6
Maui	10.3
Hawai'i	16.7

Although Kaua'i and O'ahu are much smaller than either Maui or Hawai'i in total land area, each of these older islands enjoys much greater areas of land suitable for taro irrigation. True, western Maui and the northern tip of Hawai'i do have significant areas that can be (and were) put into irrigation, but these large islands could never reach the overall capacity of O'ahu and Kaua'i. Instead, the younger islands of Maui and Hawai'i had to depend on vast regions where rainfall alone was sufficient to grow sweet potato and taro in "dryland" or rain-fed field systems. On Hawai'i, such dryland farming zones covered a staggering 212 square miles. On Maui they totaled about 54 square miles. Maui and Hawai'i were by no means without land suitable for farming; but the economies of these two large islands had to depend to a far greater extent on rain-fed cultivation, which—as any modern farmer (or commodities trader) knows—is a much riskier business than irrigation farming.

Historians, as well as anthropologists and archaeologists, have long debated the issue of "causation" in human history. To what extent do the decisive actions of great men and women determine the outcome of history, as opposed to the vagaries of nature? How do human actions compare to floods, droughts, or severe winters? Did the collapse of great civilizations depend more on the decisions made by their rulers or on the natural limitations of their environments? These are difficult questions to answer. In reality, a complex interplay between human and natural "causes" is usually involved. In Hawai'i, we can surely point to the substantial natural differences among the principal islands as one factor that helped to configure the course of events over several centuries. Better endowed with alluvial soil and flowing streams, Kaua'i and O'ahu were the first islands to be densely populated. It was these islands that the initial generations of ruling chiefs chose as their domains; they doubtless encouraged the people in their work of building up the irrigation systems. By the time of Mā'ilikūkahi, the economic capacity of O'ahu to support a large population was impressive. And as the traditions tell us, this highborn chief reorganized the land system not only to his personal advantage, but also to that of the *ali'i* class as a whole.

While Maui and Hawai'i may have been settled as early as O'ahu or Kaua'i, their populations had remained smaller, and their political influence less important, up through the end of the fourteenth century. Only in places like Waipi'o in Kohala, and in parts of western Maui such as the valleys of Wailuku and Waihe'e, was there any substantial development of irrigation. All this would soon change, however, as the potential of the vast inland slopes of these huge islands for dryland cultivation began to be opened up. Beginning in the early fifteenth century and continuing until Cook's arrival, the dryland agricultural economies of Maui and Hawai'i would come into their own. These new cultivation systems did not enjoy the waters of Kāne. They depended instead on the beneficence of a rival deity, Lono, god of thunder, lightning, and life-giving rains. Lono's sacred crop was the sweet potato, *'uala*. It was to Lono that the farmers of Hawai'i and Maui prayed for rain and an abundant crop. On these dynamic large islands, the cult of Lono grew and thrived, along with the cult of the war god, Kū. The natural resources of Maui and Hawai'i dictated that irrigation would never be sufficient to support their populations. Their farmers would have to depend on Lono's rains. And—perhaps as a result of the periodic droughts and famines that could not be averted—their leaders were drawn increasingly to engage in war and territorial conquest. As Kaua'i and O'ahu were truly the islands of Kāne, so Maui and Hawai'i would become the land of Lono and Kū, locked together in an endless embrace.

TEN · "Like Shoals of Fish"

After searching for the elusive Northwest Passage throughout the summer and fall of 1778, Captain Cook needed to replenish his ships' supplies and rest his crew. He returned south to the newly discovered Sandwich Islands. On January 17, 1779, Cook brought the *Resolution* and *Discovery* into Kealakekua Bay, a deep protected anchorage on the lee side of Hawai'i. As the sails were furled and the anchors let go, a multitude the likes of which Cook had never before witnessed in his Pacific voyages surrounded his vessels. He later wrote in his log, "I have no where in this Sea seen such a number of people assembled at one place." Thousands of Hawaiians were swimming around the ships, in Cook's words, "like shoals of fish." So many people tried to scramble up the side of the *Resolution* that she listed dangerously.

The populations of the many island groups discovered in the course of his explorations were matters of interest to Cook and his officers. Prematurely killed on the lava rocks at Ka'awaloa, Cook would never make an estimate of the Hawaiian population, as he had for other islands such as Tahiti. It fell to Lieutenant King to summarize what they had learned of this new archipelago, including the size of its indigenous population. In his handwritten shipboard journal, King began by noting that they had previously estimated the population of Tahiti at 120,000 persons. But Tahiti was a considerably smaller island; here the population seemed far greater. King continued, "I do not yet see, but we are equally as well authorized to lay down the population of this Island [Hawai'i], at 200,000." He went on then to suggest another 100,000 each for Maui, Oʻahu, and Kauaʻi, and 20,000 for the smaller, arid islands of Kahoʻolawe, Lānaʻi, and Niʻihau. In aggregate, these figures "together give half a Million for the population of these Islands." But then King threw in a note of caution, writing, "it is mere guesswork."

When the expedition returned to England, an official account was published by the British Admiralty. By then, Lieutenant King had reflected on his initial "guesswork"; he decided that a more carefully reasoned population estimate was called for. King began with the assumption that the Hawaiians lived exclusively along the coasts of the islands. He then proposed that "if the number of the inhabitants along the coast be known, the whole will be pretty accurately determined." King had a pretty good idea of the population of Kealakekua Bay, where they had anchored for weeks. He decided to base his estimate on the number of houses at Kealakekua, including the average figure of six persons to a house. This part of the coast was populated, in his view, by about 2,400 people. "If, therefore, this number be applied to the whole extent of coast round the island, deducting a quarter for the uninhabited parts, it will be found to contain one hundred and fifty thousand." King then used the same mode of calculation for the other islands of the chain, arriving at a grand total of 400,000.

King's two estimates—the first admitted "guesswork" made while at the islands, the second a carefully reasoned calculation based on several stated assumptions—are the only firsthand evidence for the overall size of the Hawaiian population at the moment of first contact. We know that King was wrong to assume that everyone lived along the coasts. Vast inland populations inhabited the fertile slopes of Hawaiʻi and Maui and lived in the deep valleys of Molokaʻi, Oʻahu, and Kauaʻi. Thus, if anything, King's second calculations are likely to have been a serious underestimation.

The Europeans who followed after Cook did not share the same penchant for careful observation. Not until 1805 did a certain George Youngson, an English carpenter who had been resident in Honolulu, jot down a population estimate of 264,160, half of what King had proposed. Two decades later, Captain James Jarves suggested just 142,050 persons across all the islands. Finally, in the years 1831–32, the Protestant missionaries who had arrived in Hawaiʻi in 1820 undertook an actual "head count" of the Hawaiian people. This census was not perfect, but it can be taken as reliable with a modest margin of error. It totaled 130,313 souls.

During the fifty years between Cook's expedition and the missionary census, the Hawaiians had been inflicted with a host of new diseases brought by the European visitors and later residents. Cook's crew certainly introduced venereal diseases, and probably respiratory ailments, if not other pathogens. While venereal disease may not kill its victims immediately (although its effect on a population with no prior exposure may have been more severe than usual), it can have a huge negative effect on women's reproductive health and fertility. The birthrate among Hawaiians plunged as a consequence of the newly introduced venereal disease. Later ships brought even more virulent diseases: dysentery, measles, tuberculosis, smallpox, and leprosy. Before Cook, the islands were free of all of these old-world scourges; consequently, Hawaiian bodies did not have antibodies or resistance against them. As we now know, such "virgin soil" epidemics can have devastating effects on indigenous populations, as they did in the New World after the Spanish conquest. Death rates may have been as high as 90 percent of the population in such cases of first contact between the Old World and the New. Thus the difference between Lieutenant King's 1779 estimate of half a million people and the missionary census listing about 130,000 fifty years later should not surprise us.

However, some twentieth-century demographers who studied and wrote about the populations of Pacific islands downplayed the impact of European diseases. Norma McArthur, an influential Australian demographer, argued that Cook, King, and other Enlightenment explorers had grossly overestimated island populations. Whereas naturalist Johan Forster, with Cook on Tahiti, had estimated the population of that island at 200,000, McArthur lowered her estimate of the contact-period population to a mere 30,000! Robert Schmitt, a recognized authority on the population of Hawai'i, wrote that King had also overestimated the islands' population. Ignoring the fact that King had assumed that no one lived in the islands' interiors, Schmitt nonetheless cut the British voyager's second, lower estimate of 400,000 down to just 300,000. In later editions of his book on Hawaiian demography, Schmitt said there could have been as few as 200,000 people when Cook arrived.

In 1989, David Stannard, a University of Hawai'i professor, published a short book titled *Before the Horror,* stirring up a hornet's nest in the staid academic word of historical demography. He revisited King's estimate and concluded that, if anything, the naval officer's assumptions had led him to seriously underestimate the true population numbers. Stannard argued that various lines of evidence—including new archaeological surveys documenting extensive inland populations—might support an estimate of as high as 800,000 people at the time of Cook's arrival in the islands. But Stannard went further, stressing the intellectual, social, and political implications of a demographic collapse on the scale that he thought had occurred. To Stannard, the introduction of foreign diseases that may have killed as many as one out of every eight Hawaiians in the first five decades after contact amounted to nothing short of a "holocaust." Stannard accused Schmitt and others who had downplayed the size of the Hawaiian demographic collapse as practicing "historical amnesia." As he wrote, "For those who bring on a holocaust, willfully or not, nothing is more desirable or sought after than historical amnesia."

Stannard was right to assert that it matters greatly to contemporary Hawaiians to know just how severe was the impact of Western contact. It also matters from a scholarly and scientific point of view that we have an accurate understanding of the demographic history of Hawaiʻi. There is not just the unresolved question of the maximum size of the archipelago's population at the moment of first contact. To understand the course of social evolution in Hawaiʻi over the eight or nine centuries from the arrival of the first voyaging canoes until Cook burst through the curtain of isolation, it is essential to know how the islands' population grew over time. Did this population grow steadily throughout this long time span? Was it still growing when Cook arrived in 1778 and 1779? Or was its demographic history more complicated, perhaps with an earlier growth phase, peaking prior to European arrival? Had the Hawaiian population already reached the "carrying capacity" of the islands? Certainly the high density of settlement over all of the lowland zones suggests that they had intensified their agriculture to a high degree. Perhaps the rate of population growth had declined as this high density was achieved. Could the Hawaiians have been in a stable balance between population and resources, the kind of "sustainability" that we all seek for our planet today?

· · ·

The ways in which human populations increase or decrease over time, how this is affected by the food supply, how growth rates are controlled by particular combinations of the birth and death rates of a population, are fundamental questions for the student of social evolution. The economic condition of any society is intimately bound up with its population, because this population not only supplies the labor force for agriculture and public works, but also determines the level of surplus upon which a political economy can be built. The eighteenth-century social philosophers such as Adam Smith and David Ricardo, who developed the first modern theories of economy and social evolution, were interested in the complex linkages among population, economy, and society. The first scholar to think deeply about demographic matters, however, was Thomas Malthus. Like Captain Cook, Malthus was a product of the Enlightenment, but he had the advantage of having been educated at Cambridge University. In 1798, Malthus published *An*

Essay on the Principle of Population. In this classic book—which was hugely influential to the later thinking of Charles Darwin—Malthus pointed out that whereas the human capacity for reproduction is exponential or geometric, our ability to increase food production is arithmetic. Hence there is a tendency over time for population growth in any particular region to exceed the capacity of the land to support the growing numbers of mouths to be fed. When this reaches a critical point, various conditions will set in to limit or reduce the population. Malthus put it this way, somewhat grimly:

> The power of population is so superior to the power of the earth to produce subsistence for man, that premature death must in some shape or other visit the human race. The vices of mankind are active and able ministers of depopulation. They are the precursors in the great army of destruction, and often finish the dreadful work themselves. But should they fail in this war of extermination, sickly seasons, epidemics, pestilence, and plague advance in terrific array, and sweep off their thousands and tens of thousands. Should success be still incomplete, gigantic inevitable famine stalks in the rear, and with one mighty blow levels the population with the food of the world.

In Malthus's view, human populations invariably grow when conditions are favorable, that is, when land and food are plentiful and not limited. But as populations increase and put pressure on land and the food supply, forces will begin to clamp down on the natural rate of reproduction. These forces, as Malthus pointed out, often include disease (which in preindustrial societies was encouraged by high population densities), but also war, especially wars for control over territory and resources. Ultimately, the great leveler of populations, always stalking in the shadows, is famine.

Six years after Malthus published his famous *Essay*, Pierre-François Verhulst was born in Brussels. A brilliant mathematician, Verhulst became intrigued by Malthus's work and the problems of population growth and limitations. In his early thirties, Verhulst published three papers on what he called "the law of population growth." Accepting Malthus's view that there was an upper limit on population size, Verhulst constructed a simple mathematical model for how a population would grow. He assumed that the limiting effects imposed by the environment would not come into play until the population was already quite large. Thus, in a country that had previously been unsettled (such as a Polynesian island), the initial expansion of population would be exponential. Verhulst's model, when plotted as a graph, produces an S-shaped curve. Population initially increases exponentially, with the graph rising steeply. As the limit is approached, the graph bends over, with population growth slowing and ultimately—in the ideal case—coming to equilibrium with the environment. Of course, that theoretical equilibrium—the point at which there was just enough food to match the absolute minimal subsistence needs of the population—would be a condition of misery. Verhulst's model got little attention at the time, but it was later rediscovered by modern population biologists. Verhulst's "law" is now known as the "logistic equation"; it is a key concept in modern demography and population biology.

In 1967, a young mathematician named Robert MacArthur and a naturalist named Edward O. Wilson who had been collecting and studying ants across the Pacific wanted to explain the patterns of plant and animal distribution across the Pacific Islands. MacArthur and Wilson realized that there was a "strategy of colonization" that successful organisms used to gain a foothold on a new island. There were also predictable changes that species went through after they had been on islands for many generations. Colonizing species typically had high reproductive rates (designated by the symbol r). In contrast, species that had evolved over a long time on an island tended to have low r rates, but they had evolved traits that maximized their carrying capacity, K. The terms r and K are key elements in Verhulst's classic logistic equation, in which population (N) is a function of the interaction between the rate of population growth and the carrying capacity of the environment. MacArthur and Wilson's model of island colonization became known as r/K selection. It too is a cornerstone of modern population biology.

Although MacArthur and Wilson were studying organisms such as ants and birds, there are good reasons to think that human populations may also have had similar population cycles. Initial human colonizers on islands would have needed high reproduction rates to keep from going extinct. And they would not have been limited (except on very small islands) by land or resources. As their populations grew larger, however, it might become necessary to limit population by various cultural means, such as the use of plants that could induce abortion, or by increasing the age at which girls married and had children. Means could also be taken to increase the food supply, by maximizing the production of gardens and fields. Eventually, however, the population would have to come into some kind of equilibrium with the island's ability to produce food. Raymond Firth, an anthropologist who studied the Polynesian island of Tikopia in the 1920s, documented just such cultural controls on human population. In fact, the Tikopia had a culturally ingrained philosophy of "zero population growth," with the ideal of one boy and one girl child for each couple. But such controls did not always work. The Tikopia sometimes had to resort to draconian methods such as infanticide and suicide voyaging to keep their population in balance with resources.

While Malthus had surely been correct in his insight that there are limits to population growth, he overlooked a crucial aspect of human culture: our ability to innovate. "Necessity is the mother of invention," the old proverb goes. In the 1960s, a Danish economist who had been studying rural agricultural populations in the Third World, Ester Boserup, came to the realization that human invention of new technologies had often raised the limits of population growth. In a slim but influential book titled *The Conditions of Agricultural Growth,* Boserup proposed that under conditions of "population pressure," humans had at times dramatically increased their agricultural output through innovation. She showed how, for example, the invention of the plow, and of irrigation, had greatly increased agricultural capacities and allowed for large population expansions. While Boserup's thesis did not contradict Malthus's fundamental principle of limits on population, it did show that such limits were not fixed, and that they could be raised over time. Modern demographers now realize that long-term population dynamics are best explained by a combination of the insights of Malthus and Boserup.

The theories and arguments of Malthus, Verhulst, MacArthur and Wilson, and Boserup give us important conceptual tools with which to think about population growth, about the vital linkages between population and environment. But to apply theory one also needs data, in this case hard numbers of people. Since there is no agreement over how many Hawaiians occupied the islands when Captain Cook first arrived, how can we hope to put real numbers on the islands' population in previous centuries?

· · ·

The early 1970s were a period of intellectual foment in Hawaiian and Polynesian archaeology. An older generation of scholars, such as Kenneth Emory, had been interested primarily in the questions of when and from where the first Polynesians had settled Hawai'i. But the study of entire settlement landscapes, including house sites and agricultural terraces, temples and fishponds, encouraged younger archaeologists to ask questions about the economic, social, political, and religious lives of the ancient Hawaiians. Rather than just reconstruct sequences of changing fishhook or adz styles, the new generation of researchers was interested in studying how these island societies evolved. I was one of those caught up in this new approach, influenced by what some called the New Archaeology. Along with other young colleagues such as Robert Hommon, Tom Riley, Patrick McCoy, and Ross Cordy, we saw ourselves as anthropologists who wanted to get at the fundamental processes driving social change over the long run of time. We were interested in history, to be sure, but we were also searching for larger patterns that might speak to the nature of human culture and society.

Doug Yen had introduced me to the writings of Ester Boserup during his graduate seminar at the University of Hawai'i in the spring of 1970. I still have my copy of her slim red-covered book, with my annotations in the margins. In Yen's seminar, we hotly debated Boserup's theory of innovation driving agricultural change. Had the development of irrigation works, such as those we were excavating in the Mākaha Valley, been a consequence of population pressure? Or had things been the other way around; had increased food production from irrigation spurred population increases? It was the chicken-and-egg conundrum, and we wondered which view was right. Meanwhile, I discovered MacArthur and Wilson's *Theory of Island Biogeography*. I admired the way in which these two biologists used models to help understand how plants and animals evolved on Pacific islands. I thought their models might also be useful for understanding how human populations had responded to island environments. Of course, one had to allow for the greater flexibility and adaptability of human culture. Humans are fundamentally different from ants or birds. Nonetheless, it seemed logical that the same basic r and K terms of Verhulst's logistic equation had to be operating in human populations occupying islands with limited land area and resources.

By 1975 I was back in Honolulu as a researcher at the Bishop Museum, after finishing my doctorate at Yale. I began to ask myself how we might trace the history of Hawaiian population over time. When the U.S. Census Bureau wants to know how many people there are in the country, it sends workers out to knock on the doors of individual households. The people that archaeologists want to count are long gone, but their houses—or at least the terraces and foundations upon which their houses once stood—are often still evident on the landscape. With the new settlement pattern approach, archaeologists in Hawai'i were beginning to get some idea of the distribution and density of ancient house sites. In the Mākaha Valley on O'ahu, Roger Green and Rob Hommon had identified small "C-shaped shelters" (low stone walls resembling a C in plan, which were the foundations for thatched shelters or huts) set among extensive dryland agricultural terraces. These seemed to have been the houses of *maka'āinana* (commoners) who worked the fields of sweet potatoes and yams. In Hālawa Valley on Moloka'i, we had mapped many terraced house sites on the colluvial slopes overlooking the irrigated taro fields. And at Lapakahi in Kohala district on Hawai'i Island, Paul Rosendahl had shown that the upland garden system was also dotted with innumerable small C-shaped shelters and house enclosures. Clearly, the Hawaiian population had densely covered the landscape at times in the past.

With our new landscape mapping—our settlement pattern approach—we could count how many ancient houses were out there. The problem was that not all of these houses had been occupied at the same time; they were not all contemporary. We needed to excavate a reasonable-size sample of these houses and date them using the radiocarbon method, so that we could put *time* into the equation. But radiocarbon dating was (and still is) expensive. Until that time, archaeologists had used their limited research dollars sparingly, obtaining dates from only those sites that were considered to be early or especially noteworthy (like the Bellows sand dune site at Waimānalo).

Help came from an unexpected source. Times were changing in Hawai'i in the 1970s, driven by larger economic forces. The old plantation economy of sugar and pineapple was on the wane, while the tourist industry was booming. Everywhere one looked a new hotel or resort was going up. The state government was funding major infrastructure developments—airports and highways in particular—to support the boom in tourism. New laws had been passed at the federal and state levels that mandated archaeological surveys for both private developments and government projects. When significant sites were discovered, "salvage excavations" were often undertaken to recover remains before they were bulldozed to make way for the latest hotel, golf course, or freeway. These developments were funded by big budgets. An unexpected consequence of the development boom was an upsurge in archaeological studies, including lots of new radiocarbon dates.

Rob Hommon was one of the first to make use of the radiocarbon dates that were beginning to flow in to the museum's archaeology lab. In his doctoral dissertation, submitted to the University of Arizona in 1976, Hommon plotted the temporal distribution of dates from coastal versus inland house sites. His graph for the coastal sites began just before A.D. 1000, climbed to a peak about 1400, and then leveled off. When he plotted the "inland" sites, the curve didn't start until about 1150 and climbed on an exponential trajectory until 1500, when it too began to bend over. Hommon's data suggested that Hawaiians had settled the coastal regions first, beginning to expand inland a few centuries later. He called this the "inland expansion hypothesis." It was a major advance in our thinking about what had happened in Hawai'i. More interestingly, both of his graphs strongly implied that Hawaiian population had not increased steadily over time. Rather, there was an earlier phase of growth, followed by a decline in the rate of increase, if not a full leveling off. It seemed as though Malthus's limits to population growth were being reflected in the Hawai'i data.

I decided to test Hommon's model with new data from the leeward side of Hawai'i Island, where a number of contract archaeology projects had generated radiocarbon dates on habitation sites. Using a "census-taking approach," I plotted out the ages of 113 house sites from the Kohala and Kona regions. The sample was still small, but better than what Hommon had been able to work with. I plotted the site frequency by hundred-year intervals on the x-axis of my chart, with the number of houses in any interval indicated by the height of the bars on the y-axis. My chart showed just a single dated house between A.D. 800 and 1000. Then the number of dated houses began to step up exponentially, until about 1600. The graph showed continued growth in the number of houses from 1600 to 1700, but the *rate* was clearly no longer exponential. Growth had slowed. The data showed a peak by 1700 and then a decline, which accelerated dramatically with European contact.

In 1984 I presented my findings in my book, *The Evolution of the Polynesian Chiefdoms*. I argued that the sequence of population growth on most Polynesian islands, and on Hawai'i in particular, had not followed a continuous exponential curve. Rather, the characteristic pattern had been the one that Pierre Verhulst had identified with his mathematical model back in the early nineteenth century: the logistic growth curve. Early Polynesians had arrived in small numbers in Hawai'i in their voyaging canoes, to encounter vast unpopulated islands. Initially, there had been no constraints on their natural reproductive propensities. Land was not limited, and resources were abundant. Moreover, there were none of the debilitating old-world diseases that would later decimate their descendants. The population grew at a rapid rate. Estimates suggest that the growth rate in the first few centuries after Hawai'i was settled may have approximated 1.2 percent to 1.8 percent annually. This would mean a doubling of the population every one to two generations. This was the r phase of growth in MacArthur and Wilson's model of island colonization.

Then, about 1500 to 1600, depending on which data set one looked at, the *rate* of population increase dramatically slowed. Our settlement pattern surveys showed us that by this time all of the most desirable valleys on the older, well-watered islands were densely settled. Moreover, the new data from areas like Lapakahi on Hawai'i suggested that this was a phase of major expansion of settlement inland and, more important, onto more marginal agricultural lands. Something seemed to be driving the expansion of Hawaiian population out of the "salubrious" core regions (as Hommon called them) into lands that had previously been passed over. But just as this major geographic expansion was taking place, the rate of population increase was also slowing. It looked like a classic case of the "Malthusian brakes" being applied. With Rosendahl's data on the intensification of the Lapakahi dryland agricultural system, one could also argue that Boserup's "innovation" was at play. The Hawaiian population was approaching its limits, as must happen on an island with circumscribed land area. At the same time, Hawaiian farmers were doing all they could to try to boost the productivity of their gardens, even as they had to move into areas where they were dependent on more seasonal rainfall, always a risky venture.

· · ·

When David Stannard published his provocative book *Before the Horror*, in 1989, it caught the attention of scholars working on Hawaiian history and archaeology. One of those was Tom Dye, then employed by the state of Hawai'i's Historic Preservation Division, where he reviewed the reports of contract archaeologists. Dye, assisted by Eric Komori, who was in charge of the division's site database, began to compile all of the radiocarbon dates obtained from these projects.

Dye and Komori's approach to ancient demography—paleodemography, as it is sometimes called—was a bit different from the one I had used earlier. They did not try to count house sites in a "census-taking" approach. Rather, they reasoned that the overall production of "anthropogenic" (human-produced) charcoal, from hearths, earth ovens, clearing agricultural fields, and so forth, should be roughly proportional to the size of the human population, given a consistent technology. More people, more fires, more charcoal. If archaeologists sampled that charcoal in a more or less random fashion, plucking samples from a hearth here, an oven there, an agricultural terrace somewhere else, then the cumulative distribution of radiocarbon dates should mirror, at least to a first approximation, the population growth curve. A plot of cumulative radiocarbon dates would not tell one the number of people at any particular time, but it should—if their assumptions were correct—provide a proxy measure of the growth rate of that population.

In 1992 Dye and Komori published the distribution curve derived from a database of 598 radiocarbon dates throughout the archipelago. The curve had a long tail extending back into the first millennium A.D. This start date was probably due to just a few samples of "old wood" and didn't necessarily mean that the islands had been populated so early. The first sign of a significant increase in the numbers of dates—and, by inference, the population—came around A.D. 1100. After that, the curve shot up exponentially. Then, about 1440, the curve peaked and began to fall off. The period from 1440 until European contact was marked by several "squiggles" in the curve, explained by technical problems in the calibration of radiocarbon dates from the last few centuries. The important take-home lesson was that Hawaiian population was marked by dramatic exponential growth between about 1100 and the mid fifteenth century, when it suddenly stabilized or declined. These results largely matched Hommon's and my earlier findings. However, they suggested that the exponential phase might have been even more marked, and that the transition to a stable or low-growth period was even more sudden.

In recent years, I have returned to tackling the problems of Hawaiian paleodemography. I decided to look first at the evidence for population growth in one of the areas that was not settled by Hawaiians until the phase of inland expansion commenced. This is the Kahikinui district along the southeastern, leeward side of Haleakalā on Maui. Kahikinui lies in the rain shadow created by that great volcanic mountain. Its annual rainfall comes mostly in the winter months from southerly, or *kona*, storms. Although its young lava soils are rich in nutrients, the limited and unpredictable rainfall makes Kahikinui a marginal region in which to practice horticulture. Yet the landscape of Kahikinui abounds in archaeological remains of house sites, temples, and agricultural plots. With my Berkeley students, I had mapped and studied this landscape for several years. We excavated many of the house sites, and accumulated a database of 168 radiocarbon dates. Once again, I was able to apply my census-taking approach to paleodemography. Within the core region of Kahikinui, an area called Kīpapa, we acquired dates from fifty-one different house sites. The earliest sites to be occupied appeared in the fifteenth century, well after the islands were first colonized by Polynesians. This five-century delay in settling Kahikinui is predictable in light of the landscape's marginal nature. One would choose to live permanently in arid Kahikinui only if other options were not available. The number of occupied houses doubled in the sixteenth century, and then dipped slightly in the seventeenth. The eighteenth century, on the eve of European contact, saw a big increase in occupied houses. Following this was the well-known crash that came with the introduction of diseases from the West.

By extrapolating these numbers from our core study area, we were able to estimate the total population of the district over time. During the fifteenth century, the districtwide population had not been more than about 400 people. But by the late eighteenth century, this population had increased by roughly an order of magnitude. Allowing for error in these estimates, my calculations suggested that the maximum population of Kahikinui had been between 3,074 and 4,096 persons. In terms of population density, this would have been an average of between 43 and 57 persons per square kilometer throughout the arable lands of the district (roughly the zone below about 900 meters in elevation). If this high a population density had been achieved in one of the most marginal regions in the islands, one could only imagine how much denser the populations of the fertile valleys and lowlands would have been.

Our "census" data from Kahikinui also gave us an opportunity to estimate the intensity of the postcontact demographic decline that came with the introduction of foreign diseases. The missionary census of 1836 had counted 447 people in the entire Kahikinui district. This figure almost exactly matched our independently derived estimate based on archaeological house counts (in this case of houses containing introduced, Western artifacts) of 445 for the mid nineteenth century. In addition to giving us greater confidence in the house-count method, these data revealed the magnitude of the population decline between the late eighteenth and mid nineteenth centuries. The population had plummeted from between 3,000 and 4,000 down to a mere 447 in just a few decades. David Stannard had been correct to challenge Schmitt and the other demographers who had downplayed the full extent of the Hawaiian "holocaust."

. . .

There is much work still to be done on the details of Hawaiian population history, especially through careful regional studies like that for Kahikinui. But we have come a long way in our understanding of the course of demographic change in the Hawaiian Islands during the period from initial settlement to contact with the West. The first Polynesians to beach their canoes on Hawaiian shores were few in number. Each canoe carried at most perhaps fifty people. Even if multiple landings occurred over the years, the immigrants from the South Pacific would not have numbered more than a few hundred in all. Yet by the time that Cook arrived off Kealakekua Bay in 1779, the Hawaiians were so numerous that their numbers astounded the seasoned explorer, appearing to him as countless "shoals of fish." Our recent estimates of eighteenth-century population densities suggest that Lieutenant King's first "guesswork" estimate of half a million people may, if anything, have been on the low side.

Our studies have shown that Hawaiian population went through two major phases. The first phase was an exponential population increase that started with the arrival of the first voyaging canoes and continued until about 1500. During most of this time, there were few constraints on population growth. At first, land was not limited. Even after the best valleys were claimed, there were still many areas adequately suited to growing taro or sweet potatoes. But gradually, and especially as the fifteenth century approached, conditions began to change. Well-watered valleys such as Kāneʻohe on Oʻahu or Wailau on Molokaʻi now had extensive irrigation works extending over their alluvial plains and slopes. Despite their high levels of production, the irrigation works were not sufficient to feed all of the new mouths. Some people moved elsewhere in search of open lands that had not previously been claimed. Such lands were mostly to be found on the vast, younger volcanic islands of Maui and Hawaiʻi. Kahikinui on leeward Maui first began to receive a trickle of farmers in the fifteenth century.

The transition to the second phase of Hawaiian demographic history may have occurred rapidly. By approximately 1500, high density levels had been achieved over all of the best agricultural lands. Marginal leeward slopes in areas like Kahikinui, and Kohala on Hawaiʻi Island, were already being converted into vast rain-fed agricultural field systems. The rate of population growth began to fall dramatically until it leveled off, closer to a steady replacement rate. Not that some areas did not continue to see local population increases: Kahikinui is a case in point. The populations of leeward Hawaiʻi Island may also have seen modest growth into the seventeenth century, as their field systems continued to be intensified. The demographic patterns of this second phase suggest a continual tension between the Malthusian brakes that threatened to clamp down as population pushed the limits of its island environment and the Boserupian innovation that had the potential to raise the ceiling. In a later chapter I will explain how the dryland farmers in Kohala and Kahikinui used all of the inventiveness they could muster to push the productivity of their rain-fed gardens. But they could only innovate so far. They could only work so hard at mulching and tending their gardens, vulnerable as these were to periodic droughts.

The demographic history of Hawaiʻi is a key to understanding how the island civilization developed. In the first few centuries, it was literally a small-scale society. With just a few hundred or a few thousand people, everyone knew one another, if not literally face-to-face then at least as related families or social groups. It is difficult to develop an elaborate hierarchy in a small-scale society in which everyone is related, tied together by bonds of kinship. But sociologists and anthropologists have long observed that once a population exceeds a certain threshold size—which may be as low as 2,500 individuals—social dynamics began to change. Hierarchy and social distance become more pervasive as societies increase in size and scale.

By the end of the sixteenth century, the populations of individual islands numbered in the tens of thousands. This large labor force had developed vast agricultural systems capable of feeding all those mouths and even yielding a considerable surplus. That surplus was available to be tapped by those who could control the social order: the hereditary chiefs, aided by their priests and warriors. The basis for a large-scale political economy was now in place. Hawaiian history was at a threshold, ripe for the emergence of a new kind of social and political organization. It is the emergence of that new kind of society—the archaic state and its institution of divine kingship—that I take up in the last part of this book.

PART THREE · THE REIGN OF THE FEATHERED GODS

ELEVEN · 'Umi the Unifier

Nestled between Hualālai volcano to the west, the cinder-cone-capped peak of Mauna Kea to the northeast, and the broad, soaring mass of Mauna Loa to the southeast, a stark cinder and lava plain occupies the high interior of Hawai'i Island. About five thousand feet above sea level, this "saddle" region is cold and forbidding. Only the hardy *pūkiawe* and stunted *'ōhi'a* bushes break the monotony of the gray-brown lava and ash. But the cold air is sparklingly clear and—when the plain is not enveloped in fog—it seems as though you could see forever. In the winter sunlight the mountains' snow-capped summits shimmer, a startling sight. This is an alpine world, far above the limits at which taro, sweet potatoes, or other Hawaiian crops could be grown. And yet the ancient Hawaiians were no strangers to this rugged mountain region. Abundant remains of trails, small stone shelters, and lava tubes and caves filled with the debris of ancient campsites abound. They were drawn here on visits from their permanent homes in the lowlands to seek nesting dark-rumped petrels *('u'au)* and other upland resources, such as fine-grained basalt rock to fashion their adzes and other tools. More important, on this vast island, the saddle between the high mountains became an important transit route. It was often much faster to cross from one side of the island via the inter-mountain saddle than to walk around its lengthy perimeter.

This windy, forbidding lava plain is the setting for one of the most unusual of Hawaiian archaeological structures, the Ahu a 'Umi (Altar of 'Umi). 'Umi was Hawai'i's most famous warrior king, the first unifier of the island's dominions, and—it must be noted—a prodigious conqueror of women's hearts. To get to the site requires an arduous daylong trek over rutted four-wheel jeep tracks, up from Kona through ranchlands and forests, passing through a succession of locked gates. I made the journey in the late 1970s, eager to see for myself this famous site. Walking around the well-constructed temple walls, built of stacked pāhoehoe lava slabs, I recalled that this unique structure had the distinction of being the first Hawaiian archaeological site to have been scientifically studied and mapped. In November 1841 the United States Exploring Expedition, under the command of Charles Wilkes, sent a party overland to explore the interior of Hawaiʻi and record some of the first observations on its volcanoes. Three of the expedition's scientists, accompanied by "eight Kanakas and two guides," set out on horseback. "After a day's travel they reached the site of the ancient temple of Kaili." Wilkes described the "main building" as a rectangle "ninety-two feet long, by seventy-one feet ten inches wide; the walls are six feet nine inches high, seven feet thick at the top, and nearly perpendicular." The party's Hawaiian guides told them that formerly the walls had been "covered with idols, and offerings were required to be brought from a great distance." They said that the temple had been built by ʻUmi, an ancient king of the island who had unified all the previously independent districts. "The three northern pyramids forming the front were originally erected by Umi," wrote Wilkes, "to represent the districts of the island he then governed; and as he conquered other districts, he obliged each of them to build a pyramid on the side of the temple."

The engraved plan and view of Ahu a ʻUmi, published in the gilt-embossed official account of the U.S. Exploring Expedition, accurately depicts the high-walled rectangular temple enclosure, with its entrance facing north, surrounded at a uniform distance by eight stone *ahu* (cairns), what Wilkes called the "pyramids." The interior of the main structure is subdivided by low walls into four rooms. These look as though they might have held the offerings brought from the distant, subjugated districts. These offerings were made up, as Wilkes commented, "generally of provisions."

I had never seen a Hawaiian *heiau* quite like this one, though many years later in the uplands of Kahikinui, Maui, I would discover a large multiroomed enclosure that may similarly have served as a collecting place for districtwide offerings. The main rectangular walled enclosure was not unlike other Hawaiian *heiau*, but the surrounding *ahu* certainly were different. I wondered whether these cairns indeed represented the island's several districts, or had some other function. In the early 1980s, A. M. DaSilva and R. K. Johnson proposed that the configuration of Ahu a 'Umi suggested that it had functioned as a kind of astronomical observatory. However, this idea has not been verified through careful archaeoastronomical observations.

Regardless of how this structure—far from most centers of population—originally functioned, it is unique in the annals of Hawaiian archaeology. More important, the Ahu a 'Umi is a tangible link back in time to one of Hawai'i's greatest kings. 'Umi, or to give his full name, 'Umi-a-Līloa, was born of a famous love affair between the great chief Līloa and a young beauty of commoner birth. The story of 'Umi-a-Līloa is perhaps the most famous single *mo'olelo* of the Hawaiian historical traditions. In this chapter I recount 'Umi's tale, based primarily on the version set down by the great scholar of Hawaiian oral traditions Abraham Fornander.

. . .

In the early sixteenth century, Līloa ruled over the northern and eastern parts of Hawai'i Island, the Kohala and Hāmākua districts. An *ali'i nui* of the highest *pi'o* rank (that is, born of a specially sanctioned incestuous union), Līloa traced his genealogy back to the ancient 'Ulu line of chiefs. He was descended from Pilika'aiea, who had come from Kahiki around the late fourteenth century with the priest Pā'ao. Līloa was in theory the paramount ruler of the entire island. But in reality he had little direct control over the leeward and southern districts of Kona, Ka'ū, and Puna, which were governed by their own chiefs. Līloa's ancestral seat was Waipi'o Valley, a great gash in the windward flank of the Kohala mountains, carved out by millennia of torrential rains and landslides. With sheer cliffs rising more than one thousand feet from the broad valley floor, the waterfall of Hi'ilawe cascades down from the cloud-draped uplands. The renegade O'ahu chiefs Mo'ikeha and 'Olopana had been the first to exploit Waipi'o's rich alluvial soils and abundant water for taro irrigation (see chapter 7). Two centuries later, the extensive irrigation works and taro fields had been rebuilt under the direction of Līloa's predecessors. As the lord of Waipi'o, Līloa administered his windward domain from the richest place on the entire island.

By Līloa's time, the beginning of the sixteenth century, the Hawaiian chiefs had instituted several measures setting them off from the commoner class, now known collectively as the *maka'āinana*. Elaborate clothing, especially capes, cloaks, and helmets adorned with yellow and red birds' feathers, along with a particular kind of neck ornament called the *lei niho palaoa*, were the distinctive symbols of high rank. Carved from whale-tooth ivory, the *lei niho palaoa* resembled a stylized tongue curving upward, supported by a necklace of finely braided human hair.

MAP 3.
Hawai'i Island with important places mentioned in the text.

The chiefs now lived separately from the commoners. Chiefly dwellings were marked by the *pūloʻuloʻu,* a staff topped with a ball of white *kapa* (barkcloth). No one without proper authorization dared trespass beyond the *pūloʻuloʻu* staff, on pain of death.

Dwelling in the royal compound called Kahaunokamaʻahala, on the black sandy dunes fronting Waipiʻo Valley, Līloa took his mother's younger sister Pinea as his sacred wife. Just when the chiefly practice of mating with close relatives began is not certain. Close marriages between siblings or half-siblings, which many cultures (our own included) would consider incestuous, began to be preferred by the high-ranked Hawaiian chiefs beginning around the fifteenth or sixteenth century. The idea was to concentrate the high-ranked bloodline, to keep it as pure as possible. In Hawaiian this practice was called *pi'o*, meaning arched, curved, or bent, with a metaphorical sense of a lineage curving back upon itself. Royal incest was not unique to Hawai'i, but was found in a number of other early state-level societies. The pharaohs of Egypt and the Inca rulers of Peru also married their own sisters. In the lush and productive valley of Waipi'o, Pinea, 'Umi's aunt and royal wife, gave birth to a son, Hākau, destined to succeed his father as ruler of the vast island of Hawai'i.

Līloa traveled periodically to the different districts to tour the many *ahupua'a* assigned to his subchiefs. After the birth of Hākau, Līloa made a journey to a place called Koholālele, not too far from the modern town of Hilo, where he dedicated a new temple, named Manini. After completing the rites of dedication, Līloa and his party continued on to Ka'āwikiwiki, where the people had gathered to play games, such as *pahe'e*, which involved throwing a slender dart of heavy wood over a prepared course.

At Ka'āwikiwiki, Līloa went to bathe in Hō'ea stream. Approaching the clear mountain water pool, he saw a striking young woman emerging from the icy water, stark naked. Her female attendant held her barkcloth wrap, as her black hair glistened against her skin. Līloa found her irresistible. She was called Akahiakuleana. She had come to Hō'ea to bathe and purify herself at the end of her menstrual period. Līloa, flush with passion, approached Akahiakuleana and spoke frankly of his desire. Whether she was equally enamored of the handsome warrior chief in his feathered cloak and helmet or he simply exercised his *droit du seigneur*, the traditions do not say. But it was not a mere one-night stand, for 'Umi stayed with Akahiakuleana for some time, until she knew that she had become pregnant.

Līloa must truly have been in love with Akahiakuleana, for he gave her the priceless symbols of his chiefship: the feathered cape, the pendant of whale-tooth ivory, his feathered helmet, and his spear of hard *kauila* wood (see color plate 4). Līloa instructed Akahiakuleana to hide these objects away safely. Should their child be a boy, she was to call him 'Umi. Līloa told Akahiakuleana to raise the youth until he reached puberty (in ancient Hawai'i, this was the time for the rites of circumcision). At that time she must bestow the chiefly symbols on 'Umi and send him to seek his father in Waipi'o. Līloa left Akahiakuleana and rejoined the other chiefs, who were more than a little surprised when Līloa arrived wearing only a loincloth of *kī* leaves and a cape of banana leaves!

Akahiakuleana did indeed give birth to a son, and dutifully named him ʻUmi. She raised him along with her commoner husband, who frequently abused the boy because ʻUmi had a habit of giving away their food and *kapa* to the children of the neighborhood. (To the Hawaiians, this behavior was a symbol of his innate *mana* and chiefly character.) One day after ʻUmi had come of age, Akahiakuleana berated her husband for having beaten him again, telling him that he was not the boy's real father. "What proof have you?" he demanded. Akahiakuleana brought forth the feathered cloak and helmet and other chiefly symbols. Her husband was deeply shaken with fear that he might now be killed for having abused a royal child.

Akahiakuleana now prepared ʻUmi to go to Waipiʻo and claim his patrimony. She told ʻUmi that he must put on the sacred garments of the high chief: the loincloth, the cloak, and the whale-tooth necklace. He must grasp the *kauila* wood spear in his hand. After making the long overland journey to Waipiʻo and descending the trail winding down the steep *pali* (cliff), he would have to swim across the river to the royal compound. Akahiakuleana cautioned ʻUmi not to try to enter via the main door (which would be closely guarded), but to climb over the fence and then enter by the rear door (which was reserved for exclusive use by the high chief himself). She told him to look for the "old man that is being guarded, with kahilis around him" and run and sit in his lap. "When he asks for your name, tell him you are ʻUmi."

Though still only a youth, ʻUmi had already adopted another boy, named ʻOmaʻokamau. Adoption—*hānai* as the Hawaiians call the practice—was widespread in ancient Polynesia. This was yet another sign that ʻUmi possessed *mana* and innate chiefly behavior. Akahiakuleana instructed ʻOmaʻokamau to accompany ʻUmi, and to carry his war club. When the two boys got to a place called Keʻahakea, they encountered some children playing at disk rolling, sliding, and running. One boy, named Piʻimaiwaʻa, skilled in running and mock fighting, was struck with the sight of ʻUmi wearing his brilliant chiefly garments. Piʻimaiwaʻa asked if he could join them on their mission to Waipiʻo, and ʻUmi agreed to adopt Piʻimaiwaʻa as well. Farther along the trail leading to Waipiʻo, at a place called Waikoʻekoʻe, the boys encountered another youth playing the game of *koʻi* on the trail—his name was also Koʻi. Again ʻUmi adopted this latest acquaintance.

By now the unusual band had reached the precipice of Koaʻekea, overlooking the wide valley of Waipiʻo. Raised in the backcountry of Hāmākua, the youths must have marveled at the majestic valley stretched out before them. More than a thousand pondfields of taro covered the valley floor, linked to the life-giving stream by ribbons of irrigation canals. High dunes of black sand defended the rich agricultural lands from the foaming breakers crashing relentlessly upon the boulder beach. On the crest of these great dunes Līloa's ancestors had established their seat of royal residences and temples.

The boys could make out Līloa's compound, Kahaunokamaʻahala, on the inland side of the beach ridge, set off by a wooden palisade. Within the compound were many separate thatched houses, including the *mua* (men's eating house); the *hale noa* ("free" house), where the high chief could mingle with both sexes; other dwellings for the royal wives; separate cookhouses for the men and women; storehouses; and other specialized structures. The royal compound overlooked a sacred taro field, Kahiki-mai-aea, which had been constructed by Līloa's great-grandfather Kahoukapu. This irrigated garden provided taro to make *poi* for the royal household. Higher up on the black sand dune rose the white barkcloth–wrapped *ʻanuʻu* tower of the feared *luakini* (war) temple of Pakaʻalana. Here the *pahu* drums had been sounded to announce the birth of the *piʻo* child Hākau. Another large *luakini*, Honuaʻula, was situated farther up the dune. Both temples were dedicated to a particular manifestation of the war god Kū, associated with the Hawaiʻi Island kings. This was Kūkāʻilimoku, or Kū the Kingdom Snatcher. The deity's physical manifestation was a human head woven of *ʻieʻie* vines, to which had been attached thousands of red feathers. Standing on the valley's rim at Koaʻekea with all this magnificence laid out before them, ʻUmi and his adopted companions might have hesitated for just a moment before putting their bold intentions into action.

· · ·

Scrambling down the steep and slippery path, ʻUmi, ʻOmaʻokamau, Piʻimaiwaʻa, and Koʻi descended into Waipiʻo. Arriving at the southern bank of the swift stream separating the king's house and the temples from the southern plain, ʻUmi told his companions to wait there. He alone would dare to breach the *kapu* barriers marking the royal compound, an act that would earn a commoner instant death. Remembering his mother's instructions, ʻUmi scaled the wooden fence. He quickly entered the rear door of the great house, thatched with thousands of perfectly selected *hala* leaves. This door was reserved for the king; the *ʻilāmuku* (executioners) were hardly expecting anyone—especially a young boy clad in royal garments—to come bounding through it. Before the startled guards could stop him, ʻUmi jumped into the lap of his father, another egregious lapse of protocol. Startled, Līloa looked at the youth on his lap, asking, *"Owai kou inoa?"* (What is your name?). Gazing up at his real father for the first time, 'Umi answered, *"O 'Umi au"* (I am 'Umi).

Līloa recognized his personal symbols of chiefship, the cloak, neck ornament, and other objects left many years before with Akahiakuleana. The high chief grasped 'Umi with his arms and pressed his nose against the boy's in the Hawaiian greeting of *honi*, exchanging their respective *hā*, the breath of life. Līloa demanded, "Where is your mother, Akahiakuleana?" 'Umi told his father that it was she who had instructed him how to come to his father. Līloa next asked the boy what had become of his war club. 'Umi explained that it was with his companions across the river. The other youths were sent for, to be brought into the king's presence.

Līloa now addressed the assembled courtiers, many of whom had come running from the other houses of the compound when they heard the cries and shouts of the *'ilāmuku*. "When we went to dedicate the temple [at Hō'ea], and one day I returned without my loin cloth and I had on a ti-leaf loin cloth, you said that I was crazy; but today you see that here is my loin cloth, my necklace of whale's tooth and my war club. I left these things for my son, who is here." Līloa then ordered the sacred *pahu* and *kā'eke* drums (made from bamboo pipes) to be sounded in Paka'alana Heiau, as befitted the birth of a sacred chief. 'Umi was taken into the *heiau* to be circumcised in the presence of Līloa's gods, transforming him into a proper *ali'i*. Līloa also elevated 'Umi's commoner adopted sons to the status of minor chiefs, *kaukau ali'i*, giving them lands in Hāmākua. Akahiakuleana was made a chiefess over the *ahupua'a* of Kealakaha.

Of course, Līloa already had a sacred royal heir, Hākau, offspring of his *pi'o* marriage to his mother's sister Pinea. Hākau was just a few years older than 'Umi. When he heard that Līloa had a new son, who was being circumcised in the *heiau*, Hākau exploded with anger. Bursting into Līloa's house, Hākau angrily confronted his father. Līloa tried to calm Hākau, saying to him, "You will be king, and he will be your man. You will have authority over him." Hākau pretended to be reconciled to his new half-brother, but in his heart he nurtured resentment and hate.

'Umi and his three companions became part of Līloa's royal court. The older warrior chiefs trained them in spear fighting, sham battles, racing, and the deadly use of the sling with rounded river pebbles, as befitted their new status as *ali'i*. Pi'imaiwa'a, in particular, excelled in these martial arts, skills that would later help him to propel 'Umi to the kingship.

After a few years, when Līloa grew old and was nearing death, he gathered the *ali'i* and courtiers to issue his final command regarding the succession of kingship. The old king lay weakly on a pile of fine mats and scented barkcloth in his *hale noa*. 'Umi was by now a young man, fully versed in all the arts and skills of an *ali'i*. As expected, Līloa told his court that the kingdom would be Hākau's to rule. But the dying king's next words must have electrified the assembled audience. To 'Umi, now known as 'Umi-a-Līloa, Līloa gave the responsibility of caring for the feared war god, the "government-snatching god," Kūkā'ilimoku. With its "skin" of brilliant birds' feathers, flashing eyes of pearl shell, and gaping mouth edged with rows of dogs' teeth, this god traced back to Līloa's own ancestor Pili. To be charged with its care was a huge responsibility, but also one that potentially gave 'Umi immense power. For 'Umi, this new charge was not wholly unexpected. His father had more than once confided in him that after his death 'Umi might need to flee Waipi'o to live in poverty in the countryside. But always, Līloa insisted, 'Umi must honor this god, who would in turn reward him. On his deathbed, Līloa uttered these words to his beloved 'Umi: *"Noho 'ilihune"* (Live humbly).

. . .

Līloa's bones were stripped of their flesh, scraped clean, and enveloped in a kind of anthropomorphic casket of woven *'ie'ie* vines, called a *kā'ai*. The *kā'ai* resembled an armless and legless torso, the head bearing pearl-shell eyes much like the image of the war god. Enshrined in this casket, Līloa's bones were placed in a special sepulchral temple named the Hale o Līloa. The burial rites concluded, Hākau redistributed the lands of his kingdom, following the ancient practice. He took up his new role as *ali'i nui*. But Hākau was not at all like his father, widely regarded as a wise and kind ruler. Hākau was haughty, with a switchblade temper. He continually abused his courtiers, verbally and physically. His wrath was especially directed at 'Umi, whom he had resented from the day that 'Umi first appeared with his backcountry companions. Hākau, reared by his mother Pinea to believe that *pi'o* chiefs such as himself were gods on earth, despised 'Umi with his commoner origins. He especially resented that Līloa had entrusted the sacred war god to 'Umi's care. 'Umi began to fear for his life. Remembering his late father's words, 'Umi decided to leave Waipi'o.

Taking with him the war god Kūkā'ilimoku, 'Umi and his companions fled through the forest to the far boundary of Hāmākua district, near Hilo. There 'Umi hid Kūkā'ilimoku in a cave at the cliff called Hōkuli. The young men posed as commoners, taking up residence in the *ahupua'a* of Waipunalei. Handsome and skilled, they quickly attracted wives among the available young women of Waipunalei. 'Umi, in fact, attracted four young women to be his cowives. 'Omaʻokamau, Piʻimaiwaʻa, and Koʻi refused to let their high chief do any menial work. The three "sons" of 'Umi would spend long days in the upland gardens tending their fields, while 'Umi and his wives rode the surf at Laupāhoehoe.

The parents of 'Umi's wives began to complain to their daughters that the strong young man was lazy and not providing for them. They especially wanted to enjoy the taste of fresh *aku* (bonito). Catching *aku* required considerable skill: the canoe had to be paddled fast while a lure of pearl shell with a bone point was trolled from the stern on a short line. This was a task that even a chief in disguise could assent to. 'Umi offered to help paddle the *aku* fishing canoe. Each day, when the canoe returned to shore, the catch was divided among the paddlers, and 'Umi received his share. Always he took some of his share secretly to his feathered war god, hidden in the cliffside cave.

The news that 'Umi had fled Waipiʻo had spread throughout the island, though no one seemed to know where the young chief and keeper of the war god had gone. Kaleiokū (his name means "The Necklace of Kū") was a priest also living at Waipunalei, the younger brother of two powerful old priests who had served Līloa in Waipiʻo. Kaleiokū noticed that ever since the young strangers had arrived, rainbows had begun to appear frequently over the cliff at Hōkuli. He also observed that the handsome young *aku* fisherman would go that way after returning from his fishing trips. Kaleiokū began to suspect that this striking young man who excelled at surfing and who walked with a chiefly bearing was none other than 'Umi. Kaleiokū devised a scheme to unmask his identity. Black pigs were known to have the power to identify chiefs. Kaleiokū procured such a pig, saying, "Here is the pig, O god, a chief-searching pig." When Kaleiokū released the pig, it ran immediately to 'Umi's feet. The priest demanded, "Are you 'Umi?" *"Ae, owau no"* (It is I), came the answer. Kaleiokū had found his chief; he swore his loyalty to 'Umi.

With Kaleiokū as his spiritual adviser and confidant, 'Umi began to plot his revenge against the evil Hākau. First, Kaleiokū ordered the construction of several large eating houses or sheds *(halau)*, so that he could enjoin travelers passing through their lands to partake of 'Umi's hospitality. This, of course, built up 'Umi's reputation among the people as a generous chief. 'Omaʻokamau and the other young men had already cleared and planted extensive upland gardens. Under Kaleiokū's direction, the people intensified their agricultural labors, raising dryland taro and sweet potatoes, and feeding the excess vegetables to large herds of pigs and dogs. All this was necessary to underwrite 'Umi's hospitality, and to support his army. And, not least of all, 'Umi—aided by his loyal companions Ko'i, 'Oma'okamau, and Pi'imaiwa'a—began to train the men of backcountry Hāmākua in the arts of war, building up an army with which to seize Waipi'o and take the kingdom from Hākau.

While Kaleiokū and 'Umi were raising up their army in the hinterlands of Hāmākua, Hākau continued his abusive ways in Waipi'o. Two elderly priests in Waipi'o, Nunu and Kakohe, the older brothers of Kaleiokū, had once been the custodians of Kūkā'ilimoku under Līloa. These priests were deeply insulted by Hākau, who had refused them a request of food, meat, and 'awa root. The priests decided to make a secret visit to find 'Umi, to judge for themselves the rumors of the young chief's hospitality. When word reached Kaleiokū that the old priests were coming, he devised a clever scheme to impress them. When Nunu and Kakohe arrived at Waipunalei, it was arranged that only 'Umi would be there to greet them (pretending to be just a commoner of the area). 'Umi would offer them outstanding hospitality. By preparing everything in advance, 'Umi amazed the old priests by having succulent roast pig and other foods ready right after the weary travelers arrived. He poured them bowls of 'awa to drink after their meal. After sleeping off the effects of the 'awa, Nunu and Kakohe awoke in the afternoon as Kaleiokū and the others were returning from the upland fields. The priests asked their younger brother, "Where is 'Umi?" Upon hearing that he was none other than the man they had assumed to be a lowly commoner, the old priests were beset with shame, for they had let this high chief serve them. To repay his great generosity, Nunu and Kakohe told 'Umi that they would see to it that the kingdom of Hawai'i would become his: "We have no riches or property to give him in return for his service; the only great property in our keeping is the whole of the island of Hawai'i; let that be our present then to the chief 'Umi."

Kaleiokū asked Nunu and Kakohe how two aged priests who had lost the favor of Hākau could promise to hand over the kingdom to 'Umi. He knew that Hākau had a large force of well-trained warriors, and would not easily be defeated. But the old priests had a clever plan. They would let 'Umi's army down the walls of Waipi'o Valley at a time when most of the chiefs and warriors had gone to the mountains to fetch 'ōhi'a wood for the haku'ōhi'a ceremony. This was a key rite in the annual rededication of the *luakini* (war) temples, at the end of the Makahiki period. Since the *haku'ōhi'a* rites marked the commencement of the period during which war could be made, a period sacred to Kū, a human sacrifice was required. What Hākau would not know was that he was the intended sacrificial victim.

After spending some time with Kaleiokū and 'Umi, the old priests Nunu and

Kakohe made their journey back to Waipiʻo. They left explicit instructions for when ʻUmi should lead his army to the valley. Kaleiokū counted down the sacred nights of the Hawaiian lunar calendar. On the night of ʻOle, during the moon's waning phase, he gave the signal for ʻUmi's army to begin its march toward Waipiʻo. After six days the force arrived at Kemamo, directly above Waipiʻo. It was now the night of Kāne, twenty-seventh in the Hawaiian lunar sequence. The following day—the day of Lono—had been declared the *kapu* day for all of the valley's people to go and seek the *ʻōhiʻa* timbers in the mountains. Only Hākau, along with his personal servant and the old priests, would remain at the royal compound.

At Kemamo, ʻUmi's men collected hard basalt stones, carefully wrapping them in *kī*-leaf bundles so that they resembled sweet potatoes ready for cooking. As the sun rose over the valley, the warriors and people of Waipiʻo climbed the mountain trail up into the forests to search out the *ʻōhiʻa* logs. Into the nearly deserted valley ʻUmi's army now descended following the same trail that ʻUmi had taken years before, when he first went to meet his father, Līloa. From the royal residence across the river, Hākau spied the shadows of people coming down the cliff, carrying the *kī*-leaf-wrapped bundles. Nunu and Kakohe reassured Hākau, saying, "They must be your own men from Hāmākua bringing you some food."

The procession continued down the cliff and across the river, until they had assembled in a circle some twelve men deep surrounding Hākau. As ʻUmi stepped forward out of the crowd, Hākau realized too late his fate. The *kī*-leaf-wrapped bundles were not offerings of sweet potatoes destined for Hākau's cookhouse, but hard stones. ʻOmaʻokamau strode over to the hated king, who now sat with his head bowed. Grabbing the king by his lower jaw, ʻOmaʻokamau pulled Hākau's face up to meet his gaze, saying, "You are killed by ʻOmaʻokamau, for ʻUmi." The mob began hurling their bundles of stones until his body was buried under a great heap. Hākau's body was then taken up to the *lele* altar of Honuaʻula *heiau*. As Kamakau recounts, when ʻUmi-a-Līloa laid his defeated half-brother's body upon the altar, "the tongue of the god came down from heaven, without the body being seen. The tongue quivered downward to the altar, accompanied by thunder and lightning, and took away all the sacrifices."

· · ·

Few in Waipiʻo mourned the passing of this evil king. Some of the warriors, however, fled back to their home districts of Kona, Kaʻū, and Puna, where they would resist ʻUmi's rule. Nonetheless, ʻUmi made the traditional division of lands, the prerogative of a new king upon his succession to power. Pi'imaiwa'a received Hāmākua district to rule over as the *ali'i-'ai-moku*. Ka'ū was given to 'Oma'okamau, and Kohala to Ko'i. As his reward for aiding 'Umi, Kaleiokū was elevated to the status of *kahuna nui*, the chief priest, as well as being made *ali'i* over Hilo. The old priests Nunu and Kakohe were also rewarded with grants of *ahupua'a* lands in Hāmākua.

While 'Umi had taken four commoner wives in Waipunalei, he was now the *ali'i nui* of the entire island. It was incumbent upon him to sire one or more potential heirs of the highest rank. This was especially important because of 'Umi's own relatively low status, owing to his mother's commoner line. Thus 'Umi took his own half-sister Kapukini-a-Līloa, with an impeccable *pi'o* pedigree, to be his royal mate. Kapukini-a- Līloa bore three sons for 'Umi: Keli'iokaloa, Kapulani, and Keawenui, two of whom would succeed him as rulers of Hawai'i Island.

Though nominally king of the entire island of Hawai'i, 'Umi at first seems to have had real authority over only the northern and windward part of the island, especially the region from Kohala to Hāmākua. 'Umi needed to subdue rebellious chiefs in the southern and leeward parts of the island. Hua'a, the chief of Puna, was one of the first to be vanquished by 'Umi, at a battle in Kea'au, south of Hilo. Next 'Umi and his followers challenged Imaikalani, ruler of Ka'ū, the land of Pele and her volcanoes. Finally, Ehunuikaimalino, the chief of leeward Kona district, was persuaded that resistance against 'Umi was futile. He ceded his district with its rich fishing grounds to 'Umi.

It may have been at this time, after the cession of Kona, that 'Umi commanded the construction of the Ahu a 'Umi on the upland saddle inland of Hualālai volcano, to commemorate his unification of the entire Hawai'i Island kingdom. 'Umi now moved his court out of the ancestral seat in rainy Waipi'o to sun-baked Kona. Several temples in Kona are supposed to have been constructed by 'Umi at this time. Although Waipi'o remained an important place, and its temples continued to be maintained, Kona would henceforth become the principal home of the rulers of Hawai'i.

At the urging of his high priest, Kaleiokū, 'Umi sent 'Oma'okamau as his emissary to Pi'ilani, king of Maui, asking permission to bring back Pi'ilani's daughter Pi'ikea as yet another royal wife for 'Umi. This was a stroke of diplomatic genius on the part of Kaleiokū, for it at once united the two largest kingdoms in the archipelago. Pi'ilani consented to the request. Pi'ikea accompanied 'Oma'okamau on his double-hulled canoe back to Hawai'i to marry 'Umi. She bore him two children, Kumalaenui-a-'Umi and Aihakoko.

With 'Umi's reign Hawai'i Island entered a period of peace and prosperity. The traditions credit 'Umi with being a renowned farmer as well as a fisherman. He not only expanded the irrigation works in Waipi'o but, after moving his seat to Kona, oversaw the expansion of the vast upland gardens on the rich volcanic slopes of Kohala, Hualālai, and Mauna Loa. As a consequence, 'Umi is sometimes called *he ali'i mahi'ai* (the farmer king). But his accomplishments extended far beyond the development of his dominion's economic base. The first to subjugate all the island's districts, 'Umi was an effective administrator. 'Umi promulgated the division of the people into distinct classes of chiefs, priests, specialists of various kinds, and general laborers. Just as Mā'ilikūkahi had done earlier on O'ahu, 'Umi brought the land system on Hawai'i into order, with a hierarchical system of *ahupua'a*, subdivided into *'ili*, and on down to the individual plots cultivated by the *maka'āinana* farmers. All of this was administered through a parallel hierarchical system of temples, of which 'Umi constructed and dedicated a great many. 'Umi never forgot his father's admonition to honor the gods through whom he ruled.

TWELVE · 'Umi's Dryland Gardens

Kohala, geologically the oldest part of Hawai'i and that great island's northern peninsula, rose from the sea as a separate volcano long before Mauna Kea and Mauna Loa began to spew forth their lavas from the hot spot on the ocean floor. Kohala's mountainous spine rises 5,480 feet above sea level, high enough to capture the moisture-laden trade winds. With most of the rain falling on the windward side of the peninsula, the leeward (west-facing) slopes have barely been sculpted by erosion. Here and there a few shallow gulches are etched into the landscape, but for the most part the terrain follows the undulations created by the lava flows that cascaded down from the volcanic rift zone. Although permanent streams are lacking on these leeward slopes, the soils are fertile, rich in phosphorus and other rock-derived nutrients. Sufficient rain blows over the mountain crest, driven by Kohala's incessant *mumuku* winds, to water all but the thirstiest crops, at least on the upper slopes. As one nears the coast, however, the rain disappears and the land turns rocky and parched.

Many cinder cones, spewed out during the final stage of volcanic activity in Kohala, dot the mountain slopes like huge pimples on the face of some sleeping giant. The Hawaiians called these cinder cones *pu'u*. Pu'u Kehena, situated at 2,517 feet above sea level, juts up from the rolling pasturelands of Parker Ranch. If you were to climb the steep side of Pu'u Kehena in the late afternoon, as I have done many times, you would be treated to a most amazing sight. Catching your breath after the steep ascent and struggling not to be blown over by the strong wind, you take your bearings. Behind you, to the east, lies the long, forest-cloaked crest of the Kohala Mountains, shrouded in cloud. A few other *puʻu* jut up here and there out of the mist. To the west, the land stretches away in a gradual slope descending to the distant silver-blue ocean. Far away to the south, the looming snow-capped summits of Mauna Kea and Mauna Loa glisten above the clouds. Your eyes focus now the vast slope below the *puʻu*, illuminated by the late afternoon sunlight. From the base of the cinder cone where you started to climb, this slope appeared to be endless pasture. You now see, to your amazement, that it is indelibly marked with the imprint of hundreds of low ridges or embankments, regularly spaced about thirty feet apart (see color plate 5). You can see these ridges clearly now because the slanting rays of the afternoon sun create low shadows on their eastern sides. If you were standing here at midday, with the sun directly overhead, the ridges would blend into the undulating green of the pasture. The regularly spaced ridges follow the contour of the land. Studying the landscape closely, you can make out other lines that cut across the ridges perpendicularly; these run up and down the slope. The ridges and the cross-slope lines subdivide the landscape into a vast reticulate grid, as though some perverse deity had pressed an immense waffle iron into the leeward side of the Kohala peninsula.

What you see before you—extending south, west, and north as far as your eye can make out—is the imprint left by centuries of intensive farming. Covering more than twenty-three continuous square miles, the leeward Kohala field system is surely one of the greatest archaeological complexes in the entire Pacific. Here the ancient Hawaiians developed a remarkable agricultural system based on the cultivation of sweet potatoes, dryland taro, sugarcane, and other crops that could be raised on the fertile volcanic slopes, watered only by the rains blown over the mountain crest on the *mumuku* winds. The low ridges are the borders of ancient fields, while the cross-cutting lines are the curbstones of trails that ran up and down the slopes, connecting the upland gardens with the coastal hamlets. For more than four centuries, Hawaiian farmers and their families worked this land, using their wooden digging sticks to clear and plant. On his tours around the island, ʻUmi took note of these fertile slopes and their rain-fed fields of sweet potato and taro. After he moved from rainy Waipiʻo Valley to the leeward side of Hawaiʻi, ʻUmi and his chiefs could directly control the dryland agricultural systems, not just in Kohala but also in Kona district. The abundant yields of these gardens provided the economic base for ʻUmi's kingdom, and for the rulers who came after him.

The first European visitor to visit and describe this part of Kohala was Archibald Menzies, surgeon and naturalist with British captain George Vancouver, in 1793.

Although he saw Kohala only from shipboard, as they coasted along the shore, Menzies could make out the productive landscape of gardens. "The country... bears every appearance of industrious cultivation by the number of small fields into which it is laid out," he wrote, "and if we might judge by the vast number of houses we saw along the shore, it is by far the most populous part we had yet seen of the island." Forty years later, Protestant missionaries in nearby Waimea observed, "The soil of Kohala is good. In different parts anything can be cultivated." But even then, in the early 1830s, this vast and productive agricultural landscape, which had supported a high population density for centuries, was on the verge of collapse. As foreign diseases devastated the people, the population crashed. There were no longer enough hands to till Kohala's fertile soils. To add to the misery, the Haole had introduced cattle and sheep, which overran the neat grid of plantations. By the early 1860s, Father Bond, a missionary stationed in Kohala, wrote despairingly of the herds of cattle, which "well nigh annihilated all possibility of cultivation." By the end of the nineteenth century, Kohala's vast dryland field system was a distant memory.

Scholars of Hawaiian history and culture forgot that Kohala had once supported a huge system of intensive farms. In the 1930s, Edward Handy of the Bishop Museum made a systematic study of Hawaiian farming, traveling around the islands and interviewing old Hawaiian planters who still had knowledge of the traditional agricultural practices. In Handy's published account, Kohala is scarcely mentioned. Handy was not even aware that a mere century earlier one of the greatest of all Hawaiian agricultural systems had flourished on those inland slopes.

In the late 1960s, archaeologists in Polynesia started to ask questions about how the Polynesians had adapted to their island ecosystems, including the ways in which they had farmed the land. At the University of Hawai'i, Stell Newman had enrolled as a graduate student under the tutelage of Professor Richard Pearson, a proponent of the new ecological approach in archaeology. Newman was beginning a second career as an archaeologist, having already been a pilot with the U.S. Air Force's Strategic Air Command in Nebraska. To help support himself in graduate school, Newman started working for a local air tour company in Hawai'i, flying tourists around in a twin-engine Cessna, giving them eye-popping views of waterfalls and volcanoes. Whenever he passed low over Kohala on the Big Island, Newman noticed the grid of lines that covered much of the Kohala uplands. He saw a similar pattern on the slopes above Kealakekua Bay, in Kona. With his eyes now informed by new training in archaeology, Newman realized that these gridwork patterns must be the vestiges of ancient Hawaiian fields.

Pearson had obtained a grant from the National Science Foundation to help train undergraduates in archaeological field methods. He suggested that they use this opportunity to carry out research for Newman's doctoral thesis. Newman checked out Kohala, where he had seen the field walls from his airplane. A ramshackle hotel in the little plantation town of Hāwī was available to accommodate the students. Consulting with the state of Hawai'i's Department of Land and Natural Resources, Pearson was told that an *ahupua'a* named Lapakahi ran from the coast to the uplands, into the area of the old field system. Permission to dig was not only available, but the state was interested in the possibility of developing a historical park out of the ruins of the coastal village site. In the summer of 1968, Newman and ten students went to work at Lapakahi to learn what they could about how the precontact Hawaiians had farmed and fished in Kohala.

With his experience in the Strategic Air Command, Newman also knew about the potential of using aerial photographs to map the endless lines of field embankments and cross-cutting trails that covered large parts of the Kohala uplands. Today, one can call up an image of virtually any place in the world using Google Earth; one forgets that such technology did not exist in the late 1960s. If you wanted air photo coverage of an area, you had to arrange for someone with a specially equipped plane to fly over it. Newman got the R. M. Towill Corporation of Honolulu to fly a plane with a large-format camera mounted in the belly of the fuselage over the Lapakahi area. When developed and printed, the 9 × 9 inch high-resolution photos showed the field lines clearly. Using enlargements of the photos, Newman created a map of the field system as it had extended over the interior parts of Lapakahi. The long-forgotten dryland gardens of Kohala had been rediscovered and brought to light by the tools of modern archaeology.

Throughout the summer of 1968, and again for two summers after that, the University of Hawai'i's archaeological field school toiled away in the *kiawe* groves of Koai'e village and in the brush-covered uplands. Another doctoral student, Paul Rosendahl, made a detailed on-the-ground map of part of the upland field system, documenting the field walls and showing that the lines crossing the field ridges were curbstone-lined trails. These trails had connected the coastal fishing settlements with the upland gardens. The trails also seemed to run along the boundaries between the *ahupua'a* as well as the smaller *'ili* land units, functioning as territorial markers in addition to facilitating communication between coastal and upland populations. Rosendahl also showed that the farmers had lived in the midst of their upland fields, at least seasonally. He found dozens of small C- and L-shaped enclosures and other stone structures that had been the foundations for thatched houses.

Excavating these, Rosendahl found hearths and earth ovens where the garden dwellers had roasted their sweet potatoes. Rosendahl sent off charcoal from these hearths for radiocarbon dating. The results showed that Hawaiians were developing the upland gardens of Kohala in the sixteenth century, roughly the time that 'Umi defeated Hākau and took control of the entire island of Hawai'i.

'Umi no doubt was aware, from touring his dominions, of the productivity of this fast-developing field system. He could see that by moving the seat of his kingdom from Waipi'o to the leeward side, he and his subchiefs, along with their land managers, could assume direct control over this economic system, enabling them to tap into the huge potential for surplus. The excess production above the immediate needs of the local farmers, whether directly in the form of sweet potatoes and taro, or as pigs raised and fattened on the tubers, would enable 'Umi and his successors to build a political economy on a scale that Polynesia had not before witnessed. This surplus could support craft specialists and retainers attached to the royal household, as well as warriors and priests. The increasing cadre of *ali'i*, rapidly emerging as a distinct class within Hawaiian society, did not produce their own food. They were sustained by the *ho'okupu*, the tribute of the common people, the farmers and fishermen. It was in 'Umi's time, from what we can deduce, that distinctions between *nā li'i*, the chiefs, and *nā kanaka*, the people, began to become formalized. As in the course of political evolution everywhere, it was developments in agriculture—and particularly the ability to produce a surplus—that made all this possible.

. . .

Julie Field and her students had just completed a long day of digging in a stone-walled house terrace situated within the ancient fields of Makiloa *ahupua'a*, a few miles south of Lapakahi. It was June 2008, and we were in the second year of a National Science Foundation–funded project to study the dynamic linkages among ancient households, population growth, and agricultural intensification in the Kohala field system. Field, who had recently completed her doctorate at the University of Hawai'i, was a postdoctoral fellow in charge of the household site excavations. It was difficult work in the Kohala uplands, exposed all day long to the blasting *mumuku* winds, which whipped up the dust and gravel as the diggers scraped at the reddish-brown sediment with their trowels. The team members wore safety goggles to keep the grit from sand-blasting their eyes, and wrapped bandanas around their heads. While Field and the students dug, Thegn Ladefoged and I spent the day using our GPS recorders to trace the system of field walls and map the locations of several temple foundations.

Ladefoged drove our four-wheel-drive pickup truck over the rutted ranch track to pick up Field and her diggers. As the truck bounced along, Field recounted the day's finds. "Mostly the same as yesterday," she reported. "We uncovered a nice fire hearth, and found some small scraps of shellfish and bone," referring to the remains of ancient meals that they had picked out of the fine-meshed sifting screens. "We did find a really unusual artifact, a worked stone with a hole in it. I've never seen anything like this before." My curiosity aroused, I asked if she could show me the stone. Field dug into her backpack, pulling out a plastic specimen bag carefully labeled with the grid unit, level, and depth at which the stone had been found.

I opened the bag and pulled out the shaftlike, hard stone object about three inches long, turning it over in my hand. I could immediately see that it had been carefully shaped. One end had two knobs or lugs; below these a perforation had been drilled from two sides through the dense basalt, the conical depressions made by the drill meeting in the center. Lacking metal tools, ancient Hawaiians found it no easy task to drill through stone. The work of making that hole must have taken long hours with a stone drill. Unfortunately, the object had broken across the perforation, so that all I held in my hand was one end of an artifact that had once been longer.

"Any idea what that is?" Field asked from the backseat of the pickup. I knew that I had seen something like this before, or at least a picture of it. The twin lugs were very distinctive. "It reminds me of a drawing in Hiroa's book on Hawaiian artifacts," I replied. "I think it might be the end of a stone club."

"Wow! A stone club! Wouldn't *that* be cool?" chimed in the student whose sharp eyes had first spied the worked stone in the sifting screen.

Back at our field house in Niuli'i village, one of my grad students had a copy of Hiroa's *Arts and Crafts of Hawai'i*, the last book written by that famous Polynesian anthropologist and director of the Bishop Museum. I thumbed the table of contents, looking for the section on "War and Weapons." There on page 458 I found the drawing that I had recalled. It was a *pīkoi*, a stone tripping club. Our specimen perfectly matched the end through which a long cord had once been attached. Hiroa's text explained that "the weighted rope was thrown at an opponent's legs to trip him." "It may be assumed," Hiroa drily observed, "that a successful throw was speedily followed up with some other weapon to complete the job." The *pīkoi* was a unique invention of the old Hawaiian warriors, not being found elsewhere in Polynesia.

It was fascinating to think that the prosaic little stone house foundation nestled within the upland fields of Makiloa, which Field and her crew were in the process of excavating, had once been the home of a man who had been not only a farmer but—when the times required—a warrior as well. While there were full-time warriors in ancient Hawai'i, many of the rank and file of Hawaiian armies also farmed or fished in times of peace. Indeed, one way to obtain a good tract of farming land was to support your chief in times of war. We know that after Kamehameha the Great conquered O'ahu Island in 1804, he gave the best irrigation lands to his warriors, displacing the old O'ahu families.

The men of Kohala had a reputation as fierce fighters and defenders of their lands. An old Hawaiian proverb speaks to the vigor and bravery of the Kohala warriors: *I 'ike 'ia no o Kohala i ka pae kō, a o ka pae kō ia kole ai ka waha*, "One can recognize Kohala by her rows of sugarcane, which can make the mouth raw when chewed." Like many Hawaiian proverbs, this is a metaphor, comparing the Kohala warriors to the stiff rows of sugarcane that grew on the ridges between the cultivated fields. The cane acted as windbreaks, to protect the vulnerable taro and sweet potato plants. Just as the cane grew straight and tough, so were the ranks of Kohala warriors.

The dryland gardens of Kohala, along with even more extensive field systems in Kona and Ka'ū to the south, became increasingly organized and managed by 'Umi and his successors. As word of their productivity spread, these rich lands became the target of rival kings on Maui Island, across the 'Alenuihāhā Channel. In the early seventeenth century, the reigning king of Maui, Kamalālāwalu, vowed to conquer the leeward region of Hawai'i. He first dispatched his son Kauhi-a-Kama to Hawai'i on a reconnaissance mission, to determine the numbers of the enemy. Kauhi-a-Kama sailed his canoe across the blustery channel, landing at Puakea in Kohala. From there he traveled along the coastline as far as Kawaihae, but saw few people en route. It did not occur to Kauhi-a-Kama that the majority of the population lived in the fertile uplands, several miles from the arid coast, which was dotted only with small fishing settlements. Because the great field system lies several miles inland, the king's son could not see the thousands of small thatched houses nestled among the fields, their edges defined by rows of waving sugarcane. Proceeding south to Kona, Kauhi-a-Kama made the same erroneous observation about the population there.

On returning to Maui, Kauhi-a-Kama reported to his father, King Kamalālāwalu, "Kohala is depopulated; the people are only at the beach." It was fatally flawed intelligence. Thinking that his army would meet little resistance, Kamalālāwalu proceeded with his plan to invade Hawai'i, disregarding the advice of the wise priest Lanikāula, who warned Kamalālāwalu that he would never defeat his Hawai'i Island rival, Lonoikamakahiki. Kamalālāwalu led his fleet of war canoes from Maui southward past Kohala, landing at Puakō. At first they met with little resistance. Emboldened, the Maui forces marched inland to the plains outside of what is today the ranching town of Waimea. They had walked into a trap, for on this wide open plain they were surrounded by the vast army of Lonoikamakahiki, including the fierce warriors of Kohala. Kamalālāwalu realized too late the true size of the Hawai'i forces. His army was routed and Kamalālāwalu himself killed on the Waimea battlefield. Only Kauhi-a-Kama and few others managed to flee and return to Maui.

Had the carved end of the stone club from the Makiloa house site seen action in the famous battle between Lonoikamakahiki and Kamalālāwalu, when the Kohala warriors defended their lands against the Maui invaders? Perhaps the *pīkoi* had broken in the heat of battle, after felling a Maui warrior. Its owner might have carried home with him the woven cord still attached to the broken tip of the shaft, as a souvenir of that day. This is certainly possible, though beyond the power of archaeology to say for certain.

· · ·

In 1995, I began to study the settlement pattern and landscape of Kahikinui, on the leeward slopes of Maui. Kahikinui, like Kohala, had been a major zone of dryland cultivation, although there the sweet potatoes were grown in the fertile swales between lava ridges. Together with my Berkeley graduate students, I was trying to understand how the Hawaiian farmers had managed to extract a living out of a seemingly harsh lava landscape. About the same time, Michael Graves at the University of Hawai'i, along with Thegn Ladefoged of Auckland University in New Zealand, began a new study of the field system in Kohala. My project in Kahikinui, and Ladefoged and Graves's work in Kohala, were beginning to provide new data on Hawaiian life and agriculture on the dryland, leeward slopes of the islands.

In 2000, the National Science Foundation put out a call for new research initiatives designed to explore "dynamically coupled human and natural systems." This short phrase perfectly captures the relationship between a human population and the environment that supports it. I thought the NSF might be interested in a project that leveraged our ongoing research on ancient Hawaiian dryland agriculture. From my office at Berkeley I telephoned Peter Vitousek over at Stanford. (Our universities may be football rivals, but in matters of research, collaboration is essential.) Vitousek, like myself a *kamaʻāina* ("child-of-the-land") of Hawaiʻi, is an ecologist who studies how nutrients cycle through Hawaiian ecosystems, helping to control forest growth. He had also seen the NSF's announcement. We agreed to submit a joint proposal to the Biocomplexity in the Environment program, proposing Hawaiʻi as a "model system" for investigating the complex relationships among agricultural populations, their farming systems, and the natural landscape over a time span of centuries.

Vitousek and I quickly assembled other members of the team that would be required for such a complex project. Graves and Ladefoged would be key partners because of their ongoing work on the Kohala agricultural field system. An expert on Hawaiian soils was essential; Vitousek suggested Oliver Chadwick of the University of California, Santa Barbara, a self-described "renegade" soil scientist who loves nothing better than sitting with his head below ground in a 3×3 foot pit, studying a soil profile. Next we approached Shripad Tuljapurkar, a demographer and population biologist. Tuljapurkar's role would be to develop computer simulation models of sweet potato production and population dynamics. We could then test these simulations against the "hard" data being generated by the archaeological surveys and excavations. Finally, Sara Hotchkiss of the University of Wisconsin would be our archaeobotanist, helping us to reconstruct the ancient vegetation of these landscapes.

Our proposal received high marks from the NSF's review panel. By January 2001 we had funds in hand for three years of intensive research. Our Hawaiʻi Biocomplexity Project, as we now called it, focused on the ways in which the Hawaiians adapted their agricultural systems to the landscape. We dug soil pits and analyzed hundreds of samples for soil nutrients, mapping out spatial patterns. Meanwhile, Tuljapurkar and his post-doc, Charlotte Lee, developed a sophisticated computer model that estimated the yield of sweet potatoes over the Kohala slopes, given parameters of soil chemistry and rainfall. In 2005 the NSF granted our team renewed support from its Human Social Dynamics program. Building on our initial results, we now turned to studying the archaeological remains of households to tease out the linkages between a rapidly growing population and the agricultural system whose output kept all those hungry mouths fed. For three summers we surveyed and dug into house sites in the *ahupuaʻa* of Makiloa, where Julie Field's team found the *pīkoi* fragment, and in Kaiholena, farther to the north.

Let me sum up what we have learned over the course of ten years of multidisciplinary research in Kahikinui and Kohala, about how the great Hawaiian dryland field systems developed over time, about the populations they nourished and fed, and about the political economy they supported. The rise of archaic states in the large eastern islands of Hawai'i and Maui, from the time of 'Umi until the arrival of Cook, was intimately bound up with these dryland gardens. It is a key part of the story of ancient Hawai'i.

The first thing we wanted to find out was how the field systems in these leeward regions were distributed over the lava slopes. The slopes vary in geological age and in their annual rainfall, crucial factors that would have affected the crops. The field systems depended entirely on rainfall to water the crops, so spatial variations in rain must have been vitally important. Oliver Chadwick's work had shown that over hundreds of thousands of years, the hard rock of Hawaiian lava gradually turns to soil. Phosphorus, an essential plant nutrient, originates as a component of the lava itself. With rainfall and water seeping into the rock, the phosphorus is gradually released, some to be taken up by plants, some forming insoluble compounds, some eroding away. The rate at which weathering proceeds is a function of rainfall. Rainfall can be as low as 10 to 15 inches per year on the leeward coasts, rising to 40 to 60 inches in the center of the Kohala field system. It soars up to an incredible 250 inches or more on the windward mountain crests. In areas of high rainfall, phosphorus and other rock-derived nutrients get almost completely leached out of the soil within a few thousand years.

Our team collected soil samples on transect lines running up and down slopes, crossing the Kohala field system, and also across the upland cultivation zone of Kahikinui on Maui. In the lab, Chadwick and his post-doc Tony Hartshorn measured the quantities of phosphorus and other elements, as well as other key soil properties of our samples. We found that on both islands, the zone of intensively cultivated land coincides with the highest concentrations of soil nutrients. The Hawaiians had figured out for themselves (through trial and error, no doubt, as they lacked soil science labs) where the soils were most productive. They then concentrated their labor on those productive areas. In general, these were zones where the soils were still relatively young, and not so high in rainfall that the key nutrients had been washed out. For Kohala, the field system turned out to correspond to a "sweet spot" that ran along the leeward mountain slopes, with rainfall between 30 and 60 inches. Below 30 inches, annual rainfall was too low to support a crop. Above 60 inches, there had been too much leaching, and phosphorus levels were depressed. As Vitousek put it one day when we were discussing our results, the Hawaiians had figured out how to "farm the rock"—not rock in the literal sense, perhaps, but young lava flows combined with just enough rain, not so much as to have leached out the nutrients that would sustain intensive cropping, year after year. Significantly, our maps showed not only that the Hawaiians had found the sweet spots, but that their dryland field systems had been pushed to the limits of those zones. By the time the field systems were at their maximum extent, they had pretty much expanded to cover all of the terrain suitable for growing sweet potatoes and dryland taro using only the water provided by Lono's rains.

Next, we wanted to find out *when* the dryland fields had been developed. In Kahikinui, I had already obtained a large number of radiocarbon dates from house sites associated with swale gardens. These suggested that people first began to move onto the leeward slopes of Haleakalā about 1400. Over succeeding centuries the population had gradually increased in size, as described in chapter 10. For Kohala, we had only a few radiocarbon dates available, mostly those obtained by Paul Rosendahl at Lapakahi more than thirty years earlier. Ladefoged and Graves decided to see whether they could date the Kohala field ridges directly. A backhoe was brought in to dig trenches through the low field embankments that were so visible in the late afternoon sunlight. After the backhoe had roughed out a trench, the archaeologists and soil scientists climbed down into the cut and began cleaning the profiles with their trowels. When seen in cross-section, the field ridges were accumulations of soil and stones, thrown up gradually as farmers tossed weeds and unwanted rocks out of their plots and into the rows of sugarcane that had once edged the fields. Under these accumulations, they could make out horizontal zones that were the original soil horizon. Sometimes these contained charcoal and ash, signs of burning, probably from initial clearing of the fields. Thirty-three samples of charcoal from under these field walls were carefully collected and sent off to Beta Analytic for AMS radiocarbon dating. One sample was a piece of carbonized sweet potato tuber, proof that this had been one of the crops grown in the fields.

The radiocarbon dates showed that the Kohala field system developed over the course of more than four centuries, much as in Kahikinui. The carbonized fragment of sweet potato gave one of the oldest dates, between 1290 and 1430. Thus Hawaiian farmers were beginning to grow their sweet potatoes on these leeward slopes as early as the fourteenth century. More dates come from a slightly later time, indicating that the field system really got going in the 1500s. This was, no doubt, the rapidly expanding and developing system of rain-fed gardens that 'Umi and his subchiefs managed and exploited for its surplus production, pushing the political economy. A final set of radiocarbon dates came from under the last field ridges to be constructed, dating after about 1650, and representing a final phase of intensive development of the field system.

Anthropologists, geographers, and economists who study pre- or nonindustrial farming around the world have long been interested in the process of agricultural intensification. Harold Brookfield, one of the leaders in this field of research, defined intensification as the addition of inputs, such as capital, labor, or skills, so as to increase the yield of a given area of land. As we saw in chapter 10, debates about the relation between population and agriculture have been going on since the Enlightenment. Malthus recognized that humans could reproduce much faster than increases in agricultural production could keep all those new mouths fed. Boserup countered that humans have an amazing propensity for innovation, frequently raising the limits on population. The "Green Revolution" of the 1960s, which raised world agricultural production significantly, is a case in point.

One way for preindustrial populations to increase their agricultural yields was simply to put more labor into their fields and gardens. By carefully weeding, tending their plants, and placing mulch around the sweet potato and taro shoots, the Hawaiian farmers found they could raise their production levels. On the older islands of Oʻahu and Kauaʻi, whose abundant irrigated taro fields produced much higher yields than the dryland rain-fed fields of Maui and Hawaiʻi, only men did the farming. On Maui and Hawaiʻi, however, nineteenth-century Hawaiian historian Samuel Kamakau informs us that women worked alongside their husbands and brothers in the gardens. The dryland field systems were made to produce high yields, but a great deal of labor was required. Still, there are limits to how hard one can push the land, without industrial-scale inputs of petrochemical fertilizers and insecticides (and these, as we are finding out, also have their limits and negative consequences). Early economists, such as David Ricardo and John Stuart Mill, recognized the "law of diminishing returns." Initially, the application of more labor to an agricultural field will produce greater returns. But gradually the amount of increased yield that can be eked out of the same piece of land will begin to decline. As Mill wrote in 1848 in his famous *Principles of Political Economy*, "land differs from the other elements of production, labour and capital, in not being susceptible of indefinite increase." He went on to note that "the produce of land increases . . . in a diminishing ratio to the increase in the labour employed."

During the late fifteenth and early sixteenth centuries, Hawaiian population soared at an exponential rate. The initial high growth rate had been spurred by the discovery of a vast new archipelago in which land seemed to be without limit. As people increased on the well-watered, fertile islands of Oʻahu and Kauaʻi, they developed large-scale irrigation systems across the alluvial valley bottoms. The high productivity of these irrigation works further encouraged population growth, in a positive feedback cycle. By the early 1400s, the most desirable lands were already becoming densely populated.

The vast leeward slopes of east Maui and of Hawai'i Island provided a new frontier for Hawaiian farmers. Moving into these previously unoccupied lands, they discovered the sweet spots of Kahikinui, Kaupō, Kohala, Kona, and Ka'ū, where they could "farm the rock" in areas where the combination of soil and rainfall was ideal. By the close of the sixteenth century, these vast leeward slopes were covered in a patchwork quilt of gardens, defined by trails marking *ahupua'a* boundaries, with long rows of sugarcane defining the field edges. It was at this point that 'Umi and the Hawai'i *ali'i* began to manage these highly productive systems intensely, drawing on the surplus production to support a burgeoning elite hierarchy and incipient state administration. As anthropologist Marshall Sahlins, a famous scholar of Polynesia, once observed, the mobilization of individual households (which are, he noted, inherently *anti*-surplus in their inclinations) for the good of a larger political economy is the key to political evolution. For an emerging political society, "the great challenge," Sahlins wrote, "lies in the intensification of labor: getting people to work more, or more people to work."

Archaeologists studying the remains of agricultural systems that ceased to function hundreds of years ago cannot go out and interview those long-dead farmers about how many hours they were spending in their fields. But the archaeological record does give us clues about the process of intensification, about how people responded to the increasing pressures that came both from growing numbers of people, and from chiefs who wanted a regular supply of tribute to support their retainers and their political aspirations. By about 1600, roughly the time that 'Umi unified Hawai'i Island, our archaeological evidence and radiocarbon dates tell us that the Kohala field system was well established. Over the next two centuries, it would be further intensified. The archaeological evidence for this lies in the grid pattern of field ridges and intersecting trails. The detailed GPS mapping that Ladefoged and Graves had done of the field system in the Makiloa and Kahuā *ahupua'a* to the south of Lapakahi clearly revealed a sequence of field and trail construction. Their work showed that during an early stage of development, the Kohala slopes were subdivided into about nine large *ahupua'a*, whose boundaries were marked by the curbstone-lined trails running from the coast to the uplands. In the favorable upland area, the space between these trails was divided into large garden plots, separated by the low field ridges or embankments, spaced as much as one hundred feet or more apart. Over time, these initially large territories came to be segmented into smaller units, as new trails and boundaries were inserted on the landscape. Each new territory was under the management of a new chief or *konohiki*, reflecting the growing numbers of *ali'i*, each of whom wanted a share in the surplus production of the fields. As the territories shrank in width, individual garden plots also became smaller and smaller, as more field ridges were inserted into the grid pattern. Eventually, by the time of Kamehameha I, leeward Kohala, which started out with nine large *ahupua'a* territories, was now subdivided into no less than thirty-five *ahupua'a*, each under the control of its own chief and land manager.

The intensification of the Kohala field system over a period of four centuries is a remarkable tribute to the skills and energy of the Hawaiian farmers. They took a landscape that was marginal for tropical horticulture and transformed it into a highly productive network of farms and homesteads that extended over twenty-three square miles. As a famous proverb says: *Le'i o Kohala i ka nuku na kānaka* (Covered is Kohala with men to the very point of land). It was this bountiful land, with its endless rows of sugarcane lining sweet potato and taro gardens, dotted everywhere with the thatched houses of the *maka'āinana*, that formed the foundation of Hawai'i's political economy.

And yet, the remarkable intensification of Kohala's dryland gardens was not without its consequences. By the eighteenth century, prior to Cook's arrival, the rain-fed field systems of leeward Hawai'i Island, as well as those of eastern Maui, were being pushed hard. As part of our Hawai'i Biocomplexity Project, we analyzed the nutrient content of the cultivated fields and compared them with soils that had been capped by the field ridges, and hence not cultivated (thus retaining the nutrient status of the original preintensification phase). The phosphorus content of the open field plots was significantly lower than that of the soil pockets under the field ridges. Sustained cultivation of the Kohala gardens over centuries, in spite of all the labor inputs in mulching and tending of the crops, had resulted in measurable declines in soil quality. No doubt the Hawaiian farmers were doing everything they could to keep up their yields. In fact, the rows of sugarcane may have played a role beyond being windbreaks. There is reason to think that the cane roots fixed nitrogen in the soils, which would have enhanced the sweet potato and taro yields.

By the eighteenth century, nonetheless, the populations of Hawai'i and Maui were no longer in a growth phase. As discussed in chapter 10, the rate of population increase leveled off after the end of the sixteenth century. The islands had been through a huge economic boom as the leeward field systems were opened up and developed. The chiefs had seen the potential of this boom, and tapped into it with the system of *ahupua'a* and *konohiki*. Surpluses produced by the thousands of *maka'āinana* households had supported the rise of a distinct *ali'i* class. But now the ability to extract surplus had become more tenuous. The farmers, men and women both working side by side, were pushing their fields to the limits. In good years, when Lono looked favorably on them and the winter rains came early, the gardens still produced enough to both feed the farmers' families and meet the demands of the *konohiki* and his chief. In other years, however, when the rains came late or hardly at all, the young vines and shoots withered in all but the most favorable locations. Hawaiian *mo'olelo* speak of periods of intense famine, when the people were reduced to going into the mountains to fetch tree ferns to eat.

By the early eighteenth century, the kings and chiefs of Hawai'i and Maui had begun to realize that their political futures did not lie in further development of their own islands' agricultural base. They had already pushed that as hard as it could bear. They began to look beyond their individual island territories, began to ponder the possibilities of conquest and expansion. The Hawai'i kings looked to the lands of eastern Maui, while their counterparts on Maui looked westward up the island chain to Moloka'i. Beyond Moloka'i were the greatest prizes of all, O'ahu and Kaua'i, rich in irrigated taro fields and "fat" with fishponds. The gardens of 'Umi had supported the rise of a new cadre of *ali'i akua*, divine kings. Their aspirations would not be held back by the rough seas of the 'Alenuihāhā Channel.

THIRTEEN · The House of Pi'ilani

As the power and prestige of the Hawaiian god-kings continued to grow, the *ali'i* increasingly relied on the emerging cadre of *kāhuna* (priests) to keep the social and political system in line. In an earlier time—in the homeland of Hawaiki—ancient Polynesian ritual had revolved around the horticultural year. There were no full-time priests in those early days. The *ariki*, leaders of the *kāinanga* groups of related kinsfolk, themselves performed the annual rites, invoking the *atua* (spirits of the ancestors). By the sixteenth century in Hawai'i, this ancestral ritual system had undergone major elaborations. A formal class of *kāhuna* now officiated over a much more complex ritual system, based on the elaborate cults of several gods.

To be sure, Hawaiian religion still retained core elements brought by voyaging ancestors from Hawaiki. The ancient Mata-fiti, a new year celebration, marked by the rising of the star cluster Mata-liki (Pleiades), remained a key element in the Hawaiian ritual cycle. Now called Makahiki, this was a four-month period coinciding with the main growing period for sweet potatoes in the dryland field systems on Hawai'i and Maui islands. However, at the close of the Makahiki, what had begun centuries earlier as simple firstfruits prestations to the gods was now an elaborate ritualized collection of tribute for the king. As in all other complex political economies, ritual had come to serve the interests of the emerging class of chiefs, *nā li'i*. The name for this tribute, *ho'okupu*, literally means "to cause to grow," reflecting its ancient origins in offerings to the deities, supplications designed to assure a bountiful crop. Collected in the name of Lono, the Hawaiian god of rain, thunder, and the sweet potato, the *hoʻokupu* consisted of an annual tax—not only of taro and sweet potatoes, pigs, and dogs, but also of other goods such as barkcloth, cordage, mats, and the precious red and yellow bird feathers used to adorn the capes and helmets of the *aliʻi*. Lono's image, the *akua loa* (literally, "long god"), consisted of a wooden shaft tipped with a carved head, to which was lashed a cross-piece from which trailed long white barkcloth streamers and the white skins of the *kaʻupu* bird. White was Lono's color. The *akua loa* was carried by a procession of priests and warriors from *ahupuaʻa* to *ahupuaʻa*, each territory rendering its *hoʻokupu* to Lono—and to the king.

Lono was one of four great gods in the Hawaiian pantheon. Just as Lono was the patron deity of the rain-fed field systems, his counterpart Kāne was the god of the streams and flowing waters that gave life to the irrigation systems. Taro was Kāne's sacred crop. Not surprisingly, where Lono was central in the Makahiki rituals of Maui and Hawaiʻi islands, dominated by their dryland field systems, the worship of Kāne was prominent on the well-watered islands of Molokaʻi, Oʻahu, and Kauaʻi. There the taro pondfields provided the main source of sustenance. Kāne's colors were yellow and red, the colors of the sunrise and sunset. The sun's daily path across the sky was the *Ala Nui o Kāne*, the "Great Road of Kāne."

Kanaloa was the third of the principal Hawaiian gods. His origins trace back to Tāngaroa of Hawaiki, about whom chief Paopao had spoken to me on the beach at Saua, in Manuʻa, as described in chapter 2. In Hawaiʻi, Kanaloa's position in the pantheon had become rather diminished. He was regarded as the patron deity of the ocean, the sea bottom, and subterranean depths. Fishermen gave offerings and prayers to Kanaloa in little shrines by the seashore. Kanaloa was also associated with death and the underworld; his color was black. But his rituals and his temples were relatively unimportant.

The fourth great deity of sixteenth-century Hawaiʻi was Kū, god of war. His origins can be found in the early Polynesian societies of Eastern Polynesia, where he is still known on certain islands as Tū. In Hawaiʻi, Kū became the patron deity of the highest chiefs and kings, who relied increasingly on their warriors and the threat—if not outright use—of force to expand their power. Kū's symbols were the red flowers of the *ʻōhiʻa* trees that grow on the high mountain ridges, the red *ʻiʻiwi* birds that feed on those flowers, and, most important, the feathers they provided. Red had always been the sacred color of chiefs in Polynesia. In Hawaiʻi, expert craftsmen in the chiefs' households now took those red feathers and tied them, by the hundreds and thousands, to woven wicker capes, cloaks, and helmets that were the special garments of the *aliʻi*. At the same time, the red feathers were also used to cover wicker images of Kū himself. One manifestation of Kū, Kūkāʻilimoku (Kū the Snatcher of Kingdoms), was the personal war god of the Hawaiʻi kings. The imagery of the king and chiefs wrapped in brilliant red feathers, carrying before them the flaring-mouthed, pearl-shell-eyed image of Kū identically adorned with the same red feathers, gave an unmistakable visual message. The king had become, as later Hawaiian scholars such as Samuel Kamakau observed, an *aliʻi akua*, a god-king.

With these developments in Hawaiian religion came significant changes in ritual practice. Each of the four main gods had his own cult, with elaborate ceremonies involving complex liturgies. A specialized priesthood had arisen to deal with these ritual complexities, a class of *kāhuna pule*, literally "prayer priests." Often junior sons of chiefly families, they were trained from a young age to memorize the long incantations and chants. Each cult had its specialist *kāhuna pule*. These were distinct from other kinds of *kāhuna*, such as the feared *kāhuna ʻanāʻanā* (sorcerers), whose terrifying prayers gave them the power of praying a man to death. There were also *kāhuna lapaʻau*, experts in the use of medicinal plants and in the diagnosis of ailments. But the most powerful *kāhuna* were those who officiated at the cults of Lono and Kū. In particular, the high priest of Kū was known as the *kahuna nui* (great priest). It was he who assisted the king in the most sacred and feared rites of all, those attending the dedication of a war temple, a *luakini*.

Only a king, an *aliʻi nui*, could build a *luakini* temple. Such temples were also known as *heiau poʻokanaka*, literally "man-head temples," in reference to the rites of human sacrifice that were performed there. Early Polynesian ritual did not involve human sacrifice. The Hawaiian traditions tell us that in the early days of Māʻilikūkahi on Oʻahu, the temple rites did not involve ritual homicide. Indeed, the *moʻolelo* attribute the introduction of human sacrifice to Pāʻao, the voyaging priest who came from Kahiki, perhaps toward the end of the fourteenth century. Pāʻao is especially associated with Hawaiʻi Island, where he set up Pilikaʻaiea as high chief over the island. Pāʻao's temples, Wahaʻula in Puna and Moʻokini in Kohala, were where he established the new worship of Kū with the rites of ritual homicide.

· · ·

The largest temple ever built in the islands, in terms of the sheer size of its footprint, lies not on Hawaiʻi Island but across the channel on eastern Maui, in the fertile district of Hāna. Sprawling over an area of almost three acres, the stone foundation of Piʻilanihale *heiau* today lies within the grounds of Kahanu Gardens,

FIGURE 9.
The massive terraced front face of Piʻilanihale temple in Hāna district, Maui Island, rises more than forty feet from the ground. The scale is evident from the coconut palm tree on the temple's summit. Photo by Thérèse Babineau.

a branch of the National Tropical Botanical Garden. Approaching the massive foundation of stacked boulders from below, one is awestruck by the imposing five-tiered façade rising nearly forty feet above ground level, all cleverly held together merely by the tight chinking of rock to rock (see figure 9). Climbing a path to the top, one arrives at the vast main terrace, flanked on two sides by complex arrangements of rooms, platforms, and pits. The entire complex is bounded on the inland side by a thick wall about eight feet high, to keep the uninitiated—the commoners—away. The terrace was constructed by filling in a swale about forty feet deep between two tongues of the ʻaʻā lava flow upon which the *heiau* sits.

In the late 1980s archaeologist Michael Kolb excavated several test pits into the terrace and associated structures at Piʻilanihale. He found that the temple was not built in a single episode, but grew over time as various additions were constructed by successive chiefs who worshipped here. Kolb estimated that the main central terrace required no less than 84,000 man-days of labor to haul and set the countless basalt boulders. Later, another 43,000 man-days were expended in building the peripheral wings. Kolb's excavations in these peripheral sectors of the complex yielded food remains and domestic artifacts that convinced him that Piʻilanihale functioned not just as a *luakini* temple, but also as the residence of a great king. The name Piʻilanihale translates as "House of Piʻilani," confirming Kolb's interpretation of a dual temple and royal residence. Piʻilani was a contemporary of Līloa on Hawaiʻi. In the traditions of Maui it was Piʻilani who first unified the two parts of the island. Piʻilanihale was the seat of Piʻilani and those who succeeded him in the line of Maui kings.

The story of Piʻilani—and of his two sons, who fought to the death over the succession to the kingship—is as central to the history of Maui as the story of Līloa and ʻUmi is to Hawaiʻi Island. In fact, the two royal houses were linked by bonds of marriage, for Piʻilani's daughter Piʻikea became one of ʻUmi's royal wives, the union arranged by the clever priest Kaleiokū as a political alliance between Hawaiʻi and Maui. Let me now recount the *moʻolelo* of Piʻilani and his sons, as it was recorded by Kamakau and Fornander in the nineteenth century.

Piʻilani ruled over Maui during the final decades of the sixteenth century, as reckoned from his genealogy. During his reign the land was mostly at peace. Fornander tells us that it was under either Piʻilani or possibly his father, Kawaokaʻōhele, that the entire island of Maui first became a unified polity. This initial unification was achieved peacefully, the Hāna chiefs acceding to the suzerainty of the Piʻilani line, which had its origins on western Maui. The ancient seat of the Piʻilani clan of chiefs was in the area called Nā Wai Ehā, the "Four Waters" of Waiheʻe, Waiehu, Wailuku, and Waikapū. These adjacent valleys, stretching inland from what is now the sprawling town of Kahului, had flowing streams feeding canals that watered extensive taro irrigation works on the alluvial flats. It was these rich taro fields that initially gave Kawaokaʻōhele and his son Piʻilani their economic base. But as on Hawaiʻi, the much more expansive, young volcanic slopes of eastern Maui were by this time being developed into vast dryland gardens. Pi'ilani must have seen the huge potential of these developing lands, with their ability to yield large quantities of *ho'okupu*.

Pi'ilani died of old age in Lahaina, on the leeward side of western Maui, a favored residence of the chiefs with its beaches, surfing, and bountiful fish. He had two high-ranking sons by his wife La'ieloheloheikawai (a sacred chiefess from O'ahu), the oldest named Lono-a-Pi'ilani (hereafter Lono), and a younger son named Kiha-a-Pi'ilani (hereafter Kiha). Kiha had been raised for the most part by his maternal kinsfolk in the royal court of O'ahu Island, whereas Lono—following the norms of patrilineal succession—had been groomed to be the heir to the Maui kingship. On his deathbed, Pi'ilani declared Lono to be the new *ali'i nui* of Maui, and commanded Kiha to live under his older brother in peace.

For the first few years of Lono's reign, all was well between the brothers. But jealousy arose between the two siblings: Lono became envious of the way in which Kiha was developing his irrigated fields in Waihe'e Valley. Usurpation is an age-old theme in Polynesian politics; Lono suspected that Kiha was secretly plotting to steal the kingdom from him. He began to humiliate Kiha, one day throwing a bowl of briny water filled with octopus into Kiha's face. Realizing that Lono was going to try to kill him, Kiha fled to Moloka'i. Lono sent warriors to pursue Kiha, but he had secretly stolen back to Maui, making his way to the dry farming lands of upland Honua'ula and Kula, on the broad slopes of Haleakalā.

Not long before Kiha had been forced to flee Maui, the *kahuna nui* called the two royal brothers to his house. When Kiha arrived he found his older brother already in the high priest's house, waiting. Kiha called out a greeting to Lono, but the latter only stared and said nothing. The priest then told them that he was going to reveal to them their goddess. The *kahuna* further explained that if one of the brothers should be overcome with fear and flee the house when the goddess appeared, that brother would become destitute. He who remained without fear would become the king of all of Maui. Lono grunted, telling the old priest to get on with it; he had no doubt in his mind which of them would flee and which would fearlessly greet the goddess.

The *kahuna* explained that after he began his *pule*, his prayer, the two chiefs would see a spider, "your deity," creep up in the rafters over the center of the earth oven that was in the floor of the house. As he began to chant in a high-pitched, quavering voice, the brothers saw a spider crawl along the rafters, until it reached a position directly over the *piko* (umbilicus) of the oven. The spider began to create a web, and seven rainbows appeared within the roof of the house, arched over the oven. The priest finished the first prayer, pausing to ask the chiefs what they had seen. Lono described the vision they had witnessed. The priest began his second prayer. As he chanted the spider dropped down and disappeared within the *imu*, the earth oven. Again the priest paused to ask what they had witnessed as he prayed. Lono replied, "She has disappeared within the *imu*." As the *kahuna* began the third prayer, the earthen mound of the *imu* began to rise and move; Lono was overcome with dread.

Concluding the third prayer, the priest grasped a corner of the barkcloth covering the *imu*. Speaking to Lono and Kiha he said, "You will now witness this goddess who has arrived for you. She is indeed a fearsome goddess, an *akua 'ino*." As the priest pulled back the covering over the *imu*, the lizard *(moʻo)* form of the goddess Kihawahine was revealed, coiled within the searing oven pit, her skin like that of the *puhiūhā* eel. Kihawahine turned her lizard head to stare directly at Lono. Terrified, Lono screamed, "*Aue!* Not a friendly god, but a malevolent *akua!*" Seized with fear, Lono ran from the priest's house.

The great lizard goddess climbed up into the rafters, her belly full of the food that had been cooked in the *imu*. She slithered through the thatch and a loud plop was heard as she landed on the ground outside. Shortly thereafter, dozens of other smaller lizards began to appear throughout the priest's house. These were the *kinolau*, the "thousand bodies" of Kihawahine, the goddess. They surrounded Kiha, jumping up on him. The priest informed Kiha that he would now see the final and true form of Kihawahine, half human and half lizard. He gave Kiha some *ʻawa* root to chew and prepare for the goddess, and also a special kind of *poi* for her to eat. As Kiha prepared the *ʻawa*, the priest hung up a sheet of *kapa ʻōlena*, barkcloth dyed with sacred yellow-red *ʻōlena* root, dividing off part of the house. When Kiha announced that the *ʻawa* was ready, the *kahuna* drew back the barkcloth curtain. Kiha gazed upon the most beautiful woman he had ever seen. As he looked upon Kihawahine in her human form, all apprehensions left him. Kihawahine turned and spoke briefly with the priest, and then disappeared from the house. The *kahuna* explained to Kiha what the goddess had just revealed—that he must flee for his life, but that he would return and, with the aid of his sister Piʻikea, conquer all of Maui from Lono, he who could not face the lizard goddess.

Now Kiha found himself in Kula, masquerading as a commoner among the simple farmers of the uplands. A famine had beset the land; the people were reduced to eating ferns and weeds. Kiha cleared a large garden plot among the ferns, large enough to plant many sweet potatoes. He then traveled to the lands of Hāmākuapoko and to Haliʻimaile, which were not parched, inducing the people there to give him an ample supply of the precious vines. When he returned with his great load of planting slips, the people of Kula began to wonder about this energetic stranger. When they noticed that rainbows (always a sign of a high chief) appeared when Kiha arrived with his vines, they realized that he was the rebel brother of the king, Lono-a-Pi'ilani. Like 'Umi on Hawai'i, Kiha demonstrated his knowledge of dryland farming, proving to the *maka'āinana* that he was fit to be their leader.

Kiha decided to leave Kula and travel around the windward side of Maui to Hāna, a district then ruled over by Ho'olaemakua, a warrior chief fiercely loyal to Lono. Kiha thought that if he could convince Ho'olaemakua to turn against Lono and support his own cause, he would have a powerful ally. Kiha possessed a large, handsome frame and, as Kamakau tells us, his eyes were "as bright as those of a *moho'ea* bird." He was an expert surfer, having learned in his youth the art of riding the *papa he'e nalu* boards on the long breakers at Waikīkī. As it happened, Ho'olaemakua had a daughter, Koleamoku, who also loved to surf. Kiha went surfing with Koleamoku. The pair became infatuated with each other, spending the evenings making love. Koleamoku was determined to have Kiha as her husband. They eloped, and Koleamoku began living with Kiha in his house at Kawaipapa. When her father Ho'olaemakua heard what had happened, he flew into a rage, for he had placed a *kapu* (taboo) on her that could be lifted only by the king, Lono-a-Pi'ilani. Ho'olaemakua disowned his daughter Koleamoku.

Koleamoku became pregnant and bore Kiha a son, whom they lovingly raised. Sensing that Ho'olaemakua's anger would have subsided, Koleamoku went to her father's house to present his grandson to him and to offer a feast of reconciliation. The Hāna chief greeted his daughter and his young grandson with great affection, but when Koleamoku presented her husband's request that they be given certain lands in Hāna, Ho'olaemakua bowed his head in silence. He told his daughter, "Your husband does not want farm lands for the two of you, but is seeking means to rebel against the kingdom." Steadfastly loyal, Ho'olaemakua refused to take his daughter's side.

When Kiha heard that his plan to gain the support of Ho'olaemakua had failed, he swore vengeance against his father-in-law. Kiha now remembered the words of the old *kahuna*, in the house where—in his *'awa*-induced trance—he had seen his lizard goddess Kihawahine. He must cross the blustery 'Alenuihāhā Channel to Hawai'i, where his sister Pi'ikea was married to 'Umi, now king of that entire island. He would seek 'Umi's assistance in gaining control over Maui.

Kiha and his retainers set sail in their canoe for Hawai'i, rounding the cape at

'Upolu and then coasting down the lee side until they came to Kona, where 'Umi was residing. They landed at the royal residence of Kamakahonu; Kiha went in search of his sister. On seeing her brother, Pi'ikea wailed with joy, for the news of how Lono had mistreated Kiha had spread to Hawai'i. 'Umi gave orders for food to be cooked for the guests. After a feast of welcoming, 'Umi inquired of Kiha, "What great purpose brought you to Hawai'i?" Kiha replied that his only purpose was vengeance against Lono. Urged on by his wife Pi'ikea, 'Umi agreed to help Kiha overthrow Lono and become the ruler of Maui.

A year was spent constructing a fleet of war canoes to transport the Hawai'i forces across the channel; clubs and spears of hardwood were prepared. 'Umi and Kiha determined to take the battle straight to Hāna, where Ho'olaemakua had refused to lend his support to Kiha. The famous old Hāna chief had prepared his fortress hill of Ka'uiki, standing sentinel over Hāna Bay (see figure 10). When the fleet of war canoes filled with Hawai'i Island warriors arrived, they were held at bay by the endless barrages of slingstones cast by the Maui warriors from their vantage point on Ka'uiki. Day after day they were unable to dislodge Ho'olaemakua's forces.

Under the cover of darkness, Pi'imaiwa'a, the famous warrior who had helped 'Umi kill Hākau and gain control over Hawai'i, crept up close to the steep entrance to the hilltop fortress. It seemed to be guarded by a huge warrior. Pi'imaiwa'a lanced his spear into the warrior, but it did not move. Climbing closer, he hit the giant with his club. It stood motionless. Pi'imaiwa'a realized that this was a dummy built of wood and wicker, to fool invaders at night so that the Maui defenders could rest. He sent word for the Hawai'i warriors to follow him up the steep ladder into the fortress. They fell upon the slumbering Maui forces. Many were killed, or leaped to their deaths off the steep cliffs encircling the hill. But in the darkness a few escaped, including Ho'olaemakua. Kiha sent Pi'imaiwa'a in search of Ho'olaemakua in the backlands of Hāna. The old warrior chief was finally hunted down at Kapipiwai, tortured, and killed. His hands were brought back to Kiha to confirm his death.

During the battle at Hāna, Lono-a-Pi'ilani had remained safely on western Maui at Wailuku, the old seat of the Pi'ilani line. When he heard that the fortress of Ka'uiki had fallen and that Ho'olaemakua had been captured and killed, he was filled with dread. Lono knew that if he were captured by Kiha he would be tortured even more severely before his body was offered up as a sacrifice to Kū on the *luakini* altar. By the time Kiha and 'Umi with the Hawai'i forces arrived at Wailuku, Lono was dead, evidently of sheer fright. Kiha desperately wanted the

FIGURE 10.
The fortress hill of Kaʻuiki in Hāna, Maui Island, scene of many famous battles recounted in Hawaiian traditions. Photo by P. V. Kirch.

corpse of his brother, to mutilate and offer up on the *heiau* platform. But Lono's loyal retainers had taken the body of their king and hidden it away, probably in a remote lava tube. As much as Kiha's men hunted, they could not find the bones of Lono-a-Piʻilani.

Kiha-a-Piʻilani was now the undisputed lord of Maui. As was the custom, he divided the districts and *ahupuaʻa* among his loyal followers and warriors. One of the sons of ʻUmi and Piʻikea, ʻAihakoko, was left on Maui to reside in the royal household. Then ʻUmi and his fleet returned to Hawaiʻi, leaving Maui under the rule of his brother-in-law. It was probably at this time that Kiha had the great terrace at Piʻilanihale *heiau* constructed, making it his royal center and principal *luakini*. The descendants of Kiha-a-Piʻilani would continue to rule over Maui in an unbroken succession until the end of the eighteenth century, when Kamehameha the Great finally took possession of the island during his conquest of the archipelago.

• • •

For more than fifteen years, I have explored the expanses of Kahikinui and Kaupō districts, which make up the southern, leeward flanks of Haleakalā on eastern Maui. Only a narrow ribbon of poorly paved or in places dirt road traverses this vast land of lava flows and deep gulches. Most of this terrain lies behind locked gates and must be accessed by four-wheel-drive tracks or on foot. The land belongs to a few cattle ranches, 'Ulupalakua Ranch, Kaupō Ranch, Nu'u Mauka Ranch, and others. Their owners have graciously granted me access over the years. The Department of Hawaiian Home Lands likewise has allowed me carry out research on its thousands of acres in Kahikinui. Only a few families live in this part of the island. Maui old-timers still refer to Kahikinui and Kaupō as "Backside," a near-literal translation of the Hawaiian term *kua'āina* (back of the land).

I have grown to love this *kua'āina* land of Kahikinui and Kaupō. It is the antithesis of the Visitors Bureau image of Hawai'i—no white sand beaches, no swaying palm trees, no waterfalls. (Well, almost none: Kaupō has a few waterfalls that flow only after heavy rains.) But this *'āina malo'o*, this arid landscape, has a beauty all its own, one that grows on you over the years. It is the beauty of a cloudless Haleakalā summit illuminated by the sunrise, of late afternoon rainbows over Ka Lae o ka 'Ilio, of the gnarled orange trunk of an old *wiliwili* tree growing out of bare lava rock, of watching a school of bottle-nosed porpoise leap in the waters off Niniali'i. But make no mistake, fieldwork here is rough: sharp lava rocks quickly grind away the soles of your boots, while the incessant winds rip the notepaper out of your hands. But the rewards have been well worth the hard work.

Precisely because it is a *kua'āina*, a backcountry lacking in water, beaches, or fertile soils, Kahikinui has been bypassed by the wave of development that swept the islands after statehood in 1959. Even cattle ranching is marginal in arid Kahikinui. Thankfully for the archaeologist, no sugarcane or pineapple plantations, no resorts, no golf courses, no suburban housing developments have intruded into this backcountry world. In large sections of all the main islands such developments have erased the signature of ancient Hawaiian land use. First the plantations plowed under vast tracts, including Līhu'e, ancient seat of the O'ahu kings. Then, as the new state's economy shifted from plantations to tourism, more of Hawai'i's past succumbed to the bulldozer's blade. Thankfully, federal and state laws passed in the 1960s began to require archaeological study of most proposed development areas. Nonetheless, it is increasingly difficult to find an entire district, a *moku*, in which an entire ancient landscape remains intact. Kahikinui is such a place.

In the course of my Kahikinui and Kaupō research, I have walked over nearly ten square miles of land, ridge by ridge, gulch by gulch, swale by swale, searching for the stone ruins that provide clues to ancient Hawaiian life in this *kua'āina*. With my students I have recorded, in sketches, notebooks, and GPS data files, more than four thousand individual stone structures. Among the nondescript walls and terraces of house sites and garden plots are a few structures that stand out as especially prominent; these are usually larger, constructed with greater finesse. They typically occupy promontories, or sit on ridges with views extending for miles. These are the foundations of the *heiau*, places of worship and sacrifice, around which the ancient inhabitants of Kahikinui and Kaupō organized their lives. In these undeveloped landscapes we have been able to study the entire *system* of *heiau*, from the smallest fishing shrines on the coast, through a range of midsize agricultural temples dedicated to Lono and Kāne, up to the largest *luakini* temples of human sacrifice where the worship of Kū was performed by Maui's kings, including the famous Kekaulike.

After a number of years of making detailed maps of dozens of these temples, I could discern systematic patterns in their architecture, their settings, and their orientations. As Kamakau tells us, "Heiaus were not alike; they were made of different kinds according to the purpose for which they were made." Special architect-priests, called *kuhikuhi pu'uone*, had the responsibility of choosing the location and laying out the foundation for a new temple. (*Kuhikuhi pu'uone* means literally "to demonstrate [with a] sand hill," perhaps because these architects used models made in the sand to formulate their designs.)

Although there are many minor variations, the temples of Kahikinui and Kaupō can be grouped into four major architectural types. First are the small enclosures, usually square or rectangular, situated near the shoreline, which served as *ko'a* (shrines) to Kanaloa and Kū'ula, gods of the fishermen. Typically these shrines are heaped with quantities of branch coral, including entire coral heads, as offerings to the gods. Some also have an elongated, phallic water-worn stone set upright, a representation of Kū'ula.

Most of the other *heiau* are located farther inland, in the main planting zone, for these were *heiau ho'o'ulu'ai*, temples for assuring the fertility of the crops. They range in size from about one hundred up to fourteen hundred square yards in area. One type consists of square enclosures, sometimes with an elevated platform in one corner. Another category has six sides, so that in plan view they look like a square with a notch removed from one corner; archaeologists call these notched *heiau*. The final major category of temples is defined by two distinct levels or courts, one higher than the other, with an elongated overall plan. I call these the elongated double-court *heiau*. The largest temples, such as Loʻaloʻa in Kaupō, which was dedicated by King Kekaulike in the early eighteenth century, often fall into this elongated double-court category. Loʻaloʻa is truly a monumental construction; its ground plan covers 4,975 square yards, and its terraced eastern face rises to a height of twenty-six feet above the ground level.

When I began to analyze the detailed maps I had made of these temples, I realized that their foundations had particular orientations. Earlier generations of archaeologists in Hawaiʻi, such as John Stokes, had thought that *heiau* were simply oriented to the contours of the land, reflecting the local topography. My systematically collected data suggested otherwise. Almost all temples have what I call an axis of orientation, indicated by one end or side of the foundation that is higher, and more prominently constructed. Often this side has multiple stepped terraces. Just as a Christian church has an axis, with the entrance at one end and the high altar under stained-glass windows at the other, so Hawaiian *heiau* were laid out with the images representing the gods at the end that had the most prominent architecture.

When I plotted the compass bearings of this axis of orientation for all of the *heiau* I had mapped onto 360-degree polar coordinate graph paper, they clustered into three clear groups. Temples in the largest group have their axes oriented between 82 and 93 degrees east of north (these are in true cardinal degrees, corrected for magnetic declination). Essentially these are east-facing. A second group was oriented to the northeast, between 64 and 73 degrees east of north. The third cluster has its axis facing roughly north, with a range between 333 and 15 degrees east of north. The east- and northeast-facing groups included *heiau* of both the square and notched types. However, most of the elongated double-court temples faced to the north (with a few exceptions, including the giant temple of Loʻaloʻa).

I asked myself if these discrete orientation patterns might provide a clue to the functions of the different *heiau*. When I turned to the writings of the nineteenth-century Hawaiian scholars Kamakau, Malo, and Kepelino, it became evident that the ancient *kāhuna* had a sophisticated knowledge of astronomy. The *kuhikuhi puʻuone* were capable of laying out temple foundations according to precise cardinal orientations. But more than that, each of the main gods of the Hawaiian pantheon was associated with a particular direction. The east was the domain of Kāne. I also observed that many of the *heiau* with an eastern axis of orientation are situated adjacent to streams or watercourses; Kāne, of course, was the deity of flowing waters. I began to suspect that these east-facing temples were Kāne's *heiau*, where his *kāhuna* would chant up the sun and make offerings of taro, his sacred plant, which thrives best when supplied with abundant water.

The temples with their axes facing the northeast are centered on the rising position of the star cluster Pleiades, which the Hawaiians call Makaliʻi (The Little Eyes). The rising of Pleiades was a crucial event in the ancient Hawaiian calendar, for it determined the onset of the Makahiki season, sacred to Lono. Toward the end of November each year, the priests of Lono would begin to scan the horizon, watching for the first visibility after sunset of the cluster of seven twinkling "eyes." The northeast-facing temples, I surmised, were those dedicated to Lono, god of rain, thunder, and the sweet potato.

What, then, of the temples oriented to the north? In Kahikinui, north is the direction of the summit of Haleakalā, rising to 10,035 feet at its highest peak. High mountains, I knew, were the realm of Kū, god of war, whence came his sacred red feathers from the birds that fed on the nectar of the red *ʻōhiʻa* blossoms. Some of the temples in this group seemed as well to be specifically pointed toward a particularly prominent, reddish cinder cone on the mountain's rim. It seemed likely that these north-facing temples, many of which are of the elongated two-court type, were dedicated to Kū. Many of them seem to be positioned so as to mark the boundaries between *ahupuaʻa* territories. They may or may not have been *heiau poʻokanaka* (temples of human sacrifice), but most likely they were all associated with prominent chiefs, and were not agricultural temples.

About the time that I was analyzing the orientations of the Kahikinui temples and their likely associations with the various gods of the Hawaiian pantheon, I began to collaborate with Warren Sharp at the Berkeley Geochronology Center, who specializes in dating corals and carbonate rocks using the uranium-series method. This follows the same basic principles as radiocarbon dating, in that uranium (U) decays over time to a by-product, thorium (Th). Corals growing in seawater take in U, which is abundant in the oceans; the U gets incorporated into their carbonate skeletons. After the coral dies, the U gradually decays to ^{230}Th. By precisely measuring the ratio between U and Th in a specimen, Warren could calculate the age of a coral. Moreover, this method is far more accurate than radiocarbon dating. The "plus or minus" error range on a U/Th coral date may be as little as ten years, and is sometimes as small as five years.

While surveying the Kahikinui *heiau*, I noticed that pieces of branch coral often occurred on their altars, where they were presumably placed as dedicatory offerings. Other pieces were frequently incorporated into the fill of stone walls, placed there during construction. I gave Warren corals that I had collected from seven different temples in Kahikinui, along with some from a temple on Molokaʻi Island. Since these samples represented a variety of temple types and orientations, I expected the dates to display a considerable age range. But when Warren and I sat down with the results from his laboratory, over drinks at the Berkeley Faculty Club, I got a shock. Instead of spanning several centuries as I had expected, the dates from all eight temples were tightly clustered. The oldest date was A.D. 1565 plus or minus eight years, while the youngest date was 1638 plus or minus six years. These corals had all been placed on the altars of the Kahikinui and one Molokaʻi *heiau* within a period of about thirty to at most sixty years. What could explain this tight age range?

Mulling over the significance of these new dates, I began to realize that the time period we were talking about—the end of the sixteenth century and beginning of the seventeenth—was precisely the period when Piʻilani and his son Kiha-a-Piʻilani first gained control over a unified Maui kingdom. The oral traditions told us that at the same time, ʻUmi on Hawaiʻi had a series of temples built throughout his dominions on Hawaiʻi Island. Had Piʻilani or his son done the same thing? Given the dates from the Kahikinui *heiau*, this seemed likely. The temple system of the Hawaiian kings and chiefs was highly organized and hierarchical. It provided the ritual basis for controlling the vast agricultural system, and for the regular collection of tribute in the form of *hoʻokupu*. The rites of Kāne and Lono helped to assure that planting and harvesting took place at the optimal times. And the temples of Kū asserted the power and rights of the *aliʻi* over their territories and the kingdom. The economic and political structures by which society was ordered were reflected in the religion with its rites and liturgies. In this, the ancient Hawaiians were no different from any other civilization.

FOURTEEN · "Like a Shark That Travels on the Land"

With 'Umi's unification of Hawai'i and Kiha-a-Pi'ilani's control over all of Maui, Hawaiian society entered a new era. These new kings were different from the Polynesian chiefs of old. Although the Hawaiki chiefs had also been the direct descendants of the ancestors, they were still related to their people as kinsmen. The new generation of Hawaiian kings, *ali'i nui*, now promoted themselves as *ali'i akua*, literally "god-kings." They were sacred, *kapu*. Their sanctity demanded that they be kept apart from the people at large, lest their *kapu* status be sullied by the commoners. Even their names reflect their exalted status: Kalanikūpule, Kalani'ōpu'u, Kalanilehua, and so on. The opening words of these names, *ka lani*, literally mean "heavenly one." The god-kings were, in the words of Kamakau, "gods, fire, heat, and raging blazes." For these kings to go abroad during the day would be dangerous, not so much for themselves, but for the people they would encounter. When they did travel, heralds marched before the kings, warning the people by shouting out *"Kapu moe, kapu moe!"* Their cries invoked the prostrating taboo, when all but the highest-ranked chiefs must lie flat on the ground, their faces pressed to the earth, until the king and his retinue had passed. To fail to heed the heralds' cry could mean sudden death. Even those few chiefs of high rank had to sit on the ground, with their torsos bared, in the presence of the *ali'i akua*.

Anthropologists call the kind of leadership that developed in Hawai'i during the late sixteenth to early seventeenth centuries, with the rise of 'Umi, Kiha-a-Pi'ilani, and their successors, divine kingship. It is a hallmark of most early states and civilizations. The late professor Bruce Trigger, in his book *Understanding Early Civilizations*, informed us that divine kingship was pervasive in all early states. Standing at the apex of society, the king "constituted the most important link between human beings and the supernatural forces on which the welfare of both society and the universe depended," Trigger wrote. Since only they could intercede with the gods, it was natural to think that these kings were the instantiation of gods on earth. In ancient Egypt, the pharaohs "were believed to be so charged with supernatural power that simply touching them or their regalia without taking ritual precautions could cause serious injury." The same was true of the *aliʻi akua* of Hawaiʻi. Hawaiian historian David Malo tells us that "when a tabu chief ate, the people in his presence must kneel, and if anyone raised his knee from the ground, he was put to death."

Trigger points out that among the ancient Maya of Mesoamerica, the "rulers wore the costumes and attributes of various gods and bore names that incorporated references to numerous deities." We find precisely the same behavior among the Hawaiian kings, whose names frequently invoked the gods. Lonoikamakahiki, for example, was a king of Hawaiʻi Island who ruled a few generations after ʻUmi. His name means "Lono of the Harvest Season," directly invoking a comparison with Lono the deity.

The most pervasive symbolism linking the Hawaiian kings with their gods was their elaborate use of feathered garments. Hawaiians perfected the art of featherwork to the highest artistic level anywhere in Polynesia, or, indeed, anywhere in the world where people used feathers to decorate their clothes and their bodies. Captain Cook and his officers (whose own grades of naval rank were carefully expressed in their uniforms) were greatly impressed with the feathered capes and cloaks of the Hawaiian chiefs and king. David Samwell, surgeon's mate on the *Discovery*, wrote admiringly in his journal, "A more rich or elegant Dress than this, perhaps the Arts of Europe have not yet been able to supply." Samwell continued his description of these chiefly garments: "Inferior Chiefs have Cloaks made of Cock's Tail feathers with a Collar of red & yellow, others of white bordered with Cocks feathers & a Collar of red & yellow. Some of the first Chiefs have long cloaks made of the fine yellow feathers of the Cocks with a Collar and borders of red and yellow feathers."

Samwell was mistaken in thinking that the brilliant feathers came from "cocks" or fowl. Rather, the red and yellow feathers were obtained from small forest birds, known to the Hawaiians as *ʻiʻiwi*, *ʻōʻō*, and *mamo* (scientists later gave these birds the names *Vestiaria coccinea*, *Moho nobilis*, and *Drepanis pacifica*). To catch the birds and procure the feathers required a great deal of patience and labor in the misty *'ōhi'a* forests of the high mountains. Thousands of feathers were required for a single small cape, hundreds of thousands for the long cloak of a king. The yellow feathered cloak of Kamehameha the Great, one of the most prized possessions of Honolulu's Bishop Museum, is estimated to contain 450,000 feathers, taken from about eighty thousand *mamo* birds!

During the annual Makahiki circuit, the chiefs and king demanded feathers as part of the *ho'okupu* (tribute) from each territory. When a sufficient number of feathers had been accumulated, the work of making a cape or cloak, called *'ahu'ula* in Hawaiian, could commence. Given the taboo nature of these garments, they were made exclusively by men. Adrienne Kaeppler of the Smithsonian Institution, who has studied Hawaiian featherwork, believes that these featherwork experts were also priests. No doubt the work of making a king's cloak was attended by special rituals. After all, they were making the garment of a god-king.

And just as the king would be covered from head to foot in his feathered helmet (see color plate 6) and long cloak (there were even special feathered loincloths *[malo]* for ritual occasions), the images of the gods were adorned in the same fashion. In particular, the images of the war god Kū, the king's special god, were made of *'ie'ie* vines woven in the shape of a human head, to which thousands of red and yellow feathers were carefully tied. The god's piercing eyes were fashioned from pearl shell, while its flaring mouth was rimmed with dog's teeth. These feathered images were carried before the king when he went into battle, and accompanied him on his tours of his dominions.

Members of the newly emerged class of Hawaiian chiefs were obsessed with genealogy. Polynesians had always placed great emphasis on keeping a record of their ancestors' names, often encoding this in poetic chant. In the world of ancient Hawaiki, membership in a particular *kāinanga* social group was validated by reciting one's genealogical connections to that group's founding ancestor. This was important because membership in a *kāinanga* was essential for rights to land, house sites, and other privileges. In seventeenth-century Hawai'i, the situation had changed radically, for the common people—the *maka'āinana*—no longer kept their genealogies. Indeed, the chiefs expressly forbade them to do so. Hawaiian society now consisted of two distinct classes of people, the chiefs *(nā li'i)* and the commoners *(nā kanaka)*. The defining hallmark of the chiefs was that they could recite their lineages, preferably back to the founding chiefs of Nānā'ulu or 'Ulu, who had arrived in the islands many generations before. The commoners knew only the names of their immediate parents and grandparents.

With this new emphasis on pedigree came a heightened interest in strategic marriage alliances, much as the royal families of Europe intermarried as a form of statecraft. Trigger informed us that marriage as a "way to construct networks of alliances uniting upper-class families" was again typical of early states. Aztec and Maya kings married their daughters to the leading nobles of tributary or neighboring states, while in Shang China the king married off his daughters to subordinate lords as a way of keeping those lords tied to him. So it was in Hawai'i, as we have seen with the marriage of Pi'ilani's daughter Pi'ikea to 'Umi, a union that proved vital to Kiha-a-Pi'ilani's efforts to overthrow his brother and gain control over all of Maui. Based on the chiefly genealogies and traditions compiled by Abraham Fornander, we can see that until roughly the end of the fifteenth century (about the time of Mā'ilikūkahi on O'ahu), marriages between chiefly lines on different islands were rare. A century later, in the time of 'Umi and Pi'ilani, such marriages began to be increasingly common. They were an important strategy used by the god-kings to form alliances between their royal houses.

High-ranking marriages were made not only to link the royal houses of different islands; they were also made within individual kingly families, to the point of incest. The practice of brothers marrying their sisters, or in some cases their half-sisters, is another trait that arose more than once among the world's early civilizations. The Inca kings of the Andes had numerous wives and concubines, but the queen was likely to be the king's own sister or half-sister. And in Egypt the pharaohs also frequently married their sisters. In Hawai'i, this practice was called *pi'o*. It became increasingly common in the two centuries prior to Captain Cook's arrival. As with the Inca and Egyptian kings, the Hawaiian kings also took multiple wives. Often a special mating was arranged with the king's sister or half-sister, with the express purpose of producing a royal heir of the purest bloodline. Such "royal incest" was the exclusive prerogative of the highest-ranked *ali'i*; lesser chiefs were not permitted such transgressions of the incest taboo.

These *pi'o* unions between brothers and sisters or half-sisters were closely linked to the rise of divine kingship. Indeed, royal incest expressly reinforced the divinity of the most exalted *ali'i*. Hawaiian scholar Lilikalā Kame'eleihiwa points out that incest was a godly practice, for in Hawaiian cosmogony the world was created through the union of Wākea and Papa, brother and sister. As she writes, "the very act of incest is proof of divinity." When a brother and sister who were both themselves of *pi'o* rank mated, their offspring was of the most sacred rank possible, called *nī'aupi'o*. William Davenport, an anthropologist who studied Hawaiian royal incest, explains that the term is a metaphor deriving from "the looping back *(pi'o)* ... of midribs of coconut leaflets *(nī'au)* onto each other." Kamakau described such marriages in which "the sister marrying her brother, and the brother his sister, this wondrous marriage *(ho'ao)* of theirs was called a *ho'ao pi'o*, an arched marriage. ... The children born of these two were gods, fire, heat, and raging blazes."

Breaking the incest taboo was not the only way in which the Hawaiian kings displayed their divinity. To them also was given the power of ritual homicide, state-sanctioned killing in the form of human sacrifices to Kū, god of war. Once again, Hawaiian civilization independently invented practices seen elsewhere in early states around the world. The Aztec, Maya, Shang, Inca, and early Mesopotamian city-states all practiced various forms of ritual killing, most often of captured prisoners of war or slaves. In Hawai'i, the introduction of this practice is sometimes credited to the priest Pā'ao, who voyaged from Kahiki. Certainly in the early centuries after the islands were settled, the first kings—such as Mā'ilikūkahi—did not kill and offer up human bodies on their temple altars. The traditions are quite clear that this practice arose later, quite possibly first on Hawai'i Island. The slaying of Hākau by 'Umi, and the offering of Hākau's corpse upon the altar of Kūkā'ilimoku in Waipi'o, is one of the first explicit references to human sacrifice in the Hawaiian traditions. After this time it appears that human sacrifice became an obligatory part of the *luakini* temple rituals. Only males could be offered up to Kū. If the sacrifice was not the body of a defeated enemy chief, a victim would be plucked from the ranks of the *kauā*, outcasts and *kapu* breakers.

Hawaiians were superb observers of nature. They saw that the ocean, like human society, has its hierarchy. Corals, sea urchins, seaweeds, and mollusks form the base stratum. Coral-grazing parrotfish and wrasses are preyed on by jacks, and so on up the food chain. At the pinnacle is the shark, who takes anything below him but is prey to no other creature (except to man, of course). Sharks are the kings of the ocean. It is not surprising, then, that the Hawaiians compared their chiefs and kings to sharks, especially the feared tiger shark with its gaping mouth and razor-sharp teeth, even able to take a man. In the eyes of his people, the king was a shark who traveled on the land. The lines of a famous Hawaiian chant sum it up:

> A shark going inland is my chief,
> A very strong shark able to devour all on land;
> A shark of very red gills is the chief,
> He has a throat to swallow the island without choking.

· · ·

If the world of the Hawaiian chiefs and kings had changed dramatically from that of their ancestors, so had the life of the common people. As we have seen, the very name by which they were called, *maka'āinana*, offers a clue to the social transformations that had taken place. The ancient Proto-Polynesian root of this word is *kāinanga*, a term that in old Hawaiki had referred to a group of people descended from a common ancestor. After some Polynesians migrated eastward to settle the Marquesas, a second word, *mata*, was prefixed to *kāinanga*, as in *mata-kāinanga*. A variant spelling of that compound word, *mata'eina'a*, is still used in the Marquesas today, meaning "tribe" or "clan." *Mata'eina'a* have proper names, usually of founding ancestors, and each occupies a specific valley or region.

When the first Polynesians arrived in Hawai'i, they brought this *mata-kāinanga* concept with them. Over time, subtle changes in the Hawaiian language (including the replacement of *k* with a glottal stop ['] and the merging of *ng* with *n*) produced the later Hawaiian form of the word: *maka'āinana*. But more than just sound changes were involved. The meaning of *maka'āinana* was now vastly different from its old Proto-Polynesian root of *kāinanga*. Gone was any reference to a lineage group, especially one that held land. Gone was the idea that membership in such a named group validated a person's right to the land. Instead, *maka'āinana* now designated the collectivity of commoners. The transformation in meaning has an ironic twist, in that a word that once meant "lineage" now stood precisely for all those who had no lineages at all. The *maka'āinana* were those who worked the land, but their right to do so was no longer validated by genealogies linking them to ancestors. They had no genealogies; these had become the exclusive prerogatives of the chiefs. And without genealogies, their right to land depended on their subordinate relationship to their chiefs.

This transformation is underscored by a second word that can also be traced back to Hawaiki. In Proto-Polynesian this word was *kāinga*; we saw in chapter 2 that it referred to smaller groups of people who resided together in a single household, sharing a specific house site and garden lands. Following the same sound changes just noted, *kāinga* became *'āina* in later Hawaiian language. As almost anyone who has spent even a little time in the islands knows, *'āina* means "land." Not a specific parcel of land, but the land in general, in an encompassing sense, as in the well-known expressions *aloha 'āina* (love for the land) or *mālama 'āina* (to take care of the land). Gone in this new meaning was any reference to a social group of persons.

The radical transformation of ancient Polynesian *kāinanga* to *maka'āinana*, and of *kāinga* to *'āina*, underscores the deep changes that took place in Hawaiian society around the sixteenth century. First, the society was now segmented into two discrete classes, the chiefs above and the common people below. The chiefs (and especially the king) were *kapu* (sacred); to them all manner of deference and ritual was due. In opposition, the people were *noa*, a word that implies the antithesis of *kapu*. But along with this institutionalization of class stratification, of a "durable inequality" to use sociologist Chuck Tilley's term for it, came important changes in the ways in which land and labor were controlled and allocated. In old Hawaiki, membership in a named lineage *(kāinanga)* and in a specific household *(kāinga)* assured one the right to farm certain lands, to live in a certain place where one's ancestors had lived. In the new Hawai'i of the early seventeenth century, the *maka'āinana* continued to work the *'āina*. But that *'āina* was no longer controlled by lineages, for these had ceased to exist for the common people. The *'āina* was divided into territories called *ahupua'a*, a system that according to the traditions had been invented back in the time of Mā'ilikūkahi on O'ahu. Each *ahupua'a* was under the control of a chief, an *ali'i*, who received it from his king, usually during a redistribution of lands that occurred on the king's accession to his rule, or after conquest in war. These *ahupua'a* were further subdivided into smaller units, called *'ili*, and the *'ili* into yet smaller parcels cultivated by individual commoner households. The *'āina* was systematically organized in a hierarchical fashion. As anthropologist Edwin Burrows once put it, a land tenure system based on "breed" or kinship had been replaced with one based on "border" or territory.

This was a very efficient system for land management, especially from the perspective of the chiefs, whose *ahupua'a* provided the surplus they required to feed themselves, their households, their craft specialists, their priests, and their retainers. The chiefs, most of them also warriors, spent much of their time at the royal courts, in Waipi'o or at Kamakahonu on Hawai'i, or at Pi'ilanihale on Maui. To manage their *ahupua'a* they appointed *konohiki*, usually junior kinsmen, as stewards over the lands and the *maka'āinana*. The *konohiki* lived in the *ahupua'a*, where they could daily keep an eye on whether the gardens were being properly looked after, the irrigation ditches cleaned, the fishpond walls repaired. They made certain that when the chief sent for food, whether it be *poi*, fish, or fatted dogs and hogs, it was dispatched to the waiting chiefly household. The *konohiki* also were responsible for seeing that the annual tribute of the *ahupua'a* was ready when the Lono priests made their annual circuit of the island during the Makahiki.

Instead of validating their rights to certain lands by virtue of descent from ancestors, the common people now found themselves in a tributary relationship with their *konohiki* and their *aliʻi*. They had the right to live in certain places, and to cultivate certain lands, or fish in certain bays, as long as they provided the *hoʻokupu* to the *aliʻi*. This tribute was owed in two forms: as outright foodstuffs or other products (including barkcloth, mats, cordage, and the highly desired birds' feathers), and as labor. When the king wanted to build a new fishpond, extend a complex of irrigated taro fields, or construct the stone foundation for a new war temple, the people provided the labor.

To further ensure the efficient production of surplus for the chiefs, a new kind of land unit was created. This was the *kōʻele*, a plot of garden land—whether irrigated or dryland—set aside for the exclusive use of the *konohiki* and the *aliʻi* above him. The people of an *ahupuaʻa* were required to work these *kōʻele* plots, but the produce that came out of these fields was reserved exclusively for the land managers and chiefs. In the Hālawa Valley on Molokaʻi, whose land records from the early nineteenth century I have studied, the largest and best taro patches on the alluvial soil of the valley bottom were *kōʻele* fields. When in 1848 the commoners of Hālawa were allowed to make claims for the parcels of land that they had worked, during the Mahele (division of lands) under King Kamehameha III, their written claims made consistent reference to the fact that they had faithfully worked the *kōʻele* fields for the *konohiki*. That is to say, they validated their right to the land by stating how they had been good tenants, always working for the chiefs, giving their labor whenever asked, always providing their tribute.

All this is not to say that the relationship between the people and the chiefs was solely one of coercive exploitation. The people benefited from an efficient economic system, and from the peace and protection that a powerful king could assure them. Kameʻeleihiwa asserts that the relations between the chiefs and the people were based on reciprocal *aloha* of each for the other. The chiefs had an obligation—a kind of *noblesse oblige*—to be *pono*. *Pono* is the Hawaiian word for that which is correct, righteous, just. A *pono* chief would respect his people and not demand more from them than they could readily give. *Pono* was the ideal. But not all chiefs and kings were *pono*, as Kamakau tells us with the example of Keliʻiokaloa, a son of ʻUmi who ruled in Kona. This *aliʻi* seized "the food of the commoners, their pigs, dogs, chickens, and other property." Other accounts of chiefly oppression are not infrequent in the traditions. No doubt there was a constant negotiation between *nā liʻi* and *nā kanaka* as to the nature of their new social contract. But a new contract it was indeed.

The change from a land tenure system organized along the lines of kinship to a territorial system organized politically must have happened many times in the course of human history. It is one of the most fundamental kinds of social and economic transformations imaginable, with all kinds of ramifications. Some years ago anthropologist Irving Goldman made an exhaustive comparative study of Polynesian societies, concluding that Hawai'i had taken the course of social and political evolution the farthest. He was especially impressed by the changes in the Hawaiian land system. In the conclusion to his book, Goldman wrote the following: "The substitution of a territorial system of subdivision for one of kinship branching has long been regarded as a major divide in human history. Apart from introducing new structural arrangements of great importance, this change has introduced a parallel change in outlook—apart from the specific imagery implied by terminology—from that of a natural to a political order. Kin groups bud, branch, and unfold. Territorial groups are created by chiefs. They express human agency. In this expression, they assert a radically new social idea."

A radical new social idea. That is indeed what had happened in Hawai'i, with its *ahupua'a* system, its new kinds of *maka'āinana* and *'āina*, its *konohiki* and *kō'ele*. After the time of 'Umi and Kiha-a-Pi'ilani, Hawaiian society had crossed a threshold, one that another famous anthropologist, Marshall Sahlins, once called "the boundary of primitive society itself." Hawaiian polities were no longer chiefdoms, they were states, ruled by divine kings. Hawai'i had discovered what it was to be a civilization.

. . .

Hawaiian society, politics, and religion were radically transformed about the end of the sixteenth century and beginning of the seventeenth. Both the Hawaiian oral traditions and the empirical evidence of archaeology provide abundant proof of this. Moreover, the kinds of changes that I have just summarized were paralleled in many other early states and civilizations, in both the Old World and the New. Why, we may well ask, did Hawai'i and these other societies all follow similar paths of social and political change? Why did these societies develop class stratification? How did chiefs become divine kings? How did they convince people that they should be allowed to marry their sisters, or that it was proper to offer up the bodies of other human beings on a temple altar? Why did people give up a land tenure system organized by kinship and agree to a new system in which their rights to land required them to owe tribute and labor to overlords?

These are questions that have puzzled philosophers, sociologists, anthropologists, historians, and archaeologists for a long time. Rousseau, Hobbes, Spencer,

Marx, Durkheim, Weber, and many others since have written tomes advancing particular theories of social change. Native Hawaiian scholar David Malo, writing the history of his people in the early nineteenth century, also asked these questions. He wrote: "As to why in ancient times a certain class of people were ennobled and made into *ali'i*, and another class into subjects *(kanaka)*, why a separation was made between chiefs and commoners, has never been explained. Perhaps in the earliest time all the people were *ali'i* and it was only after the lapse of several generations that a division was made into commoners and chiefs." One has to smile at the old scholar's humble wisdom. Perhaps we were all chiefs once.

First, we can say that the Hawaiian chiefdoms did not become states through the influence of some neighboring group. This sort of thing did happen elsewhere in the world when a chiefdom rubbed shoulders with a more advanced state, when the leaders of the former decided to emulate what they saw happening in the latter. Anthropologists call this kind of process secondary state formation. But Hawai'i was totally isolated in the vastness of the North Pacific for about five centuries after the last voyages ended between Kahiki and Hawai'i. And in any case, the societies of Kahiki, the central Polynesian homeland, were themselves chiefdoms, not states. So Hawai'i has to be an instance of what we call primary state formation, in which the processes of change were internally driven. Anthropologists call this kind of change endogenous.

We can also be sure that our quest to explain why Hawaiian society changed will not lead us to a single smoking gun. At one time, some social theorists naively thought that a single prime mover might explain the evolution of complex societies. Some proposed that war was such a prime mover, while others argued that trade was the key; yet others gave that role to population pressure. We now know that social change is complex, the outcome of many inputs. That is to say, there is no single source of causation in social evolution. In academic jargon, we say that causation is multifactorial.

One of the deepest thinkers in twentieth-century biology was the late Ernst Mayr of Harvard University, who wrote many brilliant books on science and evolution. I had the privilege of meeting him just once, over lunch at a meeting of the American Philosophical Society in Philadelphia. Mayr was then in his mid nineties, still working away on several books, and as sharp as a tack. He believed that understanding causation was a tricky business. Causation can operate at several different levels simultaneously; it is easy to get misled into thinking that just one factor is important. Mayr noted that in the inanimate world there is only the causation given by natural laws, such as gravity. But the living world is different, and this includes human societies as well as organisms. Here we must be conscious of two different kinds of causations, what Mayr called "proximate (or functional) causes" and "ultimate (or evolutionary) causes." He correctly noted that *how* questions (such as those I posed a few paragraphs earlier) are usually explained by proximate causes. *Why* questions require ultimate causes. Mayr was amused by the fact that "many famous controversies in the history of biology came about because one party considered only proximate causations and the other party considered only evolutionary ones."

Following Ernst Mayr's sage advice, we need to pay attention to these two different kinds of causation when trying to understand the radical restructuring of Hawaiian society. Let me first consider some possible ultimate causes, those which address the *why* questions. One most certainly was population growth. Chiefdoms can be fairly small in scale, with only a few hundred or perhaps a couple of thousand people. But a state requires a large labor force. With a large population, some of the people can enjoy the luxury of not producing food and other goods and can become full-time managers of those who do the work. In a very large population, one that numbers in the tens or hundreds of thousands, it is also much easier to achieve social "distance." Elites can isolate themselves by various means. And with isolation comes mystery. "Perhaps those people with the feathered cloaks *are* gods on earth," the common people may wonder to themselves, occasionally glimpsing the exalted ones from afar.

As we saw in chapter 10, the centuries just before the rise to power of 'Umi and Kiha-a-Pi'ilani were a time of tremendous population growth across the Hawaiian Islands. Population had been growing exponentially for several centuries, probably reaching a peak about the time that these kings unified their respective islands. Hawai'i may have had one hundred thousand or more persons, Maui perhaps sixty thousand. This vast population provided the labor that an ambitious king needed to expand his agricultural production, to build new and larger monuments to the glory of his gods. Thus population growth and the achievement of a large and high-density population were crucial ultimate causal factors that led to Hawai'i's radical social transformations. As a philosopher might say, this was a *necessary* precondition to change, yet in itself it does not offer a *sufficient* explanation. No prime movers, no single smoking gun.

Karl Marx drew attention to control over the "means of production" as one of the factors leading to social change. In preindustrial, agricultural states such as Hawai'i (and all other early civilizations, for that matter), this meant control over food production. We have seen how, both in the wet valleys where irrigated taro was grown and on the vast leeward slopes where sweet potato and dry taro were farmed, the Hawaiian farmers achieved a remarkable intensity of production. The inevitable tension between Malthus's population pressure and Boserup's innovation resulted in a continual tweaking of the agricultural production system. Even though the Hawaiian population had been increasing, improvements in production meant not only that the new mouths could be fed, but also that there was surplus beyond the immediate needs of the farmer households. This was especially the case while the vast dryland field systems of Hawai'i and Maui were in their early growth phases, still expanding and developing. The continual intensification of production provided a second major ultimate causation for social change, by giving the chiefs the fuel they needed to drive the engine of change.

There are, no doubt, other kinds of ultimate causations that helped to move Hawaiian society along the path from chiefdom to state. One might be found in environmental changes, such as subtle shifts in climate and rainfall patterns that may have occurred during what is called the Little Ice Age, from the late fifteenth century until about the time of European contact. We do not yet know enough about how this apparently worldwide period of climate change was reflected in local patterns of temperature and rainfall in the islands. However, it seems possible that the amount of rainfall—essential for watering the dryland field systems—may have decreased, and that the frequency of periodic droughts increased during the two centuries prior to European contact. Further research is needed to confirm this. If so, these changing climatic patterns could have heightened the stress on the agricultural infrastructure that was so essential to chiefly power. But let us move on to consider the other kind of causation that Mayr posited, proximate causation.

Anthropologists are fond of speaking about "agency." At times their writings on the topic get a bit fuzzy, but basically the concept is simple. The idea is that instead of people and their societies just responding, reacting, to external forces, people themselves are active agents of change. People have goals, desires, aspirations; they are jealous, greedy, avaricious, but also sometimes generous and even altruistic. But human agency does not just exist in a vacuum. It is mediated by culture. Hawaiian culture informs its *ali'i* that they should be *pono* (righteous), and have *aloha* for their people. At the same time, Hawaiian traditions—as taught to the chiefs from the time they could first understand what was being said to them by their elders—told them tales of junior brothers who overthrew their older siblings, or who sailed to discover new islands where they would be second to none.

Status rivalry is a deeply ingrained cultural tradition in Polynesia, and it continually influenced the actions of chiefs and directed their agency in specific ways. It was status rivalry that so intrigued Irving Goldman when he made his comparative study of the Polynesian chiefdoms. In a society where everyone is ranked by birth-order status, rivalry plays itself out between senior and junior brothers, between higher- and lower-ranked lineages, between greater and lesser polities. The political traditions and history of Hawai'i are replete with stories of status rivalry. The voyager priest Pā'ao left his home in Kahiki because of a dispute with his older brother Lonopele. 'Umi's usurpation of the kingdom from his higher-ranked half-brother Hākau was all about status rivalry, as was the fight between Lono-a-Pi'ilani and Kiha-a-Pi'ilani for the control of Maui. And as we shall see a bit later on, the eventual consolidation of the entire archipelago by Kamehameha the Great was another denouement of status rivalry. Status rivalry drove the Hawaiian *ali'i* to be more powerful, more glorious; to control more land; to make more high-ranked marriage alliances. This was agency in action.

Another aspect of agency, a proximate cause lying behind the radical social changes that swept through Hawai'i around the end of the sixteenth century, was the manipulation of ideology. The seeds for Hawaiian chiefs to become god-kings already lurked in older aspects of Polynesian theology. In Hawaiki the chiefs were regarded as the closest descendants of the ancestors, hence the best able to intercede with the supernatural to assure society's well-being. By continuing to elevate their own genealogies, and then taking the step of denying common people the right to recite genealogies at all, the *ali'i* created two classes of people. Ideology can be deliberately manipulated to serve political ends.

But it was not just ideology that was actively manipulated; it was the manifestation of ideology in the physical world. Timothy Earle and Elizabeth DeMarrais have used the idea of "materialization" to explain how people control ideology through physical objects. New systems of ideas and beliefs—such as those of kings as divine beings—get actively reinforced through the use of material symbols (feathered cloaks being a prime example), especially in ritualized public displays. Earle has written that the feathered cloak "was an ultimate power dress, signifying the sacred and potent persona of its owner." The increasingly elaborate *heiau* rituals, carried out by full-time priests on the impressive stone platforms, served to reinforce further the power and prestige of the chiefs and king. Nothing could more concretely emphasize the new order of society than the king's offering of a human body upon the altar of Kū.

Another essential aspect of agency, and of materialization, is that they tend to be quickly emulated. If a chief in Ka'ū saw that his counterpart in Kona was succeeding in convincing the people that he was more effective in acquiring *mana* from the gods because of an innovation in his temple ceremonies, the Ka'ū chief would likely adopt the same rituals. Thus innovation and change spread across political and social boundaries. Colin Renfrew, a British archaeologist and scholar of ancient states, first recognized the importance of such emulation while studying the evolution of chiefdoms in another set of islands, the Cyclades in the Aegean Sea. Renfrew called this process of proximate causation "peer polity interaction." The Hawaiian chain of islands is perfectly set up for this kind of competitive emulation. In the oral traditions we can readily see such peer polity interaction, in the relations between Hawai'i and Maui, Maui and O'ahu, and so forth. Each island's kingship was eager to keep up with new developments that it saw in its competitors, a sort of "keeping up with the Joneses" on the scale of an entire archipelago.

One could say a great deal more about how both kinds of causation—ultimate and proximate—interacted in the course of Hawaiian social and political evolution. Indeed, I have written extensively about this in my book *How Chiefs Became Kings*. For now, let us just say that the late sixteenth and early seventeenth centuries in Hawai'i were a time when a combination of forces—population, surplus production, status rivalry, peer polity interaction—came together in such a way that a radically new kind of society emerged out of the older contours of the Polynesian chiefdom. A new social contract had been forged between the *ali'i* and the *maka'āinana*. Between them, chiefs and people, they would continue over the next two centuries to work out the tensions that bound them together even as they became increasingly different in their own daily practices. The chiefs in their royal centers, training their youth in the arts of war, focused on their goals of territorial conquest, of making politically astute marriage alliances, of securing the futures of their high-ranking heirs. The people, for their part, dispersed over and farmed the *'āina*, nurturing it, striving simultaneously to raise their families while meeting their obligations to their chiefs. They were inseparably linked, *nā li'i* and *nā kanaka*, their futures bound together until the world of the foreigner would burst upon them.

FIFTEEN · The Altar of Kū

By the end of the sixteenth century, Hawai'i and Maui islands had become hotbeds of social and political change. These changes had been driven by huge population increases, fueled by the yields gained from intensified agriculture, especially in the leeward regions of Hawai'i and Maui. 'Umi on Hawai'i and Kiha on Maui built their newly expanded kingdoms on this wave of agricultural growth. But economic boom cycles do not last forever; they inevitably give way to periods of retraction or bust. Just such a transition was beginning in the mid seventeenth century, as the vast farming regions of Hawai'i and Maui were reaching their maximum extents. All of the land suitable for rain-fed cultivation of sweet potatoes and taro had by now been cleared of native forest and scrub, and brought into cultivation. The expansion phase of agricultural development had come to an end. It was still possible to turn the screws on these systems a bit harder, by increasing the labor that *maka'āinana* farmers put into weeding and tending their crops. The length of time that fields were left to lie fallow could also be shortened. But such fine-tuned intensification of production also had its limits. More important, even as the common people were pushed harder, the amount of surplus relative to the total agricultural output was beginning to decline. The *ali'i*, who by this time were dependent for their lifestyle on a formal system of taxation—the annual Makahiki tribute collection, combined with the requirement of labor in the *kō'ele* fields— were not inclined to reduce their demands for food, barkcloth, cordage, and other goods. And especially, they wanted the precious red and yellow birds' feathers.

Feathered cloaks, capes, and helmets had become key symbols that differentiated the *ali'i* from the *maka'āinana*. The feathers, and the garments adorned with them, were the basis for a new kind of chiefly wealth economy. It had not replaced the older staple economy based on agricultural production, but instead had added a new tier to the economic system.

Hawaiian scholars David Malo and Samuel Kamakau described the system of annual Makahiki tax collection, when an image of Lono was carried in a clockwise procession around the island. As the Lono image entered each *ahupua'a* territory, two poles were set in the ground; the distance between the poles was close for a small *ahupua'a*, farther apart for a large territory. The gap had to be filled, from pole to pole, with baskets of sweet potato, calabashes of *poi*, pigs, dogs, barkcloth, fishnets, fine mats, and other offerings. If the *konohiki* or *ahupua'a* manager failed to make sure that his people filled the space between the upright sticks, then the chiefs attending the Lono image would order the territory to be plundered. As Kamakau drily remarked, "Only when the keepers [of the image] were satisfied with the tribute given did they stop this plundering."

Of course, a king or high chief could become too heavy-handed in the exercise of power, pushing the people to the point of rebellion, or of rallying to the side of a rival chief and potential usurper. The Hawaiian traditions speak of such times, when an oppressive chief was overthrown and a more generous *ali'i* installed in his place. One such oppressive ruler was Keli'iokaloa. The traditions say that he seized the people's food, pigs, and dogs, cut down their coconut trees, and forced them to do onerous tasks, such as diving for sea urchins at night. Keli'iokaloa was one who was justly overthrown. With a new ruler the demands on the people might be lessened, at least for a time. But the Makahiki system of taxation was too well entrenched to be abandoned. Indeed, the entire political economy depended on it.

There was, however, another strategy that could be deployed in these times when the agricultural economy was moving from a boom toward a bust. As Malo adroitly put it, "Land was the main thing that the king and chiefs sought to gain by their prayers and worship.... They prayed also to the gods for the death of their enemies." Acquiring land—along with the people who would work the land and make it productive—became the new political strategy of the Hawaiian chiefs and kings in the seventeenth and eighteenth centuries. Whereas earlier times had been characterized by economic development, especially of the vast agricultural infrastructure, the emphasis now shifted to war with the goal of territorial expansion and conquest. The war god Kūkā'ilimoku became the most important deity in the pantheon of the Hawaiian kings, and continued his primacy until the time of Kamehameha the Great. His crested feather image, pearl-shell eyes flashing and mouth rimmed with ninety-four dog teeth, can be seen to this day in the Bishop Museum in Honolulu.

Hawaiian society of the seventeenth and eighteenth centuries revolved around the alternating cults of Lono and Kū. As discussed earlier, the year was divided seasonally into two periods, the first devoted to the rites of Lono, in whose name the taxes were collected, the second to the more terrifying rituals of Kū, in whose name war was undertaken. A peaceful period for cultivation and harvest—the Makahiki—was followed by a time for building up armies and launching campaigns of conquest against other districts and neighboring islands. An endless cycle of peace and war, Lono and Kū. The common people awaited the annual coming of Lono, bringing rain and a new harvest. But it was upon the altar of Kū that the aspirations of the Hawaiian god-kings were laid.

• • •

'Umi's feat of unifying the huge island of Hawai'i proved difficult to sustain. With 4,028 square miles of land, Hawai'i is larger than the states of Delaware and Connecticut combined. This was a very large area to try to govern when reports and commands had to be sent by human runners traveling either over earth and lava trails, or along the coasts on canoes propelled by paddlers and sails. Even with the swiftest runners, it could take the better part of a week for a message to be sent from Kohala to Ka'ū, and to receive a reply. Perhaps 'Umi himself sensed how difficult it would be to hold his vast kingdom together. On his deathbed he divided the island into two realms, giving each to one of his two *pi'o* sons born from his royal wife and half-sister, Kapukini. Keli'iokaloa received the leeward side centered on Kona, while Keawenui was given the windward, southern part of the island with Hilo at its core.

For a time the two brothers ruled their respective sections of the island independently. But Keli'iokaloa began to oppress and abuse the people of his dominions. This gave Keawenui the excuse he might already have been looking for—to make war against his brother and seize the rich agricultural zones of the leeward side. With the combined forces of Hilo, Puna, and Ka'ū under his command, Keawenui marched across the island's high central plateau, past the famous *heiau* built by his father, the Ahu a 'Umi. While these warriors descended from the uplands behind Hualālai Mountain, Keawenui's war canoes simultaneously attacked from the sea. Caught in a pincer action, Keli'iokaloa's army was routed. Keli'iokaloa was killed on a lava bed, his body offered up as a sacrifice on the altar of Kū. The entire island of Hawai'i was once again united, under Keawenui.

Keawenui's reign was peaceful, but after his death the old political fault lines ruptured again, this time in a three-way division. Keawenui's son Lonoikamakahiki, at the time still a boy, was given the southern districts of Puna and Ka'ū. Kumalaenui, a son of the great 'Umi, controlled Hilo and Hāmākua. On the western side, the important farming districts of Kona and Kohala were divided between two half-brothers, both sons of Keawenui by different high-ranking chiefesses. Kanaloa-kua'ana ruled over Kona, while his half-brother 'Umiokalani held Kohala.

These powerful *ali'i* were not satisfied with their allotments; the lure of territorial conquest began to turn their heads. The priests and counselors of Kanaloa-kua'ana, residing in Kona, applied a unique tactic to induce their king to wage war against his half-brother 'Umiokalani in Kohala. The old men went to stand before the king and urinated in his presence. Shocked at this breach of royal protocol, Kanaloa-kua'ana demanded to know why they had pissed in front of him. An old priest replied, "What was our urine like? What was its appearance, O *ali'i?*" "I do not know what your urine is like," the king answered in disgust. The old *kahuna* continued, "Your priests, counselors, and people have clear urines because they drink copiously of water. That is because you are a poor chief. The urine of the chiefs, priests, and counselors of a wealthy chief is yellow through drinking *'awa* and eating rich foods." Having been thus shamed by his old priests, admonished that he was not providing sufficiently for his people, Kanaloa-kua'ana asked what he must do. They urged him to make war against Kohala, to seize that fertile district from 'Umiokalani. After a protracted struggle, the forces of Kona prevailed in a battle at the cinder cone of Pu'u Wa'awa'a. All of Kona and Kohala were now united under Kanaloa-kua'ana.

Kanaloa-kua'ana was married to the chiefess Kaikilani-nuiali'iopuna ("The little heavenly great chiefess of Puna"), herself a descendant of the great 'Umi via Keli'iokaloa; she was thus of the highest rank. Abraham Fornander tells us that this Kaikilani was a kind of power behind the throne, to whom Keawenui had given de facto control over the island after his death. She now decided to take, as a second husband, the young chief Lonoikamakahiki, Keawenui's son who had been given the southern districts of Puna and Ka'ū. When his father had died, Lonoikamakahiki had been just a boy. Now that he had grown to manhood, Kaikilani married him, thus uniting most of the island through her joint unions with Kanaloa-kua'ana and Lonoikamakahiki. Although female *ali'i* rarely ruled overtly, they carried the high bloodlines and could wield enormous influence and power.

Lonoikamakahiki became one of Hawai'i's most famous rulers, partly thanks to the extended travels he made throughout the archipelago. He learned a great deal about the royal courts and practices on the other islands, simultaneously spreading the word of the growing power of his own Hawai'i Island kingdom. On his double-hulled canoe, Lonoikamakahiki carried the royal standard of the Hawai'i kings, a *kāhili* (staff) covered with the feathers of the rare *'ō'ō* bird. From the dual masts of his canoe hung the white skins of the *ka'upu* bird, more than a fathom in length. This king had been named for the god Lono, his name meaning "Lono of the Harvest Season." The *ka'upu* bird skins that decorated his canoe also hung from the image of Lono that was carried annually around the island to collect the Makahiki tribute. Lono was a god, but here he was manifest on Earth as a god-king.

Lonoikamakahiki first made a state visit to Maui, where he stayed in the royal court of Kamalālāwalu, grandson of the great Pi'ilani, whose daughter Pi'ikea had married 'Umi. The royal houses of Hawai'i and Maui were closely related by bonds of marriage. From Maui, Lonoikamakahiki sailed with his entourage to Moloka'i, where they stayed for a time in the deep valley of Hālawa, whose bay was a famous surfing place of the *ali'i*. Then it was on to O'ahu where Kākuhihewa was the ruling king. Lono passed the time in Kākuhihewa's court fishing, surfing, and canoe-racing. Finally, the Hawai'i king took his double-hulled canoe all the way to Kaua'i and Ni'ihau, even to the little uninhabited islet of Ka'ula, which lies beyond Ni'ihau. Lonoikamakahiki thus visited all the principal islands of the archipelago, perhaps the first chief since La'amaikahiki to do so. His grand tour completed, he turned his double canoe southeast and sailed back to Hawai'i.

Kamalālāwalu, king of Maui, had decided to make war on Hawai'i. Perhaps Lonoikamakahiki's earlier visit had piqued the Maui ruler's interest in the rich gardens of Kona and Kohala, which were far more extensive than Kamalālāwalu's own dominions. In chapter 12 I recounted the story of Kamalālāwalu's disastrous attempt to conquer Hawai'i. At first the Maui forces were victorious, capturing Kanaloa-kua'ana and torturing him by tattooing his entire body, even turning his eyelids inside out and tattooing them before killing him. On hearing of this cruelty against his close relative, Lonoikamakahiki was enraged. The superior Hawai'i forces of Lonoikamakahiki set a trap for Kamalālāwalu in the uplands of Waimea. The Maui army was almost entirely destroyed, their king speared to death on the grassy plain. This would be just the first in an escalating series of wars between Maui and Hawai'i, with sometimes one, sometimes the other as the initial aggressor. A strategy of territorial conquest was becoming the predominant theme in Hawaiian politics.

With the passing of Lonoikamakahiki, the Hawai'i Island kingdom once again splintered into independent fiefdoms. The northern part of the island was controlled by the Mahi family of *ali'i*, while the southern part around Hilo was held by the equally powerful 'Ī family. Geography again prevailed over government, the sheer size of this vast island too much for any single ruler to hold effectively. A messy succession of district chiefs ensued. Only one of these *ali'i*, Keawe'ikekahiali'iokamoku, managed to gain some degree of control over the entire leeward side of the island with its rich dryland fields, about the end of the seventeenth century. This king is remembered primarily for constructing the famous burial temple called the Hale-o-Keawe at Hōnaunau, south of the bay of Kealakekua, which housed the bones of the Hawai'i Island rulers and high chiefs.

Upon the death of Keawe'ikekahiali'iokamoku, the politics of Hawai'i were again thrown into confusion and turmoil. Several *ali'i*, including the late king's sons, vied for succession. Across the channel on Maui, a young warrior chief named Alapa'inui residing in the court of the Maui king Kekaulike heard the news from Hawai'i. Alapa'inui was the son of the great Kohala warrior chief Mahiolole and his wife Kalanikauleleiaiwi, half-sister of the deceased king Keawe. He had been residing on Maui because that island's king was married to Alapa'inui's own half-sister. The intricate bonds of royal marriages, especially given the frequent multiple unions, made for considerable intrigue between the dynastic houses of Maui and Hawai'i!

Seeing his chance, Alapa'inui rushed back to Kohala, where he quickly assembled a warrior force loyal to him. In rapid succession he defeated the two other *ali'i* (Kalanike'eaumoku and Mokulani) who had divided the island between them following the death of Keawe. Only the southern district of Ka'ū remained in some measure independent. But no sooner had Alapa'inui gained control over most of Hawai'i than he was faced with a new challenge, from none other than his brother-in-law Kekaulike, king of Maui.

Unlike Hawai'i Island, with its cyclical history of unification followed by political disintegration, Maui (see map 4) had remained a single unified kingdom after Kiha-a-Pi'ilani had unified it with 'Umi's assistance in the early years of the seventeenth century. At 1,884 square miles, Maui is a sizable island, yet more manageable than its larger neighbor to the southeast. After the ignominious defeat of Kamalālāwalu in his futile attempt to conquer Hawai'i, the Maui kingship had passed through several less-known rulers until Kekaulike came to power, in the early eighteenth century. Kekaulike moved the royal court to Kaupō, on the south-

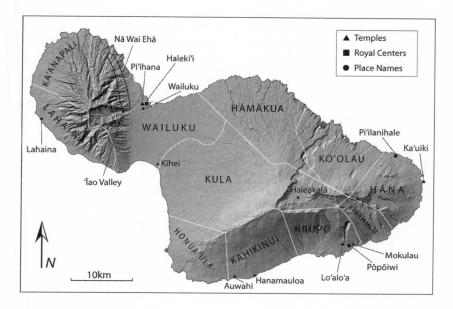

MAP 4.
Maui Island with important places mentioned in the text.

eastern side of Maui. There a great outpouring of young lavas, through the Kaupō Gap in the side of Haleakalā mountain, had created a huge fan-shaped expanse of fertile soils, called Nāholokū, "the Skirt." Covering more than ten square miles, these rich soils combined with an ideal amount of rainfall made Nāholokū perfect for sweet potato cultivation, augmented by dry taro gardens in the uplands. Kekaulike drew on the surplus production of the Kaupō fields to support his court and his army. At the time that Alapaʻinui left him to sail back to Hawaiʻi and win control of that island, Kekaulike had been busy constructing two massive war temples: Loʻaloʻa and Kānemalohemo. These are among the largest stone constructions on Maui, surpassed only by Piʻilanihale in Hāna (see chapter 13).

I have had the privilege of exploring and mapping these immense structures, which are two of the finest examples of late Hawaiian war temples. Loʻaloʻa is a terraced platform, about 375 feet long by 130 feet wide, built up of an astounding 18,000 cubic yards of volcanic rock. The western end of the platform rises only a few feet above ground level, but at its eastern end the platform towers more than 26 feet above the surrounding terrain. This eastern façade is carefully constructed in a series of four steeply descending terraces; when viewed from below it would have been an awe-inspiring sight. No doubt this was precisely Kekaulike's intention. Loʻaloʻa faces the northeast, looking out over Manawainui Gulch toward the rising position of the star cluster Makaliʻi, whose first appearance in the sky marked the onset of the Makahiki harvest and tribute collection period. This orientation suggests that Lono as well as Kū may have been worshipped here.

Kānemalohemo (also known as Pōpōiwi), closer to the shore and situated on a promontory overlooking Mokulau Bay, sprawls over an even greater area of about 3.5 acres. Today its ruins are hidden by invasive Java plum and *koa haole* trees; rank grass covers the pavements. My graduate student Alex Baer and I spent long days hacking through this green tangle with machetes, clearing the walls so that we could construct an accurate map of the multiple terraces, walls, and enclosures. At the center of the complex are two immense courtyards, one higher than the other, each paved with a thick layer of water-rounded pebbles (*ʻiliʻili* in Hawaiian). Literally tens of thousands of basketloads of these beach pebbles had been hauled up from Mokulau to cover these terraces. We were staggered to think of the labor force required. On a yet higher terrace, overlooking the broad courtyards, a series of smaller rooms and platforms were where Kekaulike and his priests conducted their sacred rites. To me the most fascinating part of this complex is the southeastern corner, which was built up with a high buttressed wall of stacked basalt boulders. The two perpendicular walls forming this corner meet at a sharp angle, forming a *V* that points directly to Hawaiʻi Island across the ʻAlenuihāhā Channel. A small room, well paved with *ʻiliʻili*, occupies the top of this corner wall. Standing there, I could visualize Kekaulike in this very place, seated on a pile of fine mats, gazing out from this rampart over his fleet of war canoes assembled in the bay below him, plotting his invasion of Hawaiʻi.

What motivated Kekaulike to attack Hawaiʻi, now controlled by his brother-in-law Alapaʻinui, is not recounted in the traditions. But attack he did, mustering his warriors and war canoes at Mokulau, lying in the shadow of the great war temples of Loʻaloʻa and Kānemalohemo. After making their offerings to Kū on the temple altars, the fleet set sail across the ʻAlenuihāhā Channel. Rounding the northern cape of Hawaiʻi at ʻUpolu, Kekaulike's fleet swept down on Kona. In the ensuing naval battle, Alapaʻinui's forces fought off the Maui invaders, forcing them to turn back. But in their retreat Kekaulike's warriors stopped at several places from Kona to Kawaihae and along the Kohala coast, raiding, killing commoners, and burning villages before crossing the channel back to Mokulau. Alapaʻinui vowed to avenge this cowardly attack on his kingdom, to take the war directly to Kekaulike in his home territory of Kaupō on Maui. In the end Alapaʻinui's war of vengeance would take the Hawai'i army all the way to O'ahu. It would be the first great interisland war between the vying kingdoms of the feathered war gods.

• • •

Although it was not yet 9:00 A.M., the hot, still air in the narrow canyon was stifling. Hopping from boulder to boulder along the dry streambed, Marshall Weisler and I made our way up to the fork where Kawela Gulch, on the southern, lee side of Moloka'i Island, divides into two branches. On either side the canyon walls rise in a series of sheer cliff faces, exposing alternating layers of lava and cinder that had once poured out of the volcanic crater at the mountain's summit. We were trying to find a way to reach the toe of a narrow ridge where the canyon forked into its two branches. From our topographic map it appeared that if we could get to that ridge, we could then make our way along its knife edge up to the plateau defined on either side by the two deeply incised canyons. We suspected that this plateau might have been a fortification mentioned in Kamakau's tradition of the great interisland war that had swept from Hawai'i through Maui all the way to O'ahu during the reign of Alapa'inui.

It was 1980, and I was overseeing an archaeological survey of Kawela *ahupua'a* on the southern Moloka'i coast. Weisler had been working at this task for a few months, concentrating on the areas most likely to be affected by a big housing development planned for the area. This morning, however, we were venturing beyond the limits of the planned development, up onto that mostly inaccessible ridge and plateau. We wanted to see if there was any remnant of the fort into which the Moloka'i warriors were supposed to have retreated. As Weisler and I continued our way up the dry streambed, I recalled what Kamakau had written of Alapa'inui's war against Maui. When Kekaulike retreated back to Mokulau in Kaupō after his marauding on Hawai'i, Alapa'inui gathered his forces together in the northern part of Kohala. There were six divisions of warriors, one from each of the island's districts, supported by a formidable fleet of war canoes. While preparations were being made to invade Maui, word spread across the channel that Alapa'inui would soon land in Mokulau with all the might of Hawai'i behind him. The Maui chiefs and warriors began to lament that their king Kekaulike had begun this war, which was about to come roaring into their home lands.

At this moment Kekaulike was seized with a violent illness, a kind of epilepsy, which the traditions call *Ka Maka Huki Lani* (Eyes Pulled Heavenward). The medicinal practitioners, *kāhuna lapa'au*, could do nothing to stop the seizures. With the king incapacitated and the enemy about to invade their shores, the council of chiefs assembled to choose a new ruler. Kamehameha-nui (not to be confused with the later Kamehameha I of Hawai'i) was a high-ranking *pi'o* son of Kekaulike by the chiefess Keku'iapoiwanui, half-sister to both Kekaulike and Alapa'inui. The council declared Kamehameha-nui to be the new king. Meanwhile, the royal court was in an uproar. The dying Kekaulike, fearing that he would become a sacrifice on Kū's altar should Alapa'inui capture him, ordered a retreat to Wailuku, ancient seat of the Maui kings on the other side of the island. In his double-hulled war canoe named Keakamilo, Kekaulike and his retainers fled around the windward Hāmākua coast. Kekaulike deserted his beloved Kaupō, and the great war temples he had built there. The sacrifices he had made on the altars of Lo'alo'a and Kānemalohemo had failed to garner Kū's support.

When Alapa'inui's war fleet reached Mokulau, they found that the Maui forces had deserted the place. Alapa'inui learned that Kamehameha-nui was now Maui's king. Kekaulike, he was told, had died at Haleki'i, a royal residence of the Maui ruling line near Wailuku. The priests had cut the flesh from his bones, which were taken by a secret party to be disposed of deep in the valley of 'Īao, so that Alapa'inui could not find and desecrate them. The new Maui king, Kamehameha-nui, was Alapa'inui's own nephew. Sailing to Kīhei, Alapa'inui met with Kamehameha-nui and with Keku'iapoiwanui, his half-sister and mother of the king, bringing an end to the hostilities.

While at Kīhei, encamped with a huge warrior force and fleet of canoes to carry them, Alapa'inui received a report that the O'ahu Island king Kapi'iohokalani had invaded Moloka'i Island to the west. Many of the Moloka'i chiefs were descendants of Keawenui of Hawai'i, and hence relatives of Alapa'inui. He heard that the Moloka'i chiefs and warriors had retreated to their mountain fortresses, only to watch helplessly while the O'ahu invaders broke down the sea walls of their fishponds and trampled their sweet potato gardens. The Hawai'i king needed no better excuse to engage his army, which was disappointed that the Maui campaign had thus far seen no real action. Alapa'inui resolved that he would commit his formidable army and naval forces to the aid of the Moloka'i defenders.

While I had been recalling this history from Kamakau's book, Weisler and I had found a rugged path out of the dry streambed up onto the boulder talus at the base of the steep ridge in Kawela, more of a goat track than a real trail, perhaps followed occasionally by a local pig hunter. Sweating profusely as the sun rose higher in the sky, we emerged from the thicket of *kiawe* and *koa haole*, picking our way up over and around the sharp lava boulders that formed the ridge edge. We proceeded carefully, pulling ourselves up one boulder at a time, careful not to trip lest one of us crash headfirst down the steep talus slope, a misstep that would end in broken bones, or worse. About halfway up the ridge, I noticed that the boulders above me were not naturally distributed, but had been stacked several courses high, to form the vertical face of a terrace about six feet high. Hauling myself up onto the level space created by the terrace, no more than about four feet wide, I looked down to see a cache of small perfectly spherical stone pebbles, each about two to three inches wide. I immediately recognized these as the armament of another era, slingstones that the Moloka'i warriors had intended to send flying with uncanny accuracy at the heads of the invading O'ahu warriors. A three-inch basalt pebble propelled from a sling by an experienced warrior was a lethal weapon. It could crack open the skull of an opponent, felling him instantly.

I had come upon a fighting stage, a terrace constructed so that the Moloka'i warriors would have a level platform from which to defend their ridge-top fortress, for the only way up was along the very knife-edge ridge we were ourselves ascending. Continuing our upward climb, Weisler and I came across several more such terraces, though we never found another cache of slingstones. Perhaps the rest had all been deployed during the battle of Kawela.

We now reached the top of the ridge. Here the land widens into a gently sloping plateau defined on either side by sheer cliffs plunging several hundred feet down into the twin branches of Kawela Gulch. A hundred yards or so ahead I could make out a stone structure looming above the stunted *kiawe* trees, its reddish-brown lava stones contrasting with the dull khaki color of the grass. A few minutes later we were exploring this ruin, which proved to be about eighty feet long east-west and about fifty feet north-south. An eastern room or chamber was defined on three sides by high, well-built stone walls, but open toward the sea, in the direction we had come. The adjacent western chamber was walled in on all four sides. We had discovered the *pu'uhonua* (place of refuge) of the Kawela people. A *pu'uhonua* was a sacred space, in which a *kapu* breaker or a defeated warrior could claim sanctuary. Under the protection of the priests and the gods, those who could make the safety of a *pu'uhonua* could not be harmed.

The Moloka'i warriors had been very clever, employing a two-tiered strategy. They put their first line of defense on the narrow ridge, armed with slingstones to be fired off from their fighting stages. But should the enemy prove too formidable, they were prepared to retreat into their sacred *pu'uhonua*, lay down their weapons, and beg forgiveness of the gods. In ancient Hawai'i, war itself was a kind of religious act, sanctified by the offering of human bodies to Kū. As a part of that religious system, the *pu'uhonua* was the ultimate mode of retreat and defense.

Penned up in their ridge-top fortress, with limited access to water and cut off from their crops on the valley floor below, the Molokaʻi defenders must have let out shouts of joy when they saw the sails of Alapaʻinui's war canoes coming to their aid across the Pailolo channel from Maui. The traditions say that the Hawaiʻi flotilla was so extensive that the canoes hauled ashore all the way from Waialua to Kaluaʻaha. The Oʻahu forces were camped to the west, with their king Kapiʻiohokalani based at Kalamaʻula. A four-day-long battle ensued, with the Hawaiʻi forces of Alapaʻinui marching westward through Kamalō to engage the Oʻahu army. The Molokaʻi defenders attacked from the hills. On the fifth day the decisive battle was fought on the coastal plain at Kawela. Kamakau described the scene: "Every able-bodied man came out of his house to fight. The Molokai forces attacked from the hills, those of Hawaii from the sea, while a great number landed from the fleet and fought on land. The battle began in the morning and lasted until afternoon. The ruling chief [king] of Oahu found himself surrounded by sea and by land and hemmed into a small space. Kapiʻiohokalani died at Kawela below Kamiloloa, and many chiefs and fighting men were slaughtered, but some escaped and sailed for Oahu." Well into the twentieth century, it was possible to see the scattered bones of the many felled warriors, eroding out of the sand dunes at Kawela.

With the Oʻahu king dead on the battlefield and his army in full rout, Alapaʻinui realized that the fertile island, the ancestral home of Māʻilikūkahi and the ancient chiefs from Kahiki, could be his for the taking. He determined that he would take this war—which had begun in his desire for revenge against Kekaulike of Maui—all the way to Oʻahu and claim it as his prize.

The Oʻahu *aliʻi* had feared exactly this possibility. Their new king, Kanahaokalani, was a mere six-year-old boy, son of the slain Kapiʻiohokalani. The Oʻahu chiefs desperately needed a seasoned leader, preferably an *aliʻi* of the highest rank. A fast canoe was dispatched to Kauaʻi, whose aristocracy was linked from ancient times to that of Oʻahu. The messengers beseeched Kauaʻi king Peleiʻōhōlani to come to the aid of threatened Oʻahu. Peleiʻōhōlani was, in fact, the brother of the recently slain Kapiʻiohokalani. Both had been fathered by the famous Oʻahu king Kualiʻi. Kualiʻi had reigned over both Oʻahu and Kauaʻi islands, leaving the Kauaʻi kingdom to Peleiʻōhōlani on his deathbed. With family honor at stake, Peleiʻōhōlani could not refuse this plea for assistance from his Oʻahu kinsmen.

Before Peleiʻōhōlani arrived, Alapaʻinui's forces had already made two attempts to land on Oʻahu, first at Waikīkī and then at Waiʻalae. But the Oʻahu defenders were stiff in their resistance. The Hawai'i forces had been repulsed at both beachheads. Alapa'inui then heard that a landing might be possible on the windward side, at an isolated place called Oneawa, between Kailua and Kāne'ohe. A successful landing was made before the O'ahu warriors had time to travel across the Ko'olau Mountains via the Pali trail.

With the Hawai'i army camped at Oneawa, skirmishing began around Kāne'ohe. No doubt another major battle would have ensued, had it not been for the wise council of Na'ili, brother of the chiefess Kamaka'imoku who was the mother of two of Alapa'inui's most skilled warrior chiefs, Kalani'ōpu'u and Keōua. Na'ili first approached Pelei'ōhōlani in his camp, telling him that the *ali'i* of Maui and Hawai'i were related to him; he could stop the war if he would go and meet with Alapa'inui. Continuing this peace-brokering, Na'ili then went to see Alapa'inui, telling the Hawai'i king that Pelei'ōhōlani was willing to meet with him. The terms of the meeting between god-kings were laid out: only those two were to meet face-to-face at a place called Naoneala'a, neither one bearing a weapon. The Hawai'i warriors were to remain in their canoes, while the O'ahu and Kaua'i forces were likewise to stay back, without arms.

Each keeping to the agreed terms, Pelei'ōhōlani and Alapa'inui met at Naoneala'a, on windward O'ahu. Samuel Kamakau dates the event to the thirteenth day of the Hawaiian month Ka'elo (roughly January), in the year 1737. In his words: "The two hosts met, splendidly dressed in cloaks of bird feathers and in helmet-shaped head coverings beautifully decorated with feathers of birds. Red feather cloaks were to be seen on all sides. Both chiefs were attired in a way to inspire admiration and awe, and the day was one of rejoicing for the end of a dreadful conflict." While the Hawai'i forces observed from their canoes and the O'ahuans watched from a distance, Pelei'ōhōlani and Alapa'inui approached each other and then embraced, pressing nose-to-nose in the *honi*, the traditional Polynesian greeting and mutual intermingling of breath. Acknowledging their distant bonds of kinship, the two god-kings agreed to put hostilities aside. It would prove to be a fragile truce.

With the war that had swept all the way from Hawai'i through Maui and Moloka'i to O'ahu concluded, Alapa'inui began the return voyage with his fleet, stopping first at Maui. He was surprised to discover that a son of the late king Kekaulike, named Kauhi'aimoku-a-Kama (hereafter Kauhi), had rebelled against Alapa'inui's nephew Kamehameha-nui. A noted warrior, Kauhi enjoyed considerable support among certain Maui chiefs. At first Alapa'inui thought that Kamehameha-nui would be able to defend his kingship, but in a bloody battle

Kamehameha-nui was forced to flee for his life. Alapaʻinui took Kamehameha-nui with him in his canoe across to Hawaiʻi, knowing that he would have to return to Maui to reinstate Kamehameha-nui by force.

After a year of preparations Alapaʻinui was ready to invade Maui and reinstall his nephew as king. This time the action was centered in the Lahaina region of western Maui, where the forces of Kauhi were assembled. Hearing that Alapaʻinui's army was about to invade, Kauhi sent an envoy to Oʻahu with gifts for Peleiʻōhōlani, who still ruled over both that island and Kauaʻi. Kauhi begged Peleiʻōhōlani to come to his aid; the latter responded with a force of 640 warriors. But the Hawaiʻi foe was formidable, with 8,440 warriors according to the traditions.

Alapaʻinui landed on Maui, driving the forces of Kauhi into mountain strongholds. In the lowlands and valleys, the Hawaiʻi warriors destroyed the taro fields and irrigation canals of their Maui opponents. When Peleiʻōhōlani and his warriors finally arrived from Oʻahu, they were prevented from joining Kauhi's rebels in the hills. A series of skirmishes and battles ensued between the Hawaiʻi and Oʻahu armies, ending in a fierce struggle at Honokawai. After major losses on both sides, Alapaʻinui and Peleiʻōhōlani once again confronted each other face-to-face, on the battlefield. Instead of fighting, they saluted each other and agreed to renew the truce they had made at Naonealaʻa on Oʻahu. Deprived of Peleiʻōhōlani's support, the rebel Kauhi could not hold out. He was captured and drowned at Alapaʻinui's orders. Kamehameha-nui was again installed as king of Maui, while Peleiʻōhōlani returned via Molokaʻi to rule over Oʻahu and Kauaʻi.

· · ·

Among the most valiant and trusted of Alapaʻinui's warrior chiefs was Kalaniʻōpuʻu, a descendant of the great ʻUmi. Kalaniʻōpuʻu's father was Kalaninuiʻimamao, a ruling chief of Kaʻū district; his mother was Kamakaʻimoku (sister of the wise counselor Naʻili who first brokered the peace between Alapaʻinui and Peleiʻōhōlani). His grandfather was Keaweʻikekahialiʻiokamoku, the king who had preceded Alapaʻinui. Kalaniʻōpuʻu and his younger half-brother Keōua Kupuāikalaninui (henceforth Keōua) had been adopted by Alapaʻinui when they were young boys, after their father was killed in the conflict following the death of Keawe, at the time that Alapaʻinui had gained the Hawaiʻi kingship. Alapaʻinui raised Kalaniʻōpuʻu and Keōua in his court, training them in the arts of war. They became his loyal generals, key supporters in the long war against Maui, Molokaʻi, and Oʻahu.

Kalaniʻōpuʻu probably never fully trusted Alapaʻinui, knowing that the latter had been responsible for the death of his father. After they had all returned to

Hawaiʻi, following the bloody campaign on Maui to reinstall Kamehameha-nui, Keōua, while living in Alapaʻinui's court at Hilo, fell ill and died. Kalaniʻōpuʻu, residing in nearby Puna at the time, heard the news and immediately suspected that Alapaʻinui had either poisoned his younger half-brother or killed him by the feared "praying to death" *(ʻanāʻanā)*. Keōua had a son, a young warrior named Kamehameha (not to be confused with Kamehameha-nui, king of Maui). Kalaniʻōpuʻu now worried that Alapaʻinui might try to kill his nephew as well. While the mourning for Keōua was going on, Kalaniʻōpuʻu made an aborted attempt to abduct Kamehameha from Alapaʻinui's court. This act opened a deep rift between the formerly loyal general and his king.

Kalaniʻōpuʻu may well have reflected that his genealogy gave him a more direct line of succession to the Hawaiʻi kingship from his revered ancestor ʻUmi-a-Līloa. Kalaniʻōpuʻu withdrew to Kaʻū, his ancestral homeland, a move that was interpreted by Alapaʻinui as open insurrection. A series of battles and skirmishes ensued, but Alapaʻinui was unable to defeat Kalaniʻōpuʻu. After a time, Alapaʻinui left him alone to rule over Kaʻū and adjacent Puna districts independently. Alapaʻinui was growing old. Soon he collected his retinue and left Hilo, moving the court to the northern part of the island, to Kawaihae. There he fell ill and, knowing that the end was coming, appointed his son Keaweʻopala as his heir.

Upon the ascension of a new king, protocol required a redistribution of the districts and *ahupuaʻa* of the kingdom among the principal chiefs and warriors. Keaweʻopala committed a serious political error by slighting several prominent *aliʻi* in this land distribution, among them a powerful warrior chief named Keʻeaumoku. Angry that he had not been allotted a proper share of the *ahupuaʻa* estates—from which all of the chiefs drew their sustenance and support—Keʻeaumoku left Keaweʻopala's court. He was soon joined by several other disaffected chiefs. Meanwhile, word of Alapaʻinui's death had reached Kalaniʻōpuʻu in Kaʻū, who decided that it was time to claim his rightful patrimony as king of the entire island. Kalaniʻōpuʻu began to move his forces northward, by land and sea, up the leeward side of the island toward Kona.

Hearing that Kalaniʻōpuʻu was approaching Kona, intending to attack Keaweʻopala, Keʻeaumoku traveled to Honomalino to meet Kalaniʻōpuʻu, offering the support of the rebels. Keaweʻopala was not about to go down without a fight. He had a battle-tested warrior named Kamohoʻula to lead his army. The two armies met north of Hōnaunau in Kona. The terrain was ʻaʻā lava, full of jagged rocks and holes that could easily trip a warrior, rendering him helpless to his opponent's club and spear. The fighting was protracted, with neither side gaining a decisive victory. Then Holoʻae, the high priest of Kalaniʻōpuʻu, declared that they must capture and kill Keaweʻopala's priest Kaʻakau, for the latter's prayers to Kū were clearly swaying the war god in his favor. Both priests had served the former king Alapaʻinui. On the battlefield, Kaʻakau was singled out by Kalaniʻōpuʻu's warriors and captured. He was cruelly killed, "baked," it is said, in an *imu* pit. Kū had his sacrifice. The tide of battle turned in favor of the southern rebels.

The year was 1754; Kalaniʻōpuʻu was now undisputed king of all of Hawaiʻi Island. He made the expected division of lands, taking care that none of his trusted warrior chiefs was slighted. As with other kings, Kalaniʻōpuʻu had numerous wives, but his most important union was with Kalola, daughter of the former Maui king Kekaulike and sister of Kamehameha-nui, who was then ruling over Maui. Given this close tie between the rulers of Hawaiʻi and Maui, one might have anticipated a period of peaceful relations between these two kingdoms. But, as Kamakau wrote, Kalaniʻōpuʻu "had one great fault; he loved war and display and had no regard for another's right over land." Ruling over the largest island in the archipelago was not enough to fulfill Kalaniʻōpuʻu's ambitions. Shortly, he would launch a series of campaigns to bring the fertile lands of eastern Maui under his control. Engaged in this war, Kalaniʻōpuʻu was fated to encounter the weather-beaten British commander James Cook, returning from his explorations of far-off New Albion. The world would never be the same, either for Cook or for Kalaniʻōpuʻu and his people.

SIXTEEN · The Return of Lono

For several years, Kalaniʻōpuʻu had been in an almost continual state of war with Maui. Ignoring his wife Kalola's pleas, Kalaniʻōpuʻu first attacked Maui while her brother Kamehameha-nui was king, about 1759. He seized Hāna and Kīpahulu districts on the island's eastern end, directly across the ʻAlenuihāhā Channel from Hawaiʻi. These well-watered, fertile lands expanded the economic base of Kalaniʻōpuʻu's kingdom. But Kalaniʻōpuʻu's conquests may well have been driven by other motives, in particular a desire for additional sources of the prized red and yellow birds' feathers. The feathers—increasingly important to the elite wealth economy—were becoming scarce on Hawaiʻi. The literally thousands of warriors and junior chiefs who filled out the ranks of the *kaukau aliʻi* (lesser nobility) were all desirous of at least a feathered shoulder cape as an emblem of their status. The highest-ranked *aliʻi* required full-length cloaks. Above Hāna and Kīpahulu the broad slopes of Haleakalā were covered in *ʻōhiʻa* forests, prime habitat of the *ʻiʻiwi* and *ʻapapane* honeyeaters whose red feathers were especially prized. By annexing these districts, Kalaniʻōpuʻu could also control the exploitation of their forest birds.

At first, Kamehameha-nui made several efforts to repel the Hawaiʻi invaders, even making a raid across the channel to Hawaiʻi Island. But Kalaniʻōpuʻu's war chief Puna controlled the strategic Kaʻuiki fortress on the cinder cone overlooking Hāna Bay; the Maui forces were unable to dislodge them. Kamehameha-nui did not have Kalaniʻōpuʻu's ambition or staying power. Eventually he gave up the fight, retreating to Wailuku, ancient seat of the Maui kings, on the far side of the island.

Seven years after Kalaniʻōpuʻu invaded Hāna, Kamehameha-nui died. The Maui kingdom now passed to Kamehameha-nui's younger brother, Kahekili-nuiʻahumanu, better known simply as Kahekili. The third child born to the great Maui king Kekaulike by his *piʻo* wife Kekuʻiapoiwanui, Kahekili was nearly fifty years old by the time he ascended to the Maui kingship. He had never expected to become king, devoting himself to the life of a warrior who supported his older brother Kamehameha-nui. Kahekili had a thin, weak voice but was fearless in combat. His favorite sport was the *lelekawa*, leaping off high cliffs into pools of water. Kahekili is reputed to have leaped off cliffs as high as 360 feet. No portraits were ever drawn of him, but his must have been a striking visage. He was tattooed solid black on one half of his entire body, from head to foot except, it was said, for the insides of his eyelids. His loyal bodyguards were similarly tattooed, known collectively as the *pahupū*, the "cut in half," from this unique style of body marking.

For about a decade after he became Maui's ruler, Kahekili tolerated the Hawaiʻi islanders in Hāna. But about 1775 Kalaniʻōpuʻu launched a further incursion, into Kaupō district, which lies to the west of Kīpahulu. Kaupō, it may be remembered, had been the favored lands of Kahekili's father Kekaulike, site of the great war temples Loʻaloʻa and Kānemalohemo. Although Kahekili resided with his royal court at Wailuku, on the windward side of Maui, Kaupō was nominally one of his dominions. He could not tolerate this bold attempt by Kalaniʻōpuʻu to seize control of yet more of eastern Maui.

Kahekili dispatched two divisions of warriors overland to confront the Hawaiʻi invaders, who, it was reported, were abusing the *makaʻāinana* of Kaupō, beating them on their heads with war clubs. The warrior chief Kāneʻolaelae led the Maui forces into battle against the Hawaiʻi army, headed by its most famous fighter, Kekūhaupiʻo. They fought hand to hand in the sweet potato fields extending inland from Ka Lae o ka ʻIlio, the Cape of the Dog. At one point, Kekūhaupiʻo's leg became entangled in the twisted sweet potato vines, tripping him. He was about to be clubbed by a Maui warrior when the young Kamehameha, nephew of Kalaniʻōpuʻu (whom the latter had tried to abduct from the court of Alapaʻinui years earlier in Hilo) saved the older warrior. This was the first battle in which Kamehameha distinguished himself. In spite of Kamehameha's bravery, the Hawaiʻi forces were badly beaten by Kahekili's warriors. Fleeing back to their canoes, Kalaniʻōpuʻu's army was forced to retreat to Hawaiʻi.

The defeat at Kaupō did not deter Kalaniʻōpuʻu. In 1776 he sailed to Maui yet again, with a force of six brigades, plus two special regiments of the best warriors.

The 'Ālapa was one of these regiments, eight hundred warriors strong, all of *ali'i* rank. These were seasoned fighters, including expert spear throwers and left-handed slingstone throwers of uncanny accuracy. The 'Ālapa landed at Keone'ō'io on Maui's southern shore, ravaging the countryside and forcing the Honua'ula people to flee into the hills. Camping at Kīhei, Kalani'ōpu'u held back the bulk of his army, sending his prized 'Ālapa over the saddle between eastern and western Maui to engage Kahekili at his royal seat in Wailuku. Kalani'ōpu'u was confident of victory, boasting to the chiefess Kalola, "My 'Ālapa have perhaps drunk of the waters of Wailuku!" But Kahekili was not to be so easily conquered. He had set a trap in the sand hills of Kahulu'u. "The fish have entered the sluice," Kahekili told his war leaders, "draw in the net." Surrounded by a far greater horde, the brilliant 'Ālapa in their feathered capes and helmets were slaughtered, corpses heaped up like firewood. Only two men escaped to relate the horrible news to Kalani'ōpu'u. Kalani'ōpu'u could only wail mournfully for his lost 'Ālapa.

Still Kalani'ōpu'u would not give up his ceaseless ambition to conquer Maui. The following day he committed his entire force, led by the fearless Kekūhaupi'o, to an overland attack against Wailuku, proceeding by way of Waikapū, along the foothills of the mountains of western Maui. But Kahekili's warriors knew the terrain better; they had positioned themselves in strategic locations before dawn. As the Hawai'i warriors marched toward Wailuku, they were met by a ferocious barrage of spears, javelins, and slingstones. The terrified Hawai'i warriors were pinned down among the creeping *'ūlei* and *'ilima* shrubs.

Kalani'ōpu'u realized that he must sue for peace or see his entire army wiped out. He first asked his wife Kalola if she would go to her brother Kahekili as a peace envoy, but Kalola demurred, replying that it was they who had started the war. The only hope, Kalola argued, was to send their son—Kahekili's nephew—Kīwala'ō to his uncle to beg for a truce. Carrying the highest bloodlines of both Hawai'i and Maui, Kīwala'ō was "endowed with the tapu of a god and covered with the colors of the rainbow," as Kamakau put it.

Dressed in a flowing cloak with a feathered helmet on his head, the young Kīwala'ō (he was a mere boy at the time) was sent to meet Kahekili, accompanied by two of Kahekili's half-brothers who were fighting on the side of Kalani'ōpu'u. As Kīwala'ō approached the battleground, the warriors of both sides were obliged by the protocol of *kapu moe* to lay down their weapons and prostrate themselves. No one dared to look at a *kapu* prince directly, even in the heat of battle. The emissaries proceeded unharmed to Wailuku, where Kahekili lay facedown on his fine mats. Kīwala'ō and Kahekili greeted with the *honi*, and wailed for each other.

Yet Kīwalaʻō was so sacred that even Kahekili could not address him directly, so he asked his two half-brothers on what errand they had come. Kahekili consented to a truce, sending a gift of sweet fish from the royal ponds of Kanaha and Mauʻoni to his sister Kalola and to his archenemy Kalaniʻōpuʻu, across the island at Kīhei. Kalaniʻōpuʻu withdrew his forces, but even this would not be the end of his ceaseless ambition against Maui.

· · ·

After Cook's ships *Resolution* and *Discovery* visited Kauaʻi and Niʻihau in January 1778, rumors began to spread eastward down the archipelago about the strange visitors in their floating *heiau* with its great white banners. Some said that the speech of the strangers was like the twittering of the ʻōʻō birds. Others observed that they blew smoke from their mouths—perhaps they were messengers from the volcano goddess Pele? Some recalled old prophecies of white men who would come, riding great dogs with long ears, to steal the land.

Kalaniʻōpuʻu first heard the news while on the windward side of eastern Maui. Despite his bitter defeat a year earlier, Kalaniʻōpuʻu was back trying to expand his grasp on eastern Maui. A man named Moho came to visit Kalaniʻōpuʻu and the other Hawaiʻi chiefs camped with him, telling them the story of the strange one-hulled giant canoes and their unusual passengers. Pailili, son of the great priest Holoʻae, declared that the leader of the strangers must be none other than the god Lono, returned from Kahiki.

At that moment, Cook's ships were just days away from returning to the islands. The *Resolution* and *Discovery*, their canvas and rigging a patchwork of repairs, had left the bone-piercing cold of New Albion in late October. Her crews were weary from months of ceaseless and futile exploration in search of the fabled Northwest Passage. Cook and his ships needed a safe harbor, a place for rest and reprovisioning. The islands they had encountered the previous year—named the Sandwich Islands by Cook—were the obvious choice. On November 26 at daybreak, land was sighted to the south-southeast. Cook wrote in his log that the island's "summit appear'd above the Clouds." Maui, with its great mountain of Haleakalā, beckoned.

Now a mere two leagues off the windward shore of Maui, the captain looked with caution upon a "steep rocky coast against which the sea broke in a dreadful surf." Houses and plantations could be discerned in the gently sloping countryside. Soon the *Resolution* and *Discovery* were surrounded by outrigger canoes eager to establish a trade. By November 27, the Hawaiian canoes were bringing out quantities of "bread fruit, [sweet] Potatoes, Tarra or eddy roots, a few plantains and small pigs." In return, the British gave them nails and iron tools.

On November 30 several large canoes came out to the *Resolution*. One of them carried a person of evident high rank, whose name Cook transcribed as "Terryaboo." This chief made Cook a present of "two or three small pigs." Lieutenant King in his journal gives us a more intimate description of Kalani'ōpu'u: "He had on a very beautiful Cap of yellow & black feathers, & a feathered Cloak which he present'd to the Captn; the Natives were very officious in their care of handing him up the side; Although not an old man, yet he was exceedingly debilitated, & every part of him shook prodigiously which we attribute to his drinking to excess the (Kava) the infusion of Pepper root. . . . At this time his whole body almost was encrustd & the Skin peal'd off in scabs; his Eyes were red & sore." Cook did not realize that this chief was actually the king of a much larger island to the southeast. For the moment, Cook was ignorant of the complex political dynamics of this newfound archipelago.

Cook learned from Kalani'ōpu'u that the island they were coasting along was called "Mow'ee" and that an even larger island now visible to windward was called "O'why'he" (Hawai'i). Continuing on past Hāna, Cook could make out the summits of its great volcanoes, capped in snow despite the subtropical latitude. Cook abandoned his plan of seeking an anchorage along leeward Maui, making for the weather coast of Hawai'i. Kalani'ōpu'u must have received with great interest the news that Cook's ships were moving toward his own island. This was the season of the Makahiki, its onset marked as always by the rising of Makali'i, the "Little Eyes," or Pleiades. Of four lunar months' duration, the Makahiki season at this moment was just opening. This was Lono's time, during which his priests would make a clockwise circuit of the island, passing through each of the hundreds of *ahupua'a* to collect the *ho'okupu* in Lono's name. Was this new Lono going to receive his *ho'okupu* from the people?

To appreciate fully the remarkable meeting of two foreign worlds that was about to unfold requires a brief digression into the Hawaiian cultural logic of the Makahiki. Cook, of course, remained ignorant of the Makahiki or what it represented to the Hawaiians, even though he had heard himself referred to as "O Rono" on the beach at Kaua'i. (The *r* and *l* sounds were interchangeable in Hawaiian.) But for the Hawaiians, chiefs and commoners alike, the Makahiki was a serious matter: source of revenue for the chiefs, and the major period of dryland cultivation and harvest for the people. The entire Makahiki season was ritually encoded, sacred to Lono. War was forbidden. Indeed, the king during this time symbolically gave the land over to Lono, in whose name the tribute was collected. Would Kalaniʻōpuʻu now have to physically yield his kingdom to this new manifestation of Lono, who had appeared from beyond the horizon? He must have pondered that question more than once.

In order to comprehend how Cook so readily played into the role of Lono—and how the Hawaiians could have mistaken his actions for those of a returning god—we need to examine briefly the ritual sequence of the Makahiki on the island of Hawaiʻi. As already noted, this was a period of four lunar months, followed by a longer period of eight lunar months when the main temple rituals of Kū were enacted. The close of one year and signal that a new Makahiki season would commence began when the king placed a rag before his principal war temple. The barkcloth rag symbolized the old year. After this, the high priest of Lono waited at night within the temple precincts, carefully scanning the heavens to the northeast, watching for the first visibility of Makaliʻi. (Pleiades is visible for six months of the year, and disappears for the other six months; it was used to recalibrate the lunar calendar.) This would occur about November 17, which was very close to when Cook's ships were first sighted off Maui. When the "Little Eyes" became visible, the king performed a second rite, the *kuapola*, breaking open green coconuts to signify life, purification, and renewal.

Scholars have called the Makahiki season both a new year's festival and a harvest festival. In fact, it was both, and more. The Makahiki occurs during the *hoʻoilo* or "wet" season of winter rains, brought by the southerly *kona* storms that dump most of the annual rainfall on the leeward sides of the islands. These winter rains were essential to the growth of the sweet potato and dryland taro plantations. The planting of sweet potatoes would have been largely completed by the time the Makahiki was declared. Thus the season itself corresponds closely to the main growing period for the crop. The fact that war was forbidden during this interval, indeed that the gardens themselves were declared to be *kapu* for a certain period, can be seen as a ritually inscribed means to assure that nothing would adversely affect the new crop. As the god of rain and thunder, Lono reigned during the Makahiki. His sacred plant was the sweet potato, principal crop of the dryland farms. It was believed that during the Makahiki Lono returned from Kahiki, the distantly remembered homeland beyond the horizon, to fertilize the land and bring the life-giving rains to the vast dryland field systems.

As we have seen, the original Makahiki had its origins long before in a simpler firstfruits offering, called Mata-fiti in ancient Polynesian times. But after ʻUmi's unification of Hawaiʻi, the Makahiki had taken on a much more important political and economic function: it was the principal ritualized form of ensuring that the surpluses produced by the common people could be systematically collected by the *aliʻi*. Two distinct phases of tribute collection occurred during the course of the Makahiki. The first is sometimes overlooked, but was clearly described by the Hawaiian scholar Kēlou Kamakau. This consisted of a kingdomwide collection of special foods, dogs in particular, as well as wealth items such as barkcloth brought specifically for the king. These were placed in front of the feathered image of the king's personal god, Kūkāʻilimoku. The king then distributed this assembled wealth to his priests, chiefs, warriors, and other favorites.

The second and much more significant phase of Lono's tribute collection was the famous circuit of the "long god" *(akua loa)*, accompanied by a shorter circuit of the "short god" *(akua poko)* through the many territories making up the kingdom. The *akua loa* consisted of a wooden shaft tipped with a carved human head. A crossbeam was attached to this shaft, from which were suspended streamers of white barkcloth and the white pelts of *kaʻupu* birds. (Note the visual similarity between these white banners hanging from a cross-piece and the white canvas of Cook's sails hanging from their square-rigged masts.) The image was brought before the king, "who cried for his love of the deity." After hanging a *niho palaoa* necklace on the image, the king placed morsels of pork, taro, and coconut pudding *(kūlolo)*, as well as *ʻawa* in the mouth of the man carrying the image. Thus properly anointed and supplicated, Lono was sent forth across the land to do his work. The long god took twenty-three days to complete his clockwise circuit around the island, whereas the short god took only four days on a counterclockwise circuit, restricting his visits to the special lands *(ʻili kūpono)* of the king. Kamakau gives a vivid description of the work of the long god as it made its circuit throughout the land:

> Much wealth was acquired by the god during this circuit of the island in the form of tribute *(hoʻokupu)* from the *mokuʻaina, kalana, ʻokana,* and *ahupuaʻa* land sections at certain places and at the boundaries of all the *ahupuaʻa*. There the wealth was presented—pigs, dogs, fowl, poi, tapa cloth, dress tapas *(ʻaʻahu)*, *ʻoloa* tapa, *paʻu* (skirts), malos, shoulder capes *(ʻahu)*, mats, *ninikea* tapa, *olona* fishnets, fishlines, feathers of the *mamo* and the *ʻōʻō* birds, finely designed mats *(ʻahu pawehe)*, pearls, ivory, iron *(meki)*, adzes, and whatever other property had been gathered by the *konohiki,* or land agent, of the *ahupuaʻa*.

Although perishable foodstuffs such as *poi* were consumed by the crowd accompanying the Makahiki circuit, the other valuables listed by Kamakau were returned to the king's storehouses.

The Makahiki season ended with a series of rites in which the king wrested symbolic control of the land back from Lono. Most dramatic of these was the *kāli'i* rite. By this time the Lono gods had returned to their principal temple, which on Hawai'i in the time of Kalani'ōpu'u was the *heiau* of Hikiau at Kealakekua Bay (Kealakekua means the Road of the God). The king set forth in a canoe to land in front of the temple. A sham battle ensued, with spears thrown at the king and/or his champion (an expert in spear dodging), who was obliged to parry them. Symbolically, the king thus reclaimed his kingdom. He then went to the temple to visit the god images, sacrificing a pig in honor of Lono, who had done the hard work of visiting the land "that belongs to us both." After this the Lono images were taken down and wrapped up, to be stored in the *mana* house of the *luakini* temple until the next year. Finally, in a concluding rite, a "tribute canoe" *(wa'a 'auhau)* filled with foods of various kinds was allowed to drift out to sea, carrying Lono back to Kahiki, whence he would return again in eight lunar cycles. With the Makahiki season ended, the king—following purification rites and the first open sea trolling for *aku* (bonito)—prepared with his high priest to begin the even more complex sequence of rituals dedicated to Kū, god of war.

. . .

Kalani'ōpu'u watched from his Hāna stronghold on Maui as Cook sailed across the channel to begin his clockwise progression off the weather coast of Hawai'i. After rounding the eastern cape the two ships followed the rugged lava shores of Ka'ū. Having appropriately arrived off Maui just as Pleiades had become visible, Cook was now enacting a maritime version of the long god's clockwise circuit of Hawai'i Island. Canoes came off daily to deliver what the Hawaiians undoubtedly regarded as *ho'okupu*, their offerings to Lono. Ignorant of the Makahiki rituals that underlay the Hawaiian side of the cultural equation, Cook viewed all this as simply good trade in provisions, with quantities of "pork, fruit and roots" coming daily onto the ships. More than food was being traded as well, for, as the expedition's surgeon, William Ellis, wrote, when the "great numbers of canoes put off as usual," many of them contained women, who spent the days belowdecks engaged in sex with the common sailors. As the clockwise progression continued, the British observed a "white flag," taken by Cook as a "signal of peace." Without doubt this was the *akua loa* with its white barkcloth streamers, being carried along by the Lono priests, keeping pace with the ships. Lono in his floating *heiau* was making his circuit around the island, receiving the *ho'okupu* on his decks. All the time, the long god was marking his progress through one *ahupua'a* territory after another.

So they progressed throughout the month of December, tacking to and fro off the Hawai'i coast, unable to find a safe harbor but keenly engaged in what the British saw as "trade," the Hawaiians as the offering of *ho'okupu*. To the Hawaiians—especially the commoners—the crucial difference was the unexpected generosity of the god himself. Unlike the long god, this new Lono bestowed reciprocal gifts of iron, nails, and beads from his seemingly endless supply in the great floating *heiau*. This was a new twist in the traditional rites, for the long god was not known for his generosity.

They rounded Ka Lae, South Point, on January 5, finding shelter in the calm, leeward waters. By January 16 they were halfway up the western coast, approaching the great bay of Kealakekua, the location of Hikiau, primary temple of Lono, and the royal residence of Kalani'ōpu'u. Lieutenant Bligh, of later *Bounty* infamy, ventured out in the ship's boat to reconnoiter the bay. On Sunday, January 17, with Bligh reporting a good anchorage and ample fresh water ashore, Cook brought the *Resolution* and *Discovery* into Kealakekua. The Hawaiians swarmed around the ships "like shoals of fish." It was a momentous occasion, for the anticipation had been building on both sides. In the words of Lieutenant King: "They express'd the greatest Joy & satisfaction, by Singg & Jumping, of our coming to Anchor, & to a more intimate & regular connection with them, nor was the Pleasure less on our side." Kamakau confirms the Hawaiian reaction: "When Captain Cook appeared they declared that his name must be Lono, for Kealakekua was the home of that deity as a man, and it was a belief of the ancients that he had gone to Kahiki and would return. They were full of joy, all the more so that these were Lono's tabu days. Their happiness knew no bounds; they leaped for joy [shouting]: 'Now shall our bones live; our *'aumakua* has come back. These are his tabu days and he has returned!'"

Kalani'ōpu'u was not among those greeting Cook at Kealakekua. He had lagged behind, anxious not to displease Lono on the god's triumphal circuit. The king would arrive shortly after, in grand style aboard his largest double-hulled canoes, accompanied by his war gods. On this morning at Kealakekua, it was one of Kalani'ōpu'u's attendants, Ko'a'a (whose name the British rendered as Koah), who would have the honor of greeting Cook and leading him up to Lono's temple of Hikiau (see figure 11). Ko'a'a and two other *ali'i* came out to the *Resolution*, dispatching the commoners who were swarming the decks. One man who dared to argue with one of the chiefs was picked up by the latter and literally thrown overboard. Approaching Cook, Ko'a'a wrapped a red barkcloth over the captain's shoulders and presented him with a small pig, repeating a "long oration or

FIGURE 11.
The main façade of Hikiau *heiau* at Kealakekua Bay, Hawai'i Island. Captain Cook was taken atop this stone platform and honored as the returning god Lono in January 1779. Photo by Thérèse Babineau.

prayer." Cook invited the chiefs to dine with him in the great cabin. Their meal finished, Cook with Lieutenant King and Ko'a'a prepared to land on the beach at Kealakekua.

As Cook stepped onto the black sand beach, he was greeted by three or four other men. These priests of Lono held wands tipped with dogs' hair. According to Lieutenant King, they "kept repeating a sentence, wherein the word Erono [E Lono] was always mention'd, this is the name by which the Captn has for some time been distinguished by the Natives." Cook was led by the priests with dogs' hair wands to the temple platform of Hikiau, a short distance away at the southern end of the beach. Each year, the long god of the Makahiki began his journey around the kingdom from Hikiau. At this temple the long god would also conclude his procession. Lieutenant King recounted the events as he witnessed them that January day:

> We were conducted to the top of the pile of stones; It was on one side raised about 8 feet from the ground. . . . I judge it to be about 20 yards broad, & more than twice that in length, the top was flat & paved with Stones: there was a stout Railing all round, on which were stuck 20 Skulls, the most of which they gave us to understand were those of Mowee [Maui] men, whom they had killd on the death of some Chief. . . . We enterd the Area on one side near the houses, & the Capt[n] was made to stop at two rude images of Wood, the faces only were Carved, & made to represent most distort'd mouths, with a long piece of Carved Wood on the head; these Images had pieces of old Cloth wrap'd round them . . . After Koah [Koʻaʻa] & a tall grave young man with a long beard whose name was Kaireekeea [Keliʻikea] repeated a few words, we were led to the end of the Area where the Scaffold was; At the foot of this were 12 Images ranged in a semicircular form, & fronting these opposite the Center figure was a corrupt'd [i.e., rotten] hog, placed on a stand, supported by post[s] 6 feet high . . .

The rotted hog was sitting on a *lele* (offering stand), which stood in front of the images of the gods. Koʻaʻa, taking Cook's hand and all the while chanting, now led him up onto the temple's *ʻanuʻu,* a wooden tower where the priests communicated with the gods, receiving their oracle. Just then a procession of ten men came around the wooden fence, carrying a freshly baked hog and a large piece of red barkcloth (red being the sacred color). Approaching the *ʻanuʻu* that Cook and Koʻaʻa had climbed up on, all ten prostrated themselves on the pavement. Keliʻikea, he with the long beard, took the red barkcloth and handed it to Koʻaʻa, who wrapped it around Cook. Koʻaʻa then took the pig and held it in his arms as he began to chant. Koʻaʻa and Keliʻikea chanted for some time, frequently in alternating stanzas. At last Koʻaʻa let the hog fall from his arms, helping Cook to descend from the *ʻanuʻu.* Koʻaʻa next addressed each of the wooden images, paying special attention to the central image: "To this he prostrated himself, & afterwards kiss'd, & desird the Capt[n] to do the same, who was quite passive, & sufferd Koah to do with him as he chose; this little Image Pareea named Koonooe-akeea, the rest they nam'd indiscriminately Kahai." This central image, which Cook agreed to kiss following Koʻaʻa's urging, was a manifestation of Kūnuiākea, "Great Widespread Kū." Koʻaʻa now led Cook and his companions to another part of the temple platform, a square space sunk down about three feet lower than the main terrace. "On one side were two wooden Images; between these the Captain was seated; Koah support'd one of his Arms, while I was made to do the same to the other." With his arms outstretched, Cook resembled the *akua loa*, the long god with its crossstaff. Having been addressed as Lono, Cook was anointed with chewed coconut meat, given *ʻawa* to drink, and hand-fed morsels of baked hog. Lieutenant King's detailed description of the ceremony on Hikiau leaves no doubt that Captain Cook had just been put through one of the essential rituals of the Makahiki, the final rite of "feeding the god" *(hānaipū)*. After what Lieutenant King called "this long, & rather tiresome ceremony, of which we could only guess at its Object & Meaning," the party descended from the temple platform back to the beach. As they walked through the village two priests with wands went before Cook, calling out "O Lono, O Lono," at which the people prostrated themselves. The *kapu moe*, prostrating taboo, was reserved for gods and god-kings.

· · ·

Cook was delighted with his reception at Kealakekua. Whether Cook fully appreciated that he had just been installed as a god is doubtful; certainly he did not comprehend the full weight that had been put upon him in the ceremony at Hikiau. But he knew that he was held in considerable awe. His crew would have no trouble provisioning the ships (not to mention that his sailors were daily satisfying their pent-up sexual desires with the *makaʻāinana* girls and women who steadily visited the ships). Cook set up an observatory in a sweet potato patch on shore (which the priests immediately made taboo), in order to take astronomical measurements necessary to fix their position precisely. Meanwhile the sailmakers were given the use of two houses near the temple. While barrels of fresh water were being filled for the ships' holds, the caulkers went to work repairing the hulls. Whenever Cook ventured ashore, he was invariably greeted by the priests with their dogs' hair wands, crying out the *kapu moe*, while others would bring offerings of pigs and sweet potatoes.

As their understanding of the situation improved (recall that many on Cook's ships spoke a kind of "pidgin Polynesian" that they had picked up in Tahiti), the British became aware that the island's true power was absent from Kealakekua. They were told that Kalaniʻōpuʻu would soon appear in the bay. On January 24, a week after they had dropped anchor, a hush fell over Kealakekua; no canoes came out to the ships. The British were told that a *kapu* had been placed on the bay because of the king's impending arrival. The next afternoon a sizable flotilla of canoes was seen coming around the northern point, landing at Kaʻawaloa, a broad lava promontory that was Kalaniʻōpuʻu's residence. At noon on January 26, three large double-hulled sailing canoes were launched from Kaʻawaloa (see color plate 7). They headed straight to the *Resolution*. John Webber captured the scene on his sketchpad. Once again, Lieutenant King recorded what transpired for posterity:

> At Noon Terreeoboo [Kalaniʻōpuʻu] in a large Canoe attended by two others set out from the Village, & paddled towards the Ships in great state. In the first Canoe was Terreeoboo, In the Second Kao with 4 Images, the third was fill'd with hogs & Vegetables, as they went along those in the Center Canoe kept Singing with much Solemnity.... Instead of going on board they came to our side, their appearance was very grand, the Chiefs standing up drest in their Cloaks & Caps, & in the Center Canoe were the busts of what we supposd their Gods made of basket work, variously coverd with red, black, white, & Yellow feathers, the Eyes represent'd by a bit of Pearl Oyster Shell with a black button, & the teeth were those of dogs, the mouths of all were strangely distorted....

The message was unmistakable. Kalaniʻōpuʻu was displaying to Lono, god of the Makahiki, the king's own feathered images of Kūkāʻilimoku, his personal war god. Kalaniʻōpuʻu and the war canoes then turned and landed on the beach, signaling that Cook/Lono should follow. Upon landing, Lieutenant King was surprised to find that Kalaniʻōpuʻu was "the same old immaciated infirm man" they had met off Maui weeks before.

Taking the long feathered cloak off his own shoulders, Kalaniʻōpuʻu wrapped it gracefully around Cook, and then placed a feathered helmet on the captain's head and put a feathered *kahili* wand in his hand. As if this were not enough, five or six more feathered cloaks, "all very beautiful, & to them of the greatest Value," were laid at Cook's feet. Collectively, those garments must have contained the feathers of literally hundreds of thousands of birds, feathers that had been offered up over the course of countless Makahiki seasons. Kalaniʻōpuʻu was giving Lono his due. After a name exchange, and the performance of yet another Lono ritual by Keliʻikea, the captain reciprocated by inviting Kalaniʻōpuʻu and the other high *aliʻi* to dine aboard the *Resolution*. In the great cabin, he presented them with gifts of trade goods.

The work of repairing and provisioning the ships progressed steadily. A small party including Midshipman George Vancouver made an exploratory foray inland, attempting to reach the "Snowy Mountain" (Mauna Loa). It proved too far away to climb, but the party did see the vast inland field system of sweet potatoes and taro that was the mainstay of Kalaniʻōpuʻu's economy. Lieutenant King wrote admiringly of the high state of cultivation extending from the village inland some three or four miles. He noted, "Walls separate their property & are made of the Stones got on clearing the Ground; but they are hid by the sugar cane being planted on each side, whose leaves or stalk make a beautiful looking hedge." They observed the closely planted sweet potatoes and taro, carefully mulched and tended by the farmers. King could not help but add, "the Tarro of these Islands is the best we have ever tasted."

On February 1, Cook made an unusual request: that they be allowed to dismantle the wooden fence surrounding Hikiau *heiau*, to use as firewood for the ship's galleys. One wonders what the Hawaiians thought of this: did Lono want to take his temple away with him? In any event, the Hawaiians helped the sailors to move the wood and even urged them to take the carved wooden images—except, that is, for the small central image of Kū. Then, on the same day, one of the older sailors, William Watman, died. Hearing this, the chiefs asked if he could be buried within the temple platform. Cook performed the burial rites as the Hawaiians looked on, who then added their own rites and threw a dead pig, coconuts, and bananas into the grave as offerings. Again, one can only ponder what cross-cultural messages were being unwittingly transmitted. Was the body of William Watman seen by the Hawaiians as a reciprocal sacrificial offering to Kū? The central image of Hikiau, which the British had not been allowed to take down, was of Kūnuiākea.

According to the protocols of the Makahiki, after the circuit of the long and short gods is completed, the king must take the land back from Lono, who is then sent away to Kahiki on his canoe. On February 2, King's journal notes that "Terreeoboo & the Chiefs became inquisitive as to the time our departing & seemd well pleas'd that it was to be soon." More hogs, sugarcane, yams, sweet potatoes, and barkcloth were brought to the beach as presents to Lono, much to the delight of the British. The ships' larders were overflowing. The next afternoon the Hawaiians entertained their visitors with boxing and wrestling matches. John Webber's exquisite drawing of the scene shows the *akua loa*, white streamers and birds' skins hanging from the cross-piece. Such games were a regular part of the Makahiki's conclusion. That night the British were surprised to see the two houses at the end of Hikiau *heiau*, where the sailmakers had done their repairs, go up in flames. Lieutenant King conjectured that this was a careless accident. In all likelihood, the Hawaiians were ritually purifying the ground on the eve of Lono's departure.

On the morning of February 4, the *Resolution* and *Discovery* hauled up anchors and sailed out of Kealakekua Bay. A number of canoes kept them company at first, and then faded astern. It was Cook's intention to explore the rest of this extensive archipelago, to map it accurately. He would then turn his ships northward for another season of exploration seeking the elusive Northwest Passage. For the next four days the ships worked their way up the leeward side of Hawai'i, encountering "baffling & Squally" winds. On February 8 they were nearly off 'Upolu Point, the island's northern tip, where strong winds deflected by the Kohala mountains can be treacherous. Encountering gale-force winds, the sailing master ordered the fore topsails to be close-reefed, but disaster struck. *Discovery*'s foremast, which had been carelessly repaired the previous year, was found to be "badly sprung." To continue north would have endangered the lives of the *Discovery*'s entire crew. Major repairs were required. Cook ordered the ships to put about and sail southward back to Kealakekua.

· · ·

To Kalani'ōpu'u's great relief, Cook/Lono had departed Kealakekua, as the Hawaiian ritual calendar dictated. Now, just a week later, the two ships were back, dropping their anchors on February 11 into the bay's calm waters. The god-king must have been horrified. He had made generous offerings to this living manifestation of Lono, bestowed on him priceless feathered cloaks and enormous quantities of pigs, yams, and sweet potatoes. Lono had been allowed free rein of his temple, even permitted to haul away its fence and images. He had been properly entertained and fêted. And then Lono had departed, as he must, so that the king could repossess the land. Lono's season was finished. The next eight months should properly be devoted to the worship of Kū. Kalani'ōpu'u was engaged in a bitter struggle with Kahekili on Maui. It was vital that his war temple be rededicated, that Kū receive his sacrifices, so that victory would be Kalani'ōpu'u's. Cook's return to Kealakekua changed everything. To Kalani'ōpu'u this act could only be interpreted as a direct challenge. Why would Lono return unless he intended to seize power, to usurp the kingship?

The British immediately noticed that everything was different. According to Lieutenant King, "Upon our first Anchoring very few of the Natives came to us. This in some measure hurt our Vanity, as we expected them to flock about us, &

to be rejoiced at our return." The carpenters set to work on the damaged mast, but ashore the formerly good relations rapidly turned sour. On February 13 a watering party was harassed by a chief. Two young midshipmen were beaten by the Hawaiians when they tried to pursue a canoe paddled by some men who had stolen items from the ship. Petty theft was now rampant, something the British had not before experienced in Kealakekua.

Waking on the morning of February 14, the officer of the watch discovered that the *Discovery*'s cutter, a large boat used to ferry crew and supplies between ship and shore, had been stolen during the night. Informed of the theft, Cook was furious. He ordered Lieutenant Rickman to blockade the bay, preventing any canoes from escaping. Then, taking with him a party of armed marines in the pinnace, launch, and small cutter of the *Resolution*, Cook set out for Ka'awaloa, royal residence of Kalani'ōpu'u. His strategy, one he had applied successfully before in his South Seas voyages, would be to take the king hostage until the stolen cutter was returned. Landing on the peninsula, Cook and the marines marched up the central alleyway to Kalani'ōpu'u's house. Meanwhile, Lieutenant Rickman's blockading party had fired at a canoe trying to cross the bay. A chief named Kalimu was killed by the musket ball. With him was the great warrior Kekūhaupi'o, who escaped and immediately made his way to Ka'awaloa to inform the king.

Arriving at the king's house, Cook asked Kalani'ōpu'u to go with him to the *Resolution*. At first the old man agreed. Dressed in his feathered cloak, Kalani'ōpu'u proceeded with Cook and the marines down the path toward the waiting boats. A crowd of several hundred *ali'i* and warriors began to gather menacingly. Kekūhaupi'o suddenly appeared with the news that Kalimu had been killed. What transpired next was later described by various eyewitnesses and others, but let us hear it from the Hawaiian point of view, as recounted by Kamakau, who wrote down the *mo'olelo* years later:

> When Ka-lola [the king's sacred wife] heard that Ka-limu was dead, shot by the strangers, she ran out of the sleeping house, threw her arms about the shoulders of Ka-lani-'opu'u and said "O heavenly one! let us go back!" Ka-lani-'opu'u turned to go back. Captain Cook tried to grasp him by the hand, but Ka-lani-mano-o-ka-ho'owaha stuck his club in the way, and Ka-lani-'opu'u was borne away by the chiefs and warriors to Maunaloia, and the fight began. Captain Cook struck Ka-lani-mano-o-ka-ho'owaha with his sword, slashing one side of his face from temple to cheek. The chief with a powerful blow of his club knocked Captain Cook down against a heap of lava rock. Captain Cook groaned with pain. Then the chief knew that he was a man and

not a god, and, that mistake ended, he struck him dead together with four other white men.

According to the marines and sailors who survived the melee, after the first fatal blow, hundreds of chiefs and warriors swarmed over Cook's body, each further dispatching him with daggers and clubs. Everyone wanted to take part in the ritual killing of the god.

Across the bay where he was keeping watch over the observatory and the carpenters, Lieutenant King heard gunfire. Looking across the bay, he could see that there was a commotion on the shore at Ka'awaloa. The cannon of the *Resolution* let loose with retorts, followed by those of the *Discovery*. King was filled with dread as he awaited Lieutenant Bligh in the cutter. Bligh brought the awful news. As King recalled, "he announcd to us the Shocking news that Captn Cook was kill'd, we saw it in his & the Sailors looks. He could only tell us that he & some Marines were killd & their bodies in the possession of the Indians." Cook, a corporal, and three marines had been felled on the lava rocks at Ka'awaloa, by Hawaiians defending their god-king Kalani'ōpu'u (see figure 12). In retaliation, the marines firing from their boats and the ships' cannon had killed seventeen Hawaiians, including five prominent *ali'i*.

A hush fell over the bay. The British went on full alert lest there be a massive attack. But the Hawaiians had no intent of further violence. Lono was dead, the threat to the kingdom had passed. About 8:00 P.M., under the cover of darkness, a canoe was seen approaching the *Resolution*. The sentries fired at her, but the men in the canoe cried out "Tinnee" (i.e., "King"). They were recognized as two of the Lono priests who had constantly attended Captain Cook whenever he had gone ashore. Permitted to board the *Resolution*, the two priests threw themselves at Lieutenant King's feet. To their horror, the officers were presented with a bundle that, when opened, "contained a piece of human flesh from the hind parts." It was all that the loyal priests could obtain of Cook's body, which had already been defleshed in order to clean the bones. And then the priests asked Lieutenant King "a Singular question," which was "when the Erono [Lono] would return . . . & what he would do to them when he return'd." Cook the man was dead, but to the priests Lono was still a living god, who should return again for the next Makahiki.

Exactly what happened to Cook's bones has been a matter of controversy. Marshall Sahlins argues that they were distributed to the king and high chiefs as war trophies. Others have suggested that at least some of the captain's bones were gathered together within a wicker *kā'ai*, a sort of anthropomorphic casket used to con-

FIGURE 12.
Kealakekua Bay as seen from the lava flats at Ka'awaloa, Hawai'i Island, at the approximate location where Captain James Cook was killed in January 1779, while attempting to take King Kalani'ōpu'u hostage. Photo by Thérèse Babineau.

tain the bones of high-ranking *ali'i*. There are reports that such a *kā'ai* containing Cook's bones was being carried around the island as late as the 1790s in the annual Makahiki procession. Visiting Hawai'i in 1823, British missionary William Ellis reported that Cook's bones had been defleshed and then "preserved in a small basket of wicker-work." This *kā'ai*, according to Ellis, was "annually carried in procession to several other heiaus, or borne by the priests round the island, to collect the offerings of the people, for the support of the worship of the god Rono." Cook the mortal expired on the lava rocks at Ka'awaloa, but his legacy persists even to this day.

SEVENTEEN · Prophecy and Sacrifice

Pele, the Hawaiian volcano goddess, makes her home in the fire pit of Halemaʻumaʻu, at the summit of Kīlauea on the southeastern flank of Hawaiʻi Island. Extending miles to the southwest of Halemaʻumaʻu, the Kaʻū Desert is an uninhabited landscape of jagged ʻaʻā and ropy pāhoehoe lava flows, cloaked in places with deposits of ash and cinder. These volcanic ashes, disgorged from Halemaʻumaʻu and often rising thousands of feet into the air, are blown by the prevailing trade winds, falling out over the Kaʻū Desert. Lying in the rain shadow of Kīlauea, the desert is arid, with only scattered clumps of stunted *ʻōhiʻa* and other native dryland plants here and there, struggling to survive in the harsh lava. When the afternoon fog sets in, obscuring the view of far-off Mauna Loa, it is easy to get disoriented and even lost in this vast wasteland.

On just such an afternoon in 1993, together with my photographer wife, Thérèse, I was searching in Kaʻū's undulating cinder dunes and swales, about six miles southwest of Halemaʻumaʻu. We were looking for fossilized human footprints preserved in volcanic ash, unique reminders that here in November 1790 Pele had destroyed nearly the entire army of Keōua. The younger son of Kalaniʻōpuʻu, Keōua ruled the southeastern half of Hawaiʻi. For some time he had been engaged in a fierce struggle with Kalaniʻōpuʻu's nephew, the warrior chief Kamehameha, for control of the entire island. After a series of indecisive battles, Keōua was marching his army from Puna to Kaʻū, by way of the trail that crossed over Kīlauea. As Keōua's army passed Pele's abode and continued along the trail

FIGURE 13.
Impression of a warrior's foot left in volcanic ash at Kīlauea, Hawai'i Island. Photo by Thérèse Babineau.

into the Ka'ū Desert, a massive explosion erupted out of Halema'uma'u. According to Kamakau, surviving witnesses said that the column of ash rose as high as the summit of Mauna Loa, more than thirteen thousand feet in the air. Then hot ashes began to fall back to earth, directly over Keōua's army. Warriors, their wives, and even their children who were traveling with them were felled by the hundreds.

Thérèse and I had found several footprints in the yellowish ash, but these were disappointingly eroded by the wind, mere depressions vaguely resembling the outline of a foot. I moved off the faintly marked trail, being careful to keep my bearings lest I become disoriented and lost in the undulating cinder hills and swales. The distinctive ash layer from the 1790 eruption is covered over in many places by later cinder falls and by windblown ash. Finally, as the afternoon sun was settling toward the distant ocean, I came across a well-preserved patch of crusty ash. My eye caught the unmistakable impression of a human foot, as sharp and detailed as if it had been made yesterday (see figure 13). Crouching down to get a closer view, I could see how the toes had gripped the scalding ash, sliding just a bit, while the heel dug in deeply. The large footprint seemed to be that of an adult male, probably a warrior. He had been running, not walking, in a desperate but presumably futile effort to escape the fatal rain of scalding ash and poisonous gas.

I called Thérèse over to have a look at the pathetic footprint. It was a poignant moment that we will never forget, like a leap out of time. We reflected on what must have been a moment of sheer horror for those who stood on this very spot two centuries before. Thérèse set up her Hasselblad camera and photographed the footprint, the slanting rays of afternoon light illuminating its crisp features. We packed up our gear as the mist began to swirl around the gnarled *ōhi'a*, intent on getting back to our car before darkness set in.

· · ·

The destruction of a large portion of Keōua's army decisively changed the course of Hawaiian history. But to appreciate the story fully, we must go back to the days of Alapa'inui, in the mid 1700s, when that great king was encamped with his army on the northern tip of Kohala, about to embark on his campaign of revenge against Kekaulike of Maui. In their company was the chiefess Keku'iapoiwa, herself of a distinguished pedigree and wife of Keōua, the younger brother of Kalani'ōpu'u (this is the Keōua who was later supposedly poisoned by Alapa'inui in Hilo). Keku'iapoiwa was on the verge of going into labor. Her baby, should it prove to be a boy, would be a chief of considerable rank. Given that Alapa'inui was himself a usurper, rumors had been flying about that there might be an attempt to take the baby's life. The *ali'i* Keawemauhili (half-brother of Kalani'ōpu'u) was said to have muttered, "Pinch off the tip of the young mulberry shoot." The allusion was to the practice of pinching off buds along the stems of the *wauke*, the paper mulberry plant, which provided the material for *kapa* (barkcloth). A young bud "pinched off" at birth could not grow up to challenge the authority of Alapa'inui.

On the night before Keku'iapoiwa went into labor, an especially bright heavenly object appeared in the sky, a special omen. This may have been an appearance of Halley's Comet in late 1758 or early 1759. (The exact birth year of Kamehameha is a matter of some dispute. Kamakau put it at 1736, but this seems too early. If he was born in 1758, Kamehameha would have been twenty-one years old at the time of Cook's return to Hawai'i Island, which fits with the latter's description of him as a young man.) Hawaiian traditions say that the chiefess gave birth during 'Ikuwā, first lunar month of the Makahiki. The night of the birth was stormy, with heavy rain, thunder, and lightning, all portents that the baby emerging from her womb was destined for greatness. Before the other chiefs could acknowledge the birth—or, perhaps more significantly, before any harm could be done to the babe—he was snatched away in the dead of night. Kamehameha, for that was the name given to the newborn, was taken by Nae'ole to be reared in seclusion at Hālawa on the windward coast of Kohala. Only when he had reached the age of five was Kamehameha returned to the court of Alapa'inui, to rejoin his father and receive the training necessary for a young *ali'i*. Kamehameha's personal instructor in martial arts was the famous warrior Kekūhaupi'o, expert in the arts of *lua*.

When Kamehameha was in his teens, his father, Keōua, died at Alapa'inui's court in Hilo. He was widely believed to have been poisoned by the king. This incident led to the attempt by Kalani'ōpu'u, fearing for Kamehameha's life, to have his nephew removed from court. As recounted in chapter 15, this was the beginning of Kalani'ōpu'u's revolt against Alapa'inui. After Kalani'ōpu'u became king of Hawai'i, Kamehameha and his mentor Kekūhaupi'o served the king in the numerous campaigns against eastern Maui.

Kamehameha was among the first *ali'i* to make contact with Captain Cook when the *Resolution* and *Discovery* appeared off windward Maui. He had gone out to the ships on the double-hulled canoe that carried Kalani'ōpu'u. Along with a few other young warrior chiefs, Kamehameha spent the night on the *Resolution*. Later, in Kealakekua Bay, Lieutenant King would describe Kamehameha thus: "Amongst [those attending Kalani'ōpu'u] was Maiha-Maiha, whose hair was now Paisted over with a brown dirty sort of Paste or Powder, & which added to as savage a looking face as I ever saw, it however by no means seemd an emblem of his disposition, which was good naturd & humorous." The paste to which King referred was slaked lime, which dyed the hair a reddish-yellow *('ehu)* color admired by the Hawaiians. King thought that Kamehameha possessed "somewhat of an overbearing spirit."

By 1780, Kalani'ōpu'u, now infirm and sensing his days were numbered, was ready to proclaim the succession to the kingship of Hawai'i. By the rules of hereditary succession, his high-ranked son Kīwala'ō, born to Kalani'ōpu'u's wife Kalola (the daughter of Maui king Kekaulike and sister of Kahekili), would be the logical heir to the kingdom. But Kalani'ōpu'u had become greatly attached to Kamehameha, who in so many ways manifested in his actions and deeds the *mana* of a true ruling chief. The old king called his court together in Waipi'o Valley, ancient seat of the Hawai'i *ali'i nui*. There, Kalani'ōpu'u made his declaration. Kīwala'ō would indeed inherit the land, and rule over the chiefs. But in a move that harked directly back to Līloa nearly two centuries earlier, Kalani'ōpu'u proclaimed that the feared war god Kūkā'ilimoku, special deity of the Hawai'i god-kings (see color plate 8), would be under the charge of Kamehameha.

Kalani'ōpu'u admonished Kamehameha to "live under" the rule of his cousin Kīwala'ō. But the entire assembly of chiefs and warriors knew well the famous history of 'Umi, to whom Līloa had entrusted the red-feathered image of Kūkā'ilimoku. They remembered how 'Umi had usurped the kingdom from Hākau. Was Kamehameha to be a new 'Umi? There must have been much wondering and speculation, there in the royal court at Waipi'o, about how this seeming reenactment of Līloa's proclamation would play out.

Soon after the proclamation at Waipi'o, news arrived of a rebellion fomented by 'Īmakakoloa, a chief of Puna. Kalani'ōpu'u moved with his army to engage 'Īmakakoloa south of Hilo. After a series of battles, the rebel forces were defeated, but 'Īmakakoloa managed to escape into the dark recesses of the Puna rain forests. For nearly a year, Puna's loyal commoners protected their defeated chief, refusing to give him up to Kalani'ōpu'u. The king, meanwhile, had moved on to the adjacent Ka'ū district, occupying his time building a *luakini* war temple named Pākini. The dedicatory sacrifice for Pākini was intended to be none other than 'Īmakakoloa.

Growing weary of waiting for his warriors to capture 'Īmakakoloa, Kalani'ōpu'u ordered them to begin a campaign of terror against the Puna commoners. Hamlet after hamlet was burned to the ground; innocent *maka'āinana* were put to death in the escalating hunt for the rebel chief. Finally, one of the Puna chief's own *kahu* (guardians)—fearing that more innocents would be slaughtered—betrayed his master.

The captured 'Īmakakoloa was taken to Kalani'ōpu'u, waiting to dedicate his new war temple of Pākini. 'Īmakakoloa was killed and for two days his body was slowly roasted over a fire of *kukui* nut shells until it was reddened and shone. His eyes were rubbed with fish oil so they resembled the terrible eyes of the man-eating *niuhi* shark. With the sacrificial body thus prepared, it was taken onto Pākini *heiau* for the rites of dedication to Kūkā'ilimoku. The roasted corpse of 'Īmakakoloa was laid out on the altar, flanked on one side by an offering of a spotted pig, and on the other by a banana stalk, symbolizing obedience.

On the temple platform, King Kalani'ōpu'u instructed Kīwala'ō to seize the sacrificial offering. The heir apparent, possibly sickened by the sight of 'Īmakakoloa's roasted body, grabbed hold of the sacrificial pig. The king then addressed Kamehameha, who was now in charge of the war god, Kūkā'ilimoku. Kamehameha immediately grasped one leg of 'Īmakakoloa's body in one hand, an arm in the other, lifting the putrid corpse up so it pressed against his own body. As the two young *ali'i* stood there, Kīwala'ō with his pig and Kamehameha with 'Īmakakoloa's corpse, the high priest Holo'ae began an incantation to Kū. Holo'ae invoked Kū of the Long Cloud, Kū of the Short Cloud, Kū of the Mountains and Forest, Kū Snatcher of the Land. The priest declared that the laws, the *kapu*, must be enforced. As the prayers and chants droned on, Kamehameha became weary with the weight of 'Īmakakoloa's corpse, finally letting it drop to the ground and falling beside it in an exhausted sleep. Kamehameha passed the night with his limbs entangled with those of 'Īmakakoloa. When he awoke, having been watched over during the night by Kekūhaupi'o and the other chiefs, Kamehameha's body shone with the grease of the sacrificial victim. This was taken as yet another omen of Kamehameha's destiny.

There was great consternation in the royal court after the dedication of Pākini *heiau*. Some said that Kamehameha had abused his privilege by taking hold of the sacrificial corpse, that it had been Kīwala'ō's right to offer up the body of 'Īmakakoloa to Kū. Others—especially the chiefs of Kohala and Kona, who secretly wanted Kamehameha to usurp the kingship—argued that Kīwala'ō had been afraid to present the human sacrifice, thereby showing that he was not fit to be ruler over all Hawai'i. Kalani'ōpu'u warned his nephew that there might be plots against his life. With Kekūhaupi'o and his kin, Kamehameha left the royal entourage in Ka'ū, heading for Kamehameha's natal land of Hālawa, on the far northern tip of Kohala. Only there would he be safe.

Kalani'ōpu'u now ensconced himself at Wai-'Ahukini, a sheltered fishing village in the lee of South Point, Ka Lae, the best trolling grounds for *'ahi* (yellowfin tuna), *mahimahi*, and other big game fish. It was the prerogative of the king to open the *'ahi* fishing season at the end of the Makahiki. But this was to be Kalani'ōpu'u's last Makahiki. He soon took ill, of what was called the *ma'i 'ōku'u* (squatting sickness), possibly a form of amoebic dysentery introduced by Cook's visit.

For many generations, the bones of the kings of Hawai'i had been interred in the Hale o Keawe, a sepulchral temple at Hōnaunau, a few miles south of Kealakekua Bay (see figure 14). Kalani'ōpu'u's dying wish had been that his remains be taken to this resting place of his ancestors. (There are indications that Kalani'ōpu'u's body was not defleshed, as was the usual case, but "embalmed." This may have

FIGURE 14.
The sepulchral temple Hale o Keawe at Hōnaunau, Hawai'i Island, where the body of Kalani'ōpu'u was deposited. The thatched house and wooden images have been reconstructed based on early historical descriptions and drawings. Photo by Thérèse Babineau.

been done by removing the organs from the body cavity and using sea salt as a preservative. If the body was kept at Wai-'Ahukini after death, that is entirely conceivable, for in an extremely dry and arid region like this a salted body could readily be preserved.) After a mourning period of several months, Kīwala'ō and those chiefs of the windward districts who supported his reign prepared to move the body of Kalani'ōpu'u north along the coast to the Hale o Keawe.

Kīwala'ō had barely reached maturity. By all indications he was rather indecisive, perhaps timid, not mean-spirited, or necessarily harboring ill intentions against Kamehameha. One gets the impression that he was simply naïve, having been pampered and endlessly fawned over as the special heir of Kalani'ōpu'u. He seems to have deferred continually to his powerful uncle Keawemauhili, younger brother of Kalani'ōpu'u and a seasoned warrior chief. Abraham Fornander called Keawemauhili "the most powerful and opulent chief on Hawaii . . . the first prince of the blood, the Doyen of the Hawaiian aristocracy." While deferring to the wish of his deceased brother that Kīwala'ō take the kingship, Keawemauhili was clearly the power behind the throne. Keawemauhili was not about to let Kamehameha or any of his supporters expand their land holdings. Indeed, he saw the customary redistribution of lands following the death of Kalani'ōpu'u as an opportunity to expand his and his supporters' dominions. After depositing the body of Kalani'ōpu'u at the sacred burial temple of Hale o Keawe, Keawemauhili would make certain that young Kīwala'ō made a land distribution that suited Keawemauhili's own interests.

The procession of canoes bearing the preserved body of Kalani'ōpu'u began to make its way up the leeward coast of Hawai'i toward Hōnaunau. Kekūhaupi'o, hearing the news, hastened to Kohala where Kamehameha was in self-imposed exile. Finding Kamehameha engaged alternately in swimming and making love, Kekūhaupi'o admonished Kamehameha to rise to his chiefly duty, telling him that when he became king the women would flock to him. Kamehameha and Kekūhaupi'o sailed south to meet Kīwala'ō and the funeral fleet bearing Kalani'ōpu'u's body. Protocol demanded that Kamehameha witness the interment of his deceased uncle's body in the sacred Hale o Keawe at Hōnaunau. More important, they were eager to hear how Kīwala'ō proposed to divide the numerous *ahupua'a* territories among the chiefs.

Kamehameha met Kīwala'ō at Hōnaunau, the sprawling royal residence and temple complex that for many generations had been a center for the Hawai'i *ali'i*. Kalani'ōpu'u's body was duly deposited in the Hale o Keawe, the special high-roofed house guarded by images of Kū and Lono. The cousins wailed together over the death of their king, performing the obligatory rites of interment. Emerging from the sepulchral house, Kīwala'ō announced the decree that all assembled had been waiting for. The new king confirmed that Kamehameha would take charge of the war god Kūkā'ilimoku, just as Kalani'ōpu'u had ordered. But Kamehameha would control only the small *ahupua'a* of Hālawa, which he had inherited from his father. Kīwala'ō claimed for himself all the lands of the kingdom, and the right to redistribute these as he saw fit. It is an understatement to say that the Kohala and Kona chiefs were dissatisfied with Kīwalaʻō's proclamation. "Unjust, unjust," they muttered among themselves. Some argued that the kingdom should be divided into two parts. The chiefs gathered around Kamehameha, urging him to war.

That evening Kamehameha, Kekūhaupiʻo, and the other high chiefs gathered with Kīwalaʻō in the royal men's house to drink ʻawa, a rite of purification after having deposited the dead king's body in the Hale o Keawe. As keeper of the war god, it was Kamehameha's prerogative to prepare the ʻawa for the new king. Kamehameha first carefully chewed the ʻawa root, placing the masticated (and thereby activated) fibrous mass in the ritual bowl. He then added the clear spring water, along with water scented with turmeric. When the infusion was ready, Kamehameha chanted a prayer to the gods, offering up the ʻawa to Kāne, Kū, and Lono, ancient deities from Kahiki. Kamehameha strained the potent brownish liquid into a coconut half-shell cup, and passed the sacred vessel to Kīwalaʻō. Unexpectedly, instead of raising the cup to his lips as protocol demanded, the young king handed the cup to a lesser chief of Maui, one of his ʻaikāne (courtly favorites). Before this low-ranking aliʻi could sully the royal ʻawa, Kekūhaupiʻo dashed the cup from his hands, spilling the ʻawa across the floor.

Angry faces illuminated by the flickering light of kukui nut torches glared at one another. The tension within the house was palpable. Too late, Kīwalaʻō realized that his careless act was a great insult to Kamehameha. Not caring that he was addressing a mighty lord and new king of the island, Kekūhaupiʻo haughtily told Kīwalaʻō that Kamehameha had not chewed the ʻawa for a commoner but for him, the aliʻi nui. Kekūhaupiʻo and Kamehameha abruptly left the men's house, heading for their canoe. They would go to Kekūhaupiʻo's homestead at Keʻei, not far to the north, to prepare for the war that was now inevitable.

On the next day Kīwalaʻō made the long-awaited apportionment of ahupuaʻa territories to the principal chiefs. Realizing that he had wronged Kamehameha, Kīwalaʻō was at first inclined to grant his cousin additional lands, but Keawemauhili dissuaded him from this idea. Instead, following Keawemauhili's advice, Kīwalaʻō not only reallocated many of the lands of the Kona and Kohala chiefs, but even short-changed some of his own close relatives.

For several days skirmishes broke out between the chiefs and warriors loyal to Kīwalaʻō who resided at Hōnaunau, and those who supported Kamehameha, now based at Kekūhaupiʻo's ancestral home at Keʻei. On the fourth day Kīwalaʻō and Kamehameha came forth to engage in combat, at a place called Mokuʻōhai. Kīwalaʻō's priest, Kalakuʻiʻaha, warned the young king with these words: "The flood tide is yours in the morning, but it will ebb in the afternoon." He urged Kīwalaʻō to postpone the fighting, but Kīwalaʻō ignored his words. At the same time, Kamehameha's priest Holoʻae gave an augury that mirrored that of his opponents: he foresaw that the morning would bring death, but that the tide of battle would turn when the sun began to descend toward the west.

Throughout the morning, the fighting went badly for Kamehameha and his supporters. Many of their warriors were felled, the bodies hauled off by Kīwalaʻō's troops to be offered at the temple of Hale o Keawe. Encouraged by this and urged on by his uncle Keawemauhili, Kīwalaʻō forgot the warning of his priest, believing that by nightfall victory would be his.

As the sun passed its zenith, two of Hawaiʻi's greatest warriors faced off on the lava fields of Mokuʻōhai. On the side of Kīwalaʻō was Ahia, a chief of Puna famous for his huge body and unparalleled command of *lua*, the art of hand-to-hand combat. Opposing him was Keʻeaumoku, an equally renowned fighter and chief from Kona. Also an expert in *lua*, Keʻeaumoku was fond of deploying the deadly *lei o mano* (shark's garland), a kind of knuckle-duster made of sennit cord wrapped around his fist, with razor-sharp sharks' teeth affixed to it.

Ahia and Keʻeaumoku circled, eyes fixed on each other. A warrior of Kīwalaʻō's distracted Keʻeaumoku for an instant. Ahia pounced, grasping Keʻeaumoku by his legs. Using a *lua* move called *iwikoʻo*, Keʻeaumoku somersaulted so that he was on Ahia's back, his feet pressed tightly against Ahia's throat while his arms held the Puna warrior's legs. Keʻeaumoku called out to Kamehameha to "catch the fish in the sluice gate." With Ahia immobilized in the *lua* hold, Kamehameha dispatched the great Puna warrior.

Kamehameha and Kekūhaupiʻo now moved off to another part of the battlefield, leaving Keʻeaumoku. Seeing their opening, several young warriors surrounded Keʻeaumoku. One, a small man named Kini, got his spear into Keʻeaumoku, who fell wounded on the lava. Kīwalaʻō, observing the action from a short distance, ran over to witness the impending demise of the Kona chief. But some of Kamehameha's party had also seen Keʻeaumoku fall and rushed to his aid. One of these, Keakuawahine, loosed a slingstone that unerringly found its way to Kīwalaʻō's temple. The king fell, not dead but stunned, a short distance from Keʻeaumoku. Ignoring the spear lodged in his body, Keʻeaumoku saw his chance. He crawled over to the prostrate body of Kīwalaʻō, and with the feared *lei o mano* of sharks' teeth, disemboweled the king. As the priests had warned, the tide would indeed ebb.

By sundown victory was Kamehameha's. Keʻeaumoku was carried from the field of battle; his wounds were treated by the *kāhuna lapaʻau* and he lived to serve Kamehameha brilliantly in future campaigns. Kīwalaʻō's corpse was offered up on the altar of Hikiau, the temple where Captain Cook had been greeted as the returning god Lono. But the battle of Mokuʻōhai was not decisive as far as the Hawaiʻi kingship was concerned. Keawemauhili, the powerful uncle who had urged Kīwalaʻō to short-change Kamehameha in the division of lands, was briefly taken captive, but managed to escape. He returned to Hilo, declaring himself the sovereign ruler of the southeastern third of the island. And after the defeat at Mokuʻōhai, Kīwalaʻō's brother Keōua had fled south with his brother's remaining warriors, setting himself up as the ruler of Kaʻū and part of Puna. Hawaiʻi was now divided into three independent and warring factions. The forces of Kamehameha, Keawemauhili, and Keōua periodically clashed over the next few years, but none was able to gain the upper hand. The standoff would persist for the several years, until Pele intervened with her violent eruption of poisonous gas and ash.

· · ·

When Cook's ships returned to England in 1780, the news that a large and fertile archipelago occupied the center of the North Pacific excited Europe. Unlike the earlier Spanish voyagers in the Pacific, the British made no attempt to conceal their discovery. The expedition's journal was promptly published, including Lieutenant Bligh's excellent chart of the islands with latitude and longitude precisely determined. While on the coast of New Albion (the Pacific Northwest), Cook's crews had traded for furs, selling them for high prices at Macau during their return voyage. Several of the men who had served with Cook realized that fortunes could be made by procuring furs in the Pacific Northwest and selling them at great profit in China. Moreover, Hawaiʻi offered a convenient watering and reprovisioning station along the route between the Northwest coast and Canton or Macau.

In 1785, English investors hoping to make a killing in the new fur trade incorporated the Sea South Company. Two ships, the *King George* under the command of Nathaniel Portlock and the *Queen Charlotte* under George Dixon, were the first vessels sent forth on this new commercial enterprise. Both captains had been members of Cook's crew, and were thus already acquainted with conditions on the Northwest coast and in Hawaiʻi. Shortly thereafter another entrepreneur, John Meares, backed by investors in Bengal, sailed in the *Nootka* from Calcutta, also bound for the fur trade. All three of these ships, and others, were in and out of Hawaiian waters between 1785 and 1788. Meares took with him a Kauaʻi chief named Kaiana, one of the first Hawaiians to venture overseas and learn the ways of the Haole. Kaiana returned to Hawai'i on the *Iphigenia*, commanded by Captain Douglas, in 1788. Kamehameha took Kaiana into his entourage, gaining from the Kaua'i chief's knowledge of firearms and the ways of foreigners.

Kamehameha and the other chiefs who had been in the company of king Kalani'ōpu'u back in 1779, when Cook's ships lay anchored in Kealakekua Bay, had been the first to be exposed to Western firepower. Although the British had little cause to use their weapons at first, after Cook attempted his fatal hostage-taking at Ka'awaloa, the Hawaiians experienced the full brunt of musket shot and cannon. At least seventeen Hawaiians had been killed in the barrage from the *Resolution* and *Discovery*. It is reputed that Kamehameha himself was injured. The efficacy of firearms and cannon were surely impressed upon the mind of this future warrior king.

Cook traded iron daggers to the Hawaiians (Kamehameha was said to have had several), but was careful not to let firearms fall into their possession. The new fur traders were not so scrupulous. They needed fresh meat and vegetables, water and firewood. The Hawaiian *ali'i*, controlling the trade to the extent they could, insisted on firearms in return. Kamehameha, in particular, was intent on ensuring that the vessels that called on his dominion be well treated. He made certain that all exchanges between his people and the Haole contributed to his mounting armory. Some years later, one sea captain characterized Kamehameha as "a magnanimous monarch, but a shrewd pork dealer . . . a grasping, trafficking savage, as shrewd and as sordid in his dealings as a white man." By the late 1780s, Kamehameha's warriors possessed a significant number of firearms.

While Kamehameha was careful to ensure peaceful relations with the fur traders when they landed on his shores, this was not always the case on the other islands, especially Maui and O'ahu, which were controlled by the old Maui war king Kahekili. In early 1790, Captain Simon Metcalfe of the trading ship *Eleanora* had a boat stolen and a sailor was killed in an incident at Honua'ula, Maui. Metcalfe took his revenge by luring several hundred innocent Hawaiians in canoes out to his ship, and then firing on them point-blank with grapeshot, killing at least one hundred. This dark affair is known as the Olowalu massacre.

Metcalfe's son Thomas was in charge of a much smaller schooner, the *Fair American*. The two vessels had become separated while on the Northwest coast, and Thomas now sailed to Hawai'i to rendezvous with his father. Six weeks after the Olowalu massacre, the *Fair American* arrived off Kona, Hawai'i. The *Eleanora* had been there a few days earlier, and Captain Metcalfe had struck one of the Kona chiefs, Kame'eiamoku, on the back with a rope. Kame'eiamoku had sworn revenge against the next Haole vessel that appeared. Surrounding the tiny *Fair American*, Kameʻeiamoku and his warriors boarded the vessel on the pretext of making trade, and then attacked the crew savagely. Thomas Metcalfe was thrown overboard by Kameʻeiamoku, and beaten by others with their canoe paddles till he drowned. The same fate met the other crew members with the exception of one Isaac Davis, who, though injured, was saved by the Hawaiians.

Kamehameha, hearing of the attack, immediately claimed the *Fair American*; it became the first foreign vessel in his war fleet. He also took Davis under his charge. From the vessel, Kamehameha secured a small cannon, which his warriors named Lopaka (the Hawaiian spelling of Robert). Lopaka was later mounted on a canoe hull with the gunner seated within, the whole being carried on stout poles by a gang of warriors, a sort of primitive tank!

Meanwhile, the *Eleanora* had dropped anchor just to the south, in Kealakekua Bay. Kamehameha feared that the senior Metcalfe, hearing of the attack on his son's schooner, would take revenge. When the *Eleanora*'s boatswain, John Young, went ashore at Kealakekua, Kamehameha ordered that Young be detained, lest he hear the news of the *Fair American* and report back to Metcalfe. The *Eleanora* waited several days for Young to return, firing cannon as a signal. Finally Metcalfe took Young for a deserter and sailed on, never learning that his son had been killed a short distance up the coast.

Kamehameha now had two skilled Haole in his entourage, Davis and Young. Neither was an educated man, but both were experienced seamen, not only knowledgeable about the workings of European ships but also capable with firearms and cannon. Davis, and especially Young, became confidants and advisers to Kamehameha, who gave them grants of land, effectively making them minor *aliʻi*. Young married a Hawaiian chiefess (Kaʻoʻanaʻeha) and set up a sort of hybrid Hawaiian-Haole household at Kawaihae, the remains of which can still be seen with its plastered lime walls. His bones are interred at the royal mausoleum in Nuʻuanu Valley, Oʻahu.

· · ·

During the 1780s, while Kamehameha was acquiring Western armaments, he continued his indecisive off-again-on-again campaigns against Keawemauhili and Keōua. The political landscape of the westerly islands was also shifting significantly. After the death of Kalaniʻōpuʻu, the half-tattooed warrior king Kahekili had recaptured the Hāna district of Maui. Kahekili then directed his expansionist aspirations toward the fertile islands of Molokaʻi and Oʻahu.

After the death of Pelei'ōhōlani, who at one time controlled the three western island of Kaua'i, O'ahu, and Moloka'i, the rule passed to his son Kumahana. This was uniformly deemed by both chiefs and commoners to be a bad choice for, in Kamakau's words, Kumahana "slept late, was stingy, penurious, deaf to the advice of others, and used to take himself off to the plains to shoot rats." There was a bloodless revolt, and at the advice of the high priest Ka'opulupulu the council of chiefs sent for Kahahana, a high chief of O'ahu descent who as an infant had been sent to the court of Kahekili and thus raised on Maui. Kahahana was still young, under the Kahekili's charge. The Maui king agreed to allow Kahahana to leave Maui and take up the rule of O'ahu, under one condition. Kahekili wanted the lands of Kualoa, an *ahupua'a* on the windward side of O'ahu long associated with the island's sacred chiefs. After Kahahana took up his new role as O'ahu's ruler, he put Kahekili's demand to the high priest Ka'opulupulu. But the priest advised Kahahana to refuse this request, saying that it would amount to giving Kahekili authority over O'ahu.

Kahekili did not immediately retaliate for this refusal of his request, but he harbored great resentment against Ka'opulupulu. Some years later, at a meeting with Kahahana on Moloka'i, Kahekili planted seeds of doubt in Kahahana's mind about the loyalty of his high priest. Believing that Ka'opulupulu was plotting against him, Kahahana evicted the priest from his court, and later had him killed. It was exactly as Kahekili had planned. With the wise high priest out of the way, Kahekili made his move. He gathered together his army, sailing by way of Lāna'i to gain an element of surprise, and landed at Waikīkī on O'ahu. Several battles were fought, but in January 1783 Kahekili won a decisive victory. Kahahana's warriors were killed or dispersed. The deposed O'ahu king spent the next two years hiding in the Ko'olau rain forests, until he was finally betrayed and killed. His body was taken to Waikīkī to be offered as a sacrifice to Kū.

The O'ahu chiefs resisted Kahekili's rule. O'ahu, it may be recalled, was the homeland of the most sacred *ali'i* of Nānā'ulu descent, tracing back to the times of Māweke. To have a Maui chief rule over them was an insult. The O'ahu chiefs laid a plot to murder Kahekili and the principal Maui chiefs, but before they could act the plot was exposed. Kahekili retaliated ruthlessly. Virtually all the O'ahu chiefs were killed, the chiefesses tortured. One of Kahekili's principal chiefs, Kalaikoa by name, had a house constructed from the bones and skulls of the murdered O'ahu *ali'i*. Kahekili was now lord of all of the islands from Kaua'i to Maui. Kahekili took up residence at Waikīkī, favored home of the O'ahu kings from the time of Mā'ilikūkahi.

FIGURE 15.
The old royal center of the Maui Island kings at Wailuku is dominated by Haleki'i *heiau*, which may have been both a residence and a temple. Photo by Thérèse Babineau.

Emboldened by his new Western advisers and armaments including the cannon Lopaka, Kamehameha decided to make his move. He still did not control all of Hawai'i, but was willing to challenge the might of Kahekili. For reasons that are not clear, his enemy Keawemauhili, ruler of the Hilo region, had made a sort of treaty with Kamehameha, and sent warriors to aid in Kamehameha's attack against Maui.

With Kahekili living at Waikīkī on O'ahu, Maui was left in the charge of Kahekili's son Kalanikūpule. Kamehameha's substantial war fleet first landed at Hāna. Skirmishes were fought along the windward side of the island, as the Maui forces retreated toward the old royal center at Wailuku (see figure 15).

Kamehameha was greatly encouraged when, during one battle, he saw the feathers of his war god Kūkāʻilimoku bristle and stand upright. When the Hawaiʻi war fleet arrived at Wailuku, the forces of Kalanikūpule retreated into the narrow valley of ʻĪao. Here the cannon Lopaka, in the hands of Isaac Davis and John Young, had a devastating effect on the concentrated Maui warriors. Tradition says that ʻĪao Stream was dammed with the bodies of Kalanikūpule's troops.

Kalanikūpule escaped by fleeing over the mountain pass from the rear of ʻĪao, descending into Olowalu Valley and thence by canoe to Molokaʻi Island. With him were the high chiefess Kalola (former wife of Kalaniʻōpuʻu and sister of Kahekili), along with her daughter Liliha Kekuʻiapoiwa, and her granddaughter Keōpūolani. Kalola, it will be recalled, was also the mother of the deceased Hawaiʻi king Kīwalaʻō, who in royal *piʻo* fashion had married his sister Kekuʻiapoiwa Liliha. After Kīwalaʻō's death at the battle of Mokuʻōhai, Kalola and her widowed daughter fled with Liliha's infant daughter Keōpūolani to Maui. Keōpūolani, now entering puberty, would be a great prize if Kamehameha could capture her. As a *nīʻaupiʻo* chiefess, she held the highest possible rank, considerably higher than Kamehameha himself. If he could take her as his royal wife, she could bear him an heir of undisputed genealogy.

When Kamehameha heard that Kalola and her sacred charge had escaped to Molokaʻi, he immediately pursued her, accompanied only by a few of his high chiefs and leaving most of his army behind on Maui. Landing on the southern coast, Kamehameha learned that Kalola and the chiefesses were camped in the uplands of Kalamaʻula. Kalola was near death. Kamehameha sent a messenger ahead, giving word that he was coming in peace, and in respect to her exalted rank. Kalola received Kamehameha, who put to her his request that after her death he be allowed to take the young chiefesses, especially Keōpūolani, under his wing. A few days later Kalola died, having given her consent. Kamehameha had his sacred prize. A few years later, Keōpūolani would bear him the heir that he desired. On the death of his father in 1819, Liholiho would become Kamehameha II.

Encouraged by his success at ʻĪao, Kamehameha intended to push on to Oʻahu to confront Kahekili directly. He sent a messenger to Kahekili at Waikīkī, bearing two stones, one white and one black. The white stone signified peace, the black one war. Kahekili's message came back loud and clear: "Tell Kamehameha," he said, "when the black tapa covers Kahekili and the black pig rests at his nose, then is the time to cast stones." The allusions are to the black funerary barkcloth that would cover the king's body after death, and to the black pigs that were killed to feed the mourners. Kahekili would not cede Oʻahu; Kamehameha would have to kill the old warrior first.

As Kamehameha pondered Kahekili's reply, word arrived from Hawaiʻi that abruptly changed Kamehameha's plans. While Kamehameha had been fighting on Maui, aided by the sons of Keawemauhili, his archenemy Keōua had attacked and killed Keawemauhili in Hilo, thus gaining control of two-thirds of the island. Keōua had then invaded the sacred valley of Waipiʻo, ripping up the taro fields, even proceeding into Kamehameha's home territory of Kohala. Oʻahu and the confrontation with Kahekili would have to wait. Keōua's challenge had to be met.

. . .

Returning to Hawaiʻi, Kamehameha landed at Kawaihae on the leeward side. Hearing that the forces of Keōua were in the uplands at Waimea, he sent his warriors to engage them, but Keōua withdrew to Hāmākua. A battle was fought at Mahiki, but Keōua again pulled his forces back, moving south toward Hilo and Puna. Rather than pursue Keōua into the latter's home turf, Kamehameha resolved to try a different strategy.

At the same time that Kamehameha had dispatched his messenger to Kahekili, he had sent the chiefess Haʻaloʻu (widow of the Maui king Kekaulike) on a parallel errand, to seek the wisdom of a certain *kahuna* named Kapoukahi, famed for reading signs and omens. Kapoukahi told Haʻaloʻu that Kamehameha must build a new *heiau*, "a great house for his god" Kūkāʻilimoku. When she asked where this house should be situated, Kapoukahi said upon the hill of Puʻukoholā, a cinder cone overlooking the bay at Kawaihae. An older temple named Mailekini already existed below the hill; the new temple would soar above it. Kamehameha now resolved to act on Kapoukahi's prophecy, to build the great *heiau* to his war god, the god he had inherited from Kalaniʻōpuʻu, which had been passed down through the generations from the time of ʻUmi.

All of the warriors and able-bodied men were set to work building the new temple atop the "Hill of the Whale," Puʻukoholā (see figure 16). Even Kamehameha himself assisted with hauling basalt boulders from the nearby beaches and canyons. When the foundation—laid from thousands of boulders and paved with fine *ʻiliʻili* gravel from the beach—was completed, the craftsmen set to work on the wooden fence, the thatched temple houses, and the *ʻanuʻu* tower.

While the *heiau* at Puʻukoholā was under construction, in the fall of 1790, Pele unleashed her deadly gas and cinder eruption from the fire pit of Halemaʻumaʻu. Keōua's army, which had been on the move from Puna to Kaʻū, was caught in its

FIGURE 16.
The stone foundations of the great war temple of Kamehameha I at Puʻukoholā, Hawaiʻi Island. Photo by Thérèse Babineau.

path. More than four hundred warriors were killed. More terrifying than the actual death of so many fighting men was the unmistakable sign that the gods had turned against Keōua. He had heard of the great temple being raised by Kamehameha at Kawaihae. Keōua harbored no doubts that his body was the intended dedicatory sacrifice.

With the *heiau* finished and only awaiting its dedication ceremony, Kamehameha sent two of his most trusted counselors, Keaweaheulu and Kamanawa, on a mission to Keōua, then camped in the uplands of Kaʻū. They were to invite Keōua to come and meet Kamehameha at Kawaihae, where the temple of Puʻukoholā now loomed over the placid bay. Arriving at Keōua's camp, Keaweaheulu and Kamanawa groveled in the dirt at the feet of the enemy king. Keōua's own advisers urged that they be killed on the spot. But the two chiefs were related by blood to Keōua; he listened to their message from Kamehameha. "I will go with you to Kawaihae," Keōua told Kamehameha's messengers.

Keōua sailed north toward Kawaihae with a small fleet of canoes and his most trusted chiefs and advisers. Passing Kailua, the party landed at Luahinewai, near Kekaha. Here Keōua went ashore to bathe ritually and garb himself in his feathered cloak and helmet, as befitted a high chief. Completing his bath, Keōua performed a shocking act: he sliced off the tip of his penis. It was the ultimate insult to Kamehameha, for in defiling his own body, his manhood, Keōua denied Kamehameha the perfect sacrificial body that the dedication rites for the new *heiau* demanded.

Keōua's double-hulled canoe rounded the point at Puakō, the imposing new temple coming into clear view. Kamehameha's war canoes covered the bay of Kawaihae. Passing them, Keōua's canoe approached the shore, where Kamehameha stood waiting. Keōua called out to his mortal enemy, "Here I am." Kamehameha answered that he should come ashore and greet him. As Keōua leaped out of his canoe, he was speared by Keʻeaumoku. The Hawaiʻi warriors fell on the others in Keōua's party; only two were spared. Keōua's body was turned over to the priests, who carried it up onto the temple platform. Puʻukoholā had received its sacrifice.

Keōua's death left Kamehameha the undisputed master of Hawaiʻi Island. He would spend the next two years preparing for the conquest of the rest of the archipelago, building his famous fleet of *peleleu* canoes and training his army. In late 1793 word arrived that Kahekili had died. The aged king's words had proved correct—not until the black cloth had covered his body would Kamehameha cast his stone. In February 1795 Kamehameha made his move. The largest war fleet the islands had ever seen crossed the ʻAlenuihāhā Channel to land at Lahaina. Maui fell with hardly any resistance. The fleet moved on to Molokaʻi, where the canoes covered the beach from Kawela to Kalamaʻula.

In April, Kamehameha moved against Oʻahu, now ruled by Kalanikūpule. Landing at Waikīkī, Kamehameha's army drove the defending forces back across the hot plains and into the deep valley of Nuʻuanu. With their superior numbers and firearms, the Hawaiʻi warriors overwhelmed the Oʻahuans, who tried to make a final stand at the head of the great cliff *(pali)* of Nuʻuanu. Some were felled by spears and gunfire, others were pushed or leaped off the cliff to their death on the rocks below. A few managed to escape by climbing up the steep ridges. Among those who escaped was Kalanikūpule. He managed to hide out in the forests for a time, but was eventually captured in 'Ewa and killed. Kalanikūpule's body was brought to Kamehameha at Waikīkī, where the corpse was offered up to his god Kūkā'ilimoku, on the *heiau* platform of Papa'ena'ena. It was the last royal sacrifice to the feathered war god.

· · ·

With Kamehameha's conquest of O'ahu, eight hundred years of Hawaiian history reached its culmination. No longer isolated from other societies and cultures, the very geographic position that had made the islands so remote now was a geopolitical asset in the expanding European-dominated world system. A convenient halfway stop between the rich fur-hunting grounds of the Pacific Northwest and the untapped markets of China, Hawai'i rapidly became a caravansary for traders, merchants, and naval vessels. Kamehameha cleverly took advantage of the Haole intruders, accommodating their needs for taro, yams, and pigs while acquiring the firearms and expertise he knew would make his army invincible.

Kamehameha's grip on the islands was not yet entirely complete. He had failed to take fair Kaua'i at the northwestern end of the island chain. His war fleet had been sundered in the stormy channel between O'ahu and Kaua'i, forced by the high seas to abandon the attack. Soon after defeating Kalanikūpule on O'ahu, before he could attempt a second invasion of Kaua'i, Kamehameha was obliged to return to Hawai'i to put down an insurrection of disloyal chiefs. The age-old patterns of rebellion and attempted usurpation died hard. But Kamehameha prevailed. In 1804 he returned to O'ahu to make Honolulu his royal seat. Kaumuali'i, the king of Kaua'i, was in 1810 induced to submit to Kamehameha's authority.

Honolulu offered a far better anchorage for the Western sailing ships that were now descending on the islands with worrisome frequency. Advised by John Young, and soon after by the worldly Spaniard Francisco de Paula Marin, Kamehameha built a fort with walls of quarried coral blocks, guarding the entrance to Honolulu harbor. Bronze cannon surmounted the ramparts, while within the fort's thick walls the king's troops drilled with an armory of muskets gained from his shrewd trading. From the fort's flagpole Kamehameha flew the flag of his new Hawaiian kingdom, a variant of the British Union Jack. The shadow of Captain Cook loomed large over this archipelago that the great explorer had opened to the world. In decades to come, the British would maintain a close alliance with the Hawaiian

FIGURE 17.
'Ahu'ena *heiau* at Kamakahonu, Kailua, Hawai'i Island, where Kamehameha kept the collection of gods he had captured from his various enemies. The thatch superstructures and images were reconstructed based on early historical period descriptions and drawings. Photo by Thérèse Babineau.

ali'i even as American traders, missionaries, and entrepreneurs made ever-deeper inroads into Hawaiian economic and political structures.

Kamehameha resided at Honolulu for more than a decade, personally overseeing the increase in trade and adroitly adapting the old Hawaiian way of governing to the new exigencies of diplomacy. In 1812, exhausted by these efforts, Kamehameha retired to his beloved island of Hawai'i. There he resided in the ancient royal compound of Kamakahonu on a point of land overlooking Kailua

Bay in Kona. In his final years Kamehameha became something of a brooding sorcerer, watching over his collection of feathered gods and carved images that he kept in the adjacent royal *heiau* of 'Ahu'ena (see figure 17). In these final years the affairs of state—and, more important, of trade—were left to his trusted chief Kalanimoku in Honolulu.

The effects of Western contact accelerated year by year. Most shocking was the high death rate resulting from exposure to old-world diseases to which the Hawaiians—having been isolated for so many centuries—had little or no resistance. Lieutenant King estimated the Hawaiian population in 1779 to be at least 400,000; it may have totaled considerably more. By 1820 that number had dwindled to perhaps 150,000. Massive economic and social changes were taking place. In 1811, several American traders began to carry shiploads of Hawaiian sandalwood, a common tree in the islands' dryland forests, to China. The Chinese markets had been hard to break open, even as the furs of the Pacific Northwest were becoming scarcer and harder to obtain. But the Chinese Mandarins coveted the scented sandalwood, which their artisans carved into fans and boxes. Yankee traders realized that they could reap huge profits by filling their ship's holds with the Hawaiian wood, cheap to obtain from the chiefs and fetching stellar prices in Canton.

In a Hawaiian society still organized by the hierarchy of chiefs and their *konohiki*, agreements to fill cargo holds with sandalwood were negotiated between the chiefs (often Kamehameha's own representative) and the ships' captains. To the common people, the *maka'āinana*, fell the hard labor of cutting the trees on the mountain ridges and hauling the wood down to the coast. The people were obligated to perform this labor as part of their *ho'okupu*, the tribute that they owed as residents of the *ahupua'a* territories. The age-old system of tributary relations between the people and the chiefs was thus subverted by the predatory capitalism of the new world system. Already wracked by disease and diminished in numbers, the *maka'āinana* suffered yet greater indignities in being forced to harvest the bounty of their forests. Meanwhile the Yankee traders made small fortunes; the chiefs enjoyed a newfound glut of imported goods from breechcloths to cutlery, all the while running up debts; and the common people got the proverbial short end of the stick. The reciprocal *aloha* that had formerly bonded chief to commoner, and the concept of *pono* that had in principle guided their society, were rapidly being eroded by the foreign ideologies.

In 1819 Kamehameha died at Kamakahonu in Kona. With him the feathered gods departed. His bones were defleshed and secreted away in the dead of night by a trusted retainer, so that they would not be desecrated by an opponent. As the proverb says, "only the morning star knows" where the bones of Kamehameha lie, even to this day.

Even as Kamehameha lay dying at Kamakahonu, the brig *Thaddeus* out of Boston was en route to Hawai'i carrying the first party of Protestant missionaries to the islands. The timing could not have been more opportune for the small gang of Congregational preachers and their wives, intent on saving heathen souls. Liholiho, heir apparent and son of Kamehameha by the *pi'o* chiefess Keōpūolani (whom Kamehameha had secured from the dying Kalola on Moloka'i), was just twenty-two years old. Ka'ahumanu, the new king's aunt and favored consort of his deceased father, stepped into the power vacuum. A formidable woman, both physically and in her personality, Ka'ahumanu would shape the destiny of the islands for decades to come. Realizing the advantages to be gained by replacing the old religious order with a new one under her direct control, Ka'ahumanu appointed Hiram Bingham, leader of the Protestant mission, her new high priest. She sent out orders to destroy the temples, to burn the images of Kū and Lono. New Christian churches arose on the ruins of the ancient *heiau*. The mission printing press churned out primers and a translation of the New Testament as rapidly as the missionaries could fathom the Hawaiian language.

By the mid nineteenth century—under the long reign of Kauikeaouli (Kamehameha III), another sacred son of Kamehameha I and brother to Liholiho—the Hawaiian kingdom became a constitutional monarchy. There was a well-organized government bureaucracy and a judiciary replete with a printed code of laws. The kingdom negotiated treaties with foreign governments that recognized the sovereignty of the Hawaiian nation. But this was the era of colonialist expansion; the islands were a great potential prize. Under pressure, Kamehameha III was induced by the missionaries and other foreign residents to end the traditional system of territorial land rights, shifting to a Western system of "fee simple" property rights. By the 1870s, vast tracts of land had been acquired by an expanding class of white sugar planters. Mostly of American origin, these businessmen chafed under onerous tariffs imposed on Hawaiian sugar by the United States government. They also harbored racist views of the Hawaiian people, views on the "natural inferiority of brown-skinned peoples" so typical of the Victorian era. Convinced that the islands would be better off as a territory of the United States, these Haole business leaders engineered a coup in 1893, overthrowing the last Hawaiian monarch, Queen Liliu'okalani. In 1898, Hawai'i was officially annexed by the United States as the territory of Hawai'i. The long history of Hawai'i's experiment with kingship came to a close.

EPILOGUE · Hawai'i in World History

If history teaches us deeper truths about ourselves, then the history of Hawai'i has a significance that transcends time and place. What transpired in Hawai'i before Cook pulled aside the veil of isolation that had encircled the islands for eight centuries was a unique experiment in social evolution. Although the Hawaiian Islands might seem to be just an obscure and remote corner of the globe, whose past could be taken as a mere curiosity on the grander stage of world history, their deep history holds the very civilizing process up to a mirror. The history of Hawai'i is a microcosm of the development of civil society. It is a history in which complexity, hierarchy, stratification, economic specialization, state religion, and everything that constitutes what anthropologists call "complex societies" emerged out of a simpler and more fundamental kind of social formation.

About the close of the tenth century, when Europe was only beginning to stir out of the depths of its Dark Ages, a small group of Polynesians, no more than a few related families, set forth on an unprecedented voyage of exploration and discovery. Sailing in their double-hulled canoes where no humans had ever ventured, across more than 2,500 miles of untamed ocean, they discovered the most remote archipelago on Earth. They were not new to the game of discovery and colonization. Their ancestors had perfected this art over generations. They came equipped with the biotic resources—crops and domestic animals—to settle a new land. They shared a legacy of hard-earned knowledge about how to adapt to islands. But the island chain they discovered while chasing the flightways of the golden plover was unlike anything their immediate ancestors had encountered. Hawai'i encompassed eight islands with more than 6,400 square miles among them. The newcomers' root and tuber crops thrived in the subtropical climate. The reefs teemed with edible marine life. The coastal plains and valleys offered a seeming infinity of land over which to spread out, to cultivate, to build a new society based on the founding principles they brought with them.

No omnipotent social scientist could have designed a better experiment in social evolution. Take a small group of people, with a competent knowledge of their environment and a strong collective will to succeed, and place them on a remote group of islands rich in resources. Isolate them, through millennia of voyaging away from the Asiatic continent, from virtually all of the diseases and afflictions of the Old World, so that they enjoyed a degree of health probably not seen elsewhere in the contemporary world. As a consequence, their demographic potential to grow and expand was limited only by the innate reproductive capacity of *Homo sapiens*. After a century or two in which they continued to maintain sporadic contacts with their homeland, they cut those ties to the motherland completely. Then, in a world utterly unto itself, let the course of social evolution run over another four hundred years.

This "experiment" might have been designed by an omnipotent social scientist, but in fact it was simply the contingent outcome of a small group of people and generations of their offspring interacting together and with the resource-rich environment that they had the good fortune to discover and settle. What ensued was a remarkable replay of the histories of other societies in similarly favorable conditions throughout both the Old World and the New. The history of Hawai'i parallels similar histories across the globe: in the valley of the Nile, along the banks of the Tigris and Euphrates, in the Yellow River valley, in the highlands and tropical rain forests of Mesoamerica, and in the Andes. In a mere eight centuries, the descendants of the first voyagers to arrive in their double-hulled canoes at Hawai'i invented nearly all of the same essential social, political, and religious institutions characteristic of the first archaic states found elsewhere in the world. Class stratification, divine kingship, royal incest, human sacrifice, intensive irrigated and dryland farming, a territorial system of land tenure, elite control of staple surplus, a wealth economy based on prestige goods, formal priesthoods, and cults of fertility and war—all of these were independently invented by the Hawaiians. These were the same traits fundamental to all early civilizations.

Yet there is one essential difference between Hawai'i and other early archaic states. All of the other primary states that emerged out of simpler societies in various parts of both the Old World and the New did so thousands of years ago. Consequently, their histories can only be inferred from archaeological vestiges. No Enlightenment voyagers, no ethnographers, were present to describe how these early societies functioned. But in remote Hawai'i, a group of island states developed in total isolation over the same time period during which Europe made its transition from the Dark Ages to the Enlightenment. Thus at the close of the eighteenth century, Enlightenment explorers encountered the last pristine civilization to have emerged on Earth. Captain Cook literally came face-to-face with Kalani'ōpu'u, the divine ruler of a society that had had no external contacts for half a millennium. The British captain knew an aristocratic society when he met one, for the Admiralty itself was steeped in hereditary rank, and British navy captains ruled their ships with unquestioned authority. King George III, to whom Cook had been introduced at Buckingham Palace, basked in the fading aura of divine kingship, even as the "divine right of kings" would soon be tested in revolutions across Europe. But in far-off Hawai'i, divine kingship was still ascendant. The late eighteenth-century voyagers to Hawai'i documented for posterity the previously cloistered world they had stumbled upon in their battered and leaky ships. Their journals and logs, pencil sketches and lithographs, remain a priceless record of the last pristine civilization to have arisen through an independent course of cultural evolution. Unaware of their own impact on this society, the British visitors described it dispassionately at the fateful moment of "first contact."

I would argue that Hawai'i is at once both unique and typical on the stage of world history. It was unique in being the last pristine archaic state to arise on Earth. Unique in that the processes underlying its economic, social, and political transformation were entirely endogenous, self-contained. Unique in that it was the only pristine civilization to have been observed and recorded by Europeans. Unique in that it has its own indigenous historical record of how its society came to be. At the same time, Hawai'i is a microcosm that illuminates common recurring themes in the larger history of human societies. It was typical in that over the course of eight centuries a social structure based on kinship was transformed into one of divine kingship. Typical in the light of a demographic history that reflects the Malthusian tendency toward population limits as well as the Boserupian potential for innovation and intensification. Typical in the ways that a small segment of elites took a system of land tenure rooted in kinship and residence and transformed it into a territorial hierarchy based on obligatory tribute and labor. Typical in the sense of a religious system that evolved from small-scale ancestor worship to institutionalized cults of agriculture and war, maintained by formal priesthoods whose rites were enacted upon monuments that still inspire awe. Unique yet typical, that is the compelling nature of Hawaiian history.

· · ·

Social scientists have argued long and hard about the underlying causes that drive social evolution. Some, such as the American Museum's brilliant ethnologist Robert Carneiro, believe that a phenomenon called circumscription was at the heart of early state formation. As Carneiro sees it, adjacent and competing chiefdoms expanded within a limited zone of prime agricultural resources, inevitably coming into competition with one another, which led to warfare and territorial conquest. Ultimately, the largest and most aggressive polity, the one that swallowed up its neighbors, became a state. Other researchers, such as Charles Spencer and Else Redmond, who have studied the rise of states in Mesoamerica, also stress the central role of territorial expansion, combined with the extraction of tribute from conquered vassal territories. But the rise of states is not just about territorial expansion and conquest, important as such factors may have been. Colin Renfrew of Cambridge University, who studied another group of islands—the Cyclades of the Aegean Sea—concluded that interaction and emulation among elites in adjacent societies was a powerful force driving social change. He called this theory peer polity interaction. Renfrew's theory is a kind of "keeping up with the Joneses" model of how leaders copy one another's actions and strategies.

These models hardly exhaust the possible causal mechanisms or driving forces ("prime movers" they are sometimes called) for social evolution. Trade between social groups can be a compelling force for economic and social change, as William Rathje once argued in an attempt to explain the rise of the Maya civilization. "Praise the gods and pass the metates" was the way he epitomized it. Others—such as David Webster, who also specializes in studying the Maya—think that war played the most significant role in state emergence. Most archaeologists and other social scientists who study the evolution of chiefdom and state societies reject any single causal explanation. Timothy Earle, who made a comparative analysis of ancient societies in Hawaiʻi, Peru, and Denmark, argues that elites in all of these societies used three distinct kinds of power strategies. The first was control over economic production, especially agriculture. Next came the use of military power. But not to be ignored was control over ideology, especially as manifested in religion. Earle and his students put forward the idea of "materialization" to encapsulate the ways in which chiefs and other leaders exercise power through public presentation and display, especially in ostentatious ceremonies and rituals. One need only reflect for a moment on how the British royal family depends to this day on a materialization strategy (changing of the guard at Buckingham palace, royal rituals in Parliament and Westminster Abbey) to maintain a tenuous hold on a monarchical system long outdated. Indeed, one is reminded of Karl Marx's famous summing up of the importance of ideology and materialization when he wrote, "Religion is the opiate of the masses."

Two of the most respected archaeologists who have tackled the problem of state formation, the husband-and-wife team of Kent Flannery and Joyce Marcus at the University of Michigan, have argued for a multicausal approach to social evolution. In a beautifully documented study of the rise of Zapotec civilization in Mexico's Oaxaca Valley over eight millennia, Marcus and Flannery apply what they call action theory. Action theory (sometimes referred to as agency or practice theory by anthropologists) highlights the importance of individual human actors in the course of history, while at the same time recognizing that people operate within specific environmental, demographic, economic, and social contexts and constraints.

The history of Hawai'i's island civilization offers a parallel case of social evolution, shorter in duration than that of Oaxaca, but no less interesting. What does the experiment of Hawaiian social evolution tell us about how archaic states, along with their most salient feature, divine kingship, emerged from simpler kinds of societies? How does Hawaiian history help us to sort through the different theories of state emergence put forward by archaeologists and social scientists? In what ways was the evolution of archaic states in isolated Hawai'i similar to or different from the evolution of a complex society in Oaxaca? Or like that in any of the other places where archaic states first developed in the New World and the Old? Let me briefly address those questions by summing up the ways in which Hawaiian society was transformed over eight centuries.

I am convinced that understanding and explaining social evolution requires a commitment to multicausality. In popular science writing, physics and chemistry are often referred to as the "hard" sciences. Yet pinpointing causality in a physical or chemical system is so much easier than in biological—or especially social—systems. It is the social sciences that are truly the "hard" sciences! Causes are multiple, complexly interwoven, and historically contingent, not at all like explaining the rotation of a planet around its star, or even the behavior of electrons around their nucleus. In chapter 14 I described how I have found inspiration in the writings of the late Ernst Mayr, Harvard's great evolutionary biologist. Mayr cautioned that complex historical phenomena usually require both ultimate and proximate causations. In seeking to explain the rise of archaic states and divine kingship, whether in Hawai'i or elsewhere, ultimate causations include such factors as Carneiro's circumscription, population growth and pressure, or the capture of surplus from intensified agriculture. Proximate causations are those associated with individual human agency, including Renfrew's peer polity emulation among elites, and the application of power through materialization as described by Earle. Ultimate causations tend to work slowly, on a time scale of decades to centuries. Proximate causations can be very fast, as when a charismatic leader convinces his followers to take up the worship of a new cult.

As described in part 1 of this book, the early Polynesians who settled Hawaiki and developed their distinctive island society there were already sedentary agriculturalists. The Polynesians of ancient Hawaiki also possessed a social system marked by differences in rank, by heritable asymmetry in which these differences are passed down through family lines. That is to say, the early Polynesians already occupied the general phase of cultural evolution that anthropologists call the chiefdom. Yet ancestral Polynesian chiefdoms were still relatively small in scale; the differences between hereditary chiefs and commoners were subtle. Most important, the ancient Hawaiki people organized themselves into lineages that traced descent back to common ancestors. These lineages were the basis on which they held rights to land and property.

When the first canoes hauled up on a Hawaiian beach about A.D. 1000, their crews carried with them a variant of this ancestral Polynesian culture. At first the new settlers did not change much, other than to adapt their practices of fishing, hunting, and farming to local conditions. In the subtropical climate of the Hawaiian Islands, some of the tropical crop plants they brought with them did better than others. But most of the fish and shellfish in Hawaiian waters were the same or similar to those they had known in the home islands. Some of the plants and trees were different, but many were the same. All that was required of these first colonists was minor adaptations in technology and the way they used their resources.

The newly discovered Hawaiian Islands were vastly larger than the Marquesan homeland, indeed, larger than any other group of islands the Polynesians were to discover in their voyages (except for New Zealand, which, being temperate, was not very favorable to most of the Polynesian crops). At first, this new archipelago must have seemed almost infinitely large to the new settlers. And although some regions on Maui and Hawai'i were forbidding with their arid lava slopes, there were ample areas offering ideal combinations of fertile soil, fresh water, and adjacent coral reefs abounding in marine life. It was as if any limits or restrictions that might put a check on either human reproduction or economic growth were lifted. With only a few hundred people to spread out over hundreds of square miles of fertile land, the classic Malthusian brakes simply did not apply.

Content in their fertile islands, protected from the ravages of old-world diseases by their extreme isolation, the Hawaiian population grew, and grew fast. Within two centuries after Polynesian arrival, the most desirable valleys all had permanent settlements. Another two hundred years on, by about 1400, the islands of Kauaʻi and Oʻahu, along with the best parts of Maui and northern Hawaiʻi, were densely settled. This surge in population, which expanded exponentially, is the first ultimate cause underlying the changes that would dramatically transform Hawaiian society. It was not just the high rate of population increase, but the large and dense populations on individual islands that were essential to social evolution. Chiefdoms can exist with just two or three thousand people, but states require a much larger critical mass. By the fifteenth century, Oʻahu and Kauaʻi islands probably had populations in the range of fifty thousand persons or more. These multitudes provided the labor force necessary to underwrite substantial public works, especially the irrigation terraces and canals.

It was in the realm of agriculture and the staple economy that Hawaiian society showed the first signs of moving from a chiefdom toward a state. The rapid increase in population necessitated that more land be brought under cultivation. Of course, the converse was just as true: increases in agricultural production meant not only that more people could be accommodated within a particular tract of land, but also that more labor (in terms of more bodies working in the fields) was required. The linkage between population and agriculture is like the old chicken-and-egg conundrum. More mouths require more food, but producing more food necessitates more field hands. The two are inseparably linked. Systems modelers call this a positive feedback loop.

Between the thirteenth century and the fifteenth, major irrigation works were constructed throughout Oʻahu and on windward Molokaʻi; presumably the same was true for Kauaʻi, which is even richer in alluvial lands suited to irrigation. These irrigation systems probably started out as small-scale projects carried out by single households or groups of related farmers. But over time their scale increased. Entire valley floodplains were brought under terraced management. Mile-long irrigation canals diverted stream flow into hundreds and even thousands of pondfields, each a highly productive microecosystem for growing taro, the prized Hawaiian staple. While these complex systems do not necessarily require a hierarchical administration to run them, there are benefits to centralized management (such as making sure that communal tasks like canal repair are carried out, and ensuring the equitable distribution of water during droughts or times of low stream flow). Centralized management has the advantage of seeing the system as a whole, and of using that advantage to capture that portion of the total production or yield which exceeds the needs of individual households. The difference between what farming households need to survive and reproduce, and what the intensive irrigation works are capable of producing, is what we call surplus. And no complex society—certainly no state—can exist without tapping into and controlling a surplus. (In our own society, that surplus—the production over and above individual households needs—is what the state extracts through taxation.)

The Oʻahu chiefs first gained control over surplus by taking over the management of the irrigation works. In so doing they pushed Hawaiian society along the transition from chiefdom to state. The *moʻolelo* of the Oʻahu king Māʻilikūkahi is especially revealing, for it was under his direction in the fifteenth century that the hierarchical system of *ahupuaʻa* and *ʻili* land units was created. It is possible, of course, that Māʻilikūkahi was given credit for a sweeping reform that had been in process for some generations. It may simply be that the system reached its formal conclusion under his reign. Regardless of how it happened, with the *ahupuaʻa* system and its corresponding *konohiki* (land managers) in place, the Hawaiian *aliʻi* possessed what every state needs: a formal means both for managing intensive agriculture (including the use of collective labor) and for the regular collection of surplus in the form of tribute (i.e., taxes). With these changes the old social organization of Hawaiki was gone forever. No longer did the kinsfolk of a *mata-kāinanga* lineage make collective decisions about land use and allocation. No longer did they share the collective fruits of their labors among themselves exclusively. Now they were all *makaʻāinana*, commoners who worked the land under the supervision of the *konohiki*, supporting the *aliʻi* with the fruits of their labors. And *aliʻi* were no longer just titular heads of lineages, but a separate category of persons who did not grow their own food. They became full-time leaders and managers. The old Polynesian society was segmented vertically by groups of kinsfolk who traced common descent from ancestors. The new Hawaiian society was divided horizontally, into a class of *aliʻi* above and a class of *makaʻāinana* below. Over succeeding generations, this distinction between the chiefs above and the people below would become more rigid and marked.

In contrast to the arguments that aggression and war play a major role in state formation, virtually all of these initial changes in Hawaiian society seem to have occurred peacefully. True, the tradition of Kalaunuiohua, who marauded up the archipelago from Hawai'i to Kaua'i, does suggest a limited degree of interisland competition, but this seems to have been an anomaly. To the extent that there was fighting, it was usually between chiefly rivals within the same island, attempting to seize power or usurp the kingship. Not until later in Hawaiian history, after the main islands had been consolidated into single kingdoms, did war and territorial conquest become driving factors in Hawaiian social and political evolution.

Had the Hawaiian Islands consisted merely of the geologically older islands of Kaua'i, O'ahu, and Moloka'i, the course of social evolution might have slowed at this point, with the emergence of these irrigation-based kingdoms. Irrigation is remarkable for its tendency toward what the anthropologist Clifford Geertz once called involution, the ability to absorb greater and greater numbers of people as the highly productive pondfields are pushed harder and harder. The island of Bali, which Geertz studied, is a classic case in point. The Balinese societies were simple kingdoms, "theatre states" Geertz called them, invoking their particular use of ritual performance to validate and legitimate their social hierarchy. But the Balinese elites were not divine kings. They did not marry their sisters and half-sisters, did not offer up human sacrifices, were not regarded by the people as living gods on earth. What pushed Hawaiian social evolution further, beyond the irrigation kingdoms of a Bali? The answers, I believe, are to be found in the large, volcanically youthful islands of Maui and Hawai'i. Those large islands were the setting for the next stage in Hawaiian history.

The innovations that had first been developed on O'ahu and Kaua'i, the restructuring of the land system and implementation of an agro-managerial structure coupled to an efficient apparatus for surplus collection, were quickly adopted by the *ali'i* of Maui and Hawai'i. On these much larger islands, however, the potential for irrigation was much more limited. But the vast leeward slopes of eastern Maui and of Hawai'i could be made highly productive through intensive rain-fed farming, especially of sweet potatoes, and of dryland taro in areas where rainfall was slightly higher. Maui and Hawai'i together have more than 270 square miles of land suited to intensive dryland farming, surpassing the total of 54 square miles amenable to irrigation on O'ahu and Kaua'i. The Hawai'i and Maui chiefs could see that their futures lay in developing the extensive rain-fed plantations.

As the O'ahu irrigation works reached a peak of construction in the late fifteenth and early sixteenth centuries, the leeward field systems of Hawai'i and Maui were just entering a great phase of expansion. 'Umi, whose famous story I recounted in chapter 11, was the first Hawai'i Island king to move the royal seat from Waipiʻo Valley (the best irrigation locality on the island) to the leeward district of Kona. From Kona, ʻUmi could watch closely over and control the dryland gardens that were now expanding over tens of square miles. ʻUmi is known in the Hawaiian traditions as a farming king. His power sprang from the surplus that could be extracted from those miles and miles of sweet potato farms, from the pigs that could be fattened on the rich tubers, and from the other products of his dominions.

ʻUmi came to power at a crucial juncture in the long-term economic cycle of Hawaiʻi. The leeward field systems of Kohala and Kona were still expanding. Surplus production was increasing, the economy was growing. This gave ʻUmi tremendous advantages that would not accrue to later rulers, when the expansion phase had ended, when the rate of surplus production began to decline. ʻUmi rode the crest of an economic boom, and he appears to have done so wisely. With ʻUmi we also get a glimpse of other causal factors beginning to work in Hawaiian social evolution, factors of the kind that Ernst Mayr called proximate and that involve human agency. One of these was ʻUmi's strategic use of marriage alliances to consolidate power. Advised by his wise priest Kaleiokū, ʻUmi married the high-ranking daughter of Maui king Piʻilani. But he also had other wives taken from *aliʻi* ranks throughout Hawaiʻi Island. Perhaps because his own mother was a commoner and he had suffered from this stigma under the rule of his half-brother Hākau, ʻUmi understood well the importance of rank and bloodlines. It was about his time that *piʻo* marriages between close royal siblings started to become a regular practice, on both Hawaiʻi and Maui.

Linked by royal marriage, the ruling houses of Hawaiʻi and Maui offer a classic example of Renfrew's theory of peer polity interaction. Practices taken up in one royal court were quickly copied in the other. This extended to the ways in which religion and the materialization of ideology were used to legitimate the new kingship order. Tū and Rongo were ancestral deities venerated throughout Eastern Polynesia, so we know that they were a part of the religious ideology of the first Polynesians to settle Hawaiʻi. By the early seventeenth century on Hawaiʻi and Maui, however, newly innovated cults of Kū and Lono (the Hawaiian variants of Tū and Rongo) had become central to the power strategies of ʻUmi, Kiha-a-Piʻilani, and their successors. Elevated to a status far above mere ancestors, Kū and Lono became the principal gods of the kingdom. Lono represented rain-fed agriculture and the sweet potato, ultimate source of the king's power. It was in Lono's name that the king collected the annual tribute from his dominions. Kū became the deity of human sacrifice, representing the king's divine right to exercise force. The red-feathered image of Kūkāʻilimoku (Kū the Snatcher of Kingdoms) became the personal deity of the king. The Hawaiʻi and Maui kings began to construct large stone temple platforms upon which the increasingly elaborate rites of Lono and Kū were performed, each by full-time priesthoods specializing in the individual cults. The displays were designed to be awe-inspiring and intimidating. This was the power strategy of materialization, of pomp and circumstance and ritualized display, all of which only served further to widen the social chasm between the king, his nobles, and the common people.

But the materialization of social distance did not end with the elaboration of the Kū and Lono cults on their imposing temple platforms. It was especially promulgated in the dress and bodily practices of the *aliʻi*. *Tapu* (taboo), an age-old concept in Polynesia, was elaborated to isolate and protect the high-ranking elite, to emphasize their distinctiveness and godlike qualities. We do not know exactly when the practice of the *kapu moe*, the prostrating taboo, was first instituted in Hawaiʻi. ʻUmi's story, however, does mention the existence of the *pūloʻuloʻu*, the special staff set up in front of the king as a symbolic barrier. As in royal courts everywhere, the Hawaiian kings were protected and fêted, venerated, and—no doubt—manipulated.

Most notable was the efflorescence of the arts of featherwork. The capes, cloaks, helmets, *lei*, and *kāhili* came to be the material symbols par excellence of the Hawaiian ruling class. In promoting these arts, and in using the feathered garments as a means not only to signal their elite status but, perhaps more important, to distinguish among themselves fine gradations in rank and power, the *aliʻi* created a new wealth economy on top of the older staple economy. Red and yellow feathers made up a significant part of the annual tribute from each chief's territory. The circulation of feathers and featherwork became increasingly essential to the social relations of the *aliʻi* class. Hawaiian social evolution had entered a hypertrophic stage.

The creation of a wealth economy in feathers and featherwork did not—indeed, could not—supplant the fundamental staple economy of sweet potatoes, taro, pigs, and dogs on which everyone, from commoner to king, depended for their daily sustenance. But as we trace Hawaiian history into the late seventeenth and early eighteenth centuries, especially on the politically dynamic islands of Hawaiʻi and Maui, every indication is that this staple economy was no longer growing. Every effort was being made to intensify production, but the limits of intensification were rapidly approaching. Absolute surplus was probably on the decline, meaning that the chiefs risked the wrath of the common people if their tribute demands were too harsh. The reason for this decline in the economic cycle is fairly simple. The dryland or rain-fed field systems could be pushed only so far without the kinds of artificial fertilizers or genetically engineered crops that modern-day agribusiness has pioneered. To be sure, Hawaiian farmers had perfected the arts of manual horticulture. They used mulch to protect their plants from evapotranspiration, and terraces to reduce soil erosion. They had learned to plant closely spaced rows of sugarcane between their fields, protecting fragile leaves from winds and possibly fixing nitrogen in the soils. But all of these techniques had their limits. And by the eighteenth century, those limits were all too evident to the land managers. The problems became especially acute in years when rainfall was less than required to water the crops on the leeward slopes. In El Niño years, the leeward sides of the Hawaiian Islands tend to be hit by droughts. Several successive years of such drought could lead to misery and famine among the large and dense populations who depended on the leeward field systems. In fact, the Hawaiian traditions recount a number of such famines.

It was in this final stage of the Hawaiian experiment in social evolution that intersocietal conflict, aggression, war, and territorial conflict truly came into play as major driving forces. The Hawaiian traditions offer detailed accounts of the various invasions, counterinvasions, and attempted conquests that became increasingly frequent from the later seventeenth century and on into the eighteenth. Maui and Hawai'i entered into an almost perpetual state of war. But the Maui kings also coveted Moloka'i and had their eyes on O'ahu and Kaua'i beyond. For the Hawai'i and Maui kings, territorial conquest provided a way out of the trap of declining production in their dryland agricultural base. In particular, they were keen to gain control of the rich irrigation lands on Moloka'i, O'ahu, and ultimately Kaua'i. The wealth economy in feathers and featherwork also played a role in the heightened pattern of aggression, as the *ali'i* sought new sources of the prized red and yellow feathers.

This is where Hawai'i's unique experiment in social evolution had arrived when Captain James Cook broke through the centuries-long period of isolation. Three archaic states now divided the islands among them, each headed by a divine *ali'i akua* (god-king). One state was centered on Hawai'i but controlled parts of Maui, a second ruled from western Maui but incorporated Moloka'i, and a third encompassed O'ahu and Kaua'i. The agricultural economies supporting these states were truly impressive, both the irrigation works that dominated the western islands and the vast rain-fed field systems of the younger eastern islands. Cook and other early visitors remarked frequently on these marvels of intensive agriculture. Even though the economy was no longer in an expansion phase, as it had been two to four centuries earlier, it was still more than adequate to support a population totaling at least half a million people, perhaps more. The vast majority of these people made up the ranks of the *maka'āinana*. Also known as *nā kanaka* (the people), they were ruled over and governed by *nā li'i*, the chiefs. These two had become distinct classes of persons, each going about their daily lives in distinctive ways, yet bound up in a common system of rights, privileges, and obligations. Hawaiian society had changed immensely over the eight centuries since the first voyagers arrived, yet in some respects it remained much the same. People still fished, planted their crops, cooked their food in earth ovens, made barkcloth, thatched their houses, and in countless other details went about their daily lives much as their ancestors had done. At the same time, the ancient lineage system was long gone. To work the land now required that one make regular payments of tribute. Now the highest-ranked nobles were so sacred that they could not even be gazed at without fear of death. If a Polynesian time traveler from A.D. 1000 were to have arrived in the Hawai'i of A.D. 1777, he or she would—at first glance—have found the scene wholly familiar, if only to discover soon enough in unsettling ways the extent to which it was totally foreign.

Social change did not end with Cook's opening of Hawai'i to the outside world. Change would continue, faster than ever, but now the events of history would be heavily influenced by the forces of the capitalist world system that quickly descended upon the islands. Many times I have been asked by someone in an audience after lecturing on Hawaiian history, What would have happened had Cook not opened up the islands in 1778? What if the islands had remained cloistered for another two, three, four hundred years? I can only speculate about something that was not to be. But I think that the process of territorial conquest and political consolidation would likely have continued, eventually bringing the islands under a single unified ruler. This might have taken considerably longer than it did for Kamehameha, whose conquest of the entire archipelago was achieved with the aid of European armaments. And it is entirely likely that unification would not have been continuous, that the archipelago would have cycled through a series of unifications interrupted by ruptures into independent, warring factions.

More interesting is to ask whether two characteristics of many other early states that are lacking in Hawai'i—urbanism and writing—would eventually have emerged. Urban centers, I think, would not have developed, mostly owing to the nature of the Hawaiian staple economy. Taro, sweet potato, and the other crops grown by the Hawaiians are mostly not amenable to long-term storage. Towns and cities need a food supply that can be transported to them and stored for considerable periods. They are based on a magazine economy. It is no coincidence that most early urban centers were built around granaries. The Hawaiian kings and nobles had developed sizable royal centers, but these were limited to the residences of the *aliʻi* and their retainers, along with those of the priests alongside the temples. Such royal centers would probably have continued to grow in size and grandeur, but the people at large would likely have remained dispersed over the landscape, tending to their farms and gardens. As for writing, this I think would ultimately have been invented by the Hawaiians. The control of information is extremely important to a state. Hawaiian elites were interested in all kinds of information, from their genealogies and political histories to the details of landholdings and tribute. At the time of contact, such information was the purview of specialists who were charged with memorizing these details and recounting them when called on. When an alphabet was introduced by the missionaries in 1821, the Hawaiians seized upon it and within a few years had become a more literate people than the population of the United States! Thus I think it very likely that in due course they would have invented some kind of writing system for themselves.

Such speculations aside, the story of what transpired in this most isolated of Earth's archipelagoes after it was discovered by an intrepid band of canoe-borne explorers remains one of the great sagas of world history. It is a history fascinating both in its own details and for its own sake, and also for what it tells us about the dynamic potential of human society. We are a very plastic species, we humans. Our cultures and our social organizations are capable of great flexibility and adaptation. Societies never remain static; they are always changing, albeit some faster and others slower. Yet when examined over the broad sweep of history, we tend to travel along the same roads. Egalitarianism gives way to stratification and hierarchy, time and again. A select few come to dominate over the many. And those few exercise power in the same predictable ways, through control of the means of production, through manipulation of ideology and religion, and ultimately through the application of force. Meanwhile the vast working population that supports a state or a civilization does not necessarily accept these changes passively. Though dominated from above, they frequently develop means of resistance, often subtle, sometimes overt. While the details may vary, similar patterns emerge over and over through the course of social evolution. Discovering these great patterns and trends helps us to understand what is universal about human society. The history of Hawaiʻi, that most isolated of all archipelagoes, reaches beyond its unique aspects to help inform us about these great social truths.

ALPHABETICAL LIST OF HAWAIIAN HISTORICAL PERSONS

AKAHIAKULEANA   Mother of 'Umi by Līloa, ruling chief of Hawai'i.

ALAPA'INUI   Son of the Kohala warrior chief Mahiolole and his wife Kalanikauleleiaiwi, half-sister of Hawai'i Island king Keawe'ikekahiali'iokamoku. Alapa'inui became king over most of Hawai'i Island and repulsed an attempted invasion by Maui king Kekaulike.

EHUNUIKAIMALINO   Ruling chief of Kona district on Hawai'i Island in the time of 'Umi.

HA'ALO'U   Chiefess and widow of Maui king Kekaulike who was sent by Kamehameha to receive the oracle of the prophet Kapoukahi.

HAKA   A descendant of Māweke and ruling chief of O'ahu in the fifteenth century. An evil ruler, he was overthrown and replaced by Mā'ilikūkahi.

HĀKAU   Son of Līloa with Pinea, he succeeded Līloa as ruling chief of Hawai'i in the mid sixteenth century. Hākau was deposed by his half-brother 'Umi.

HIKAPOLOA   An early ruling chief of Kohala district, Hawai'i Island, and grandfather of Lu'ukia.

HOLO'AE   High priest of Hawai'i Island during the reign of King Kalani'ōpu'u.

HO'OIPOIKAMALANAI   A daughter of the Kaua'i ruling chief Puna, wife of Mo'ikeha, and mother of Kila.

HO'OLAEMAKUA   Ruling chief of Hāna district of Maui Island in the early seventeenth century. He was killed by Pi'imaiwa'a of Hawai'i during the war for succession to the Maui kingship between Lono-a-Pi'ilani and Kiha-a-Pi'ilani.

HUAPOULEILEI   Ruling chief of Waialua district on Oʻahu in the fifteenth century who fought against Kalaunuiohua of Hawaiʻi.

IMAIKALANI   Ruling chief of Kaʻū and a famous warrior, he opposed ʻUmi but was killed by Piʻimaiwaʻa.

ʻĪMAKAKOLOA   A ruling chief of Puna district, Hawaiʻi Island, who rebelled against Kalaniʻōpuʻu, and whose body was offered up by Kamehameha as a sacrifice on the temple of Pākini.

KAʻAHUMANU   Favorite wife of Kamehameha I, and *kuhina nui* (premier) during the reigns of Kamehameha II and Kamehameha III. She was born about 1773 and died in 1832.

KAHAHANA   Succeeded Kumahana as king of Oʻahu in the late eighteenth century, until he was deposed and killed by Maui king Kahekili.

KAHAʻI-A-HOʻOKAMALIʻI   A grandson of Moʻikeha, he made one of the last round-trip voyages to Kahiki, returning with the first breadfruit to be introduced to Hawaiʻi.

KAHEKILI   Also known as Kahekili-nuiʻahumanu, he was the younger son of Kekaulike, and succeeded his older brother Kamehameha-nui as king of Maui. He died on Oʻahu, which he had conquered, in 1794.

KAIKILANI-NUIALIʻIOPUNA   A high-ranking chiefess descended from ʻUmi and married to both Kanaloa-kuaʻana and Lonoikamakahiki.

KAKOHE   A priest of Waipiʻo who aided ʻUmi in overthrowing Hākau.

KĀKUHIHEWA   The king of Oʻahu in the early seventeenth century. His reign was noted as a time of great prosperity on the island.

KALĀKAUA, DAVID   Born in 1836, Kalākaua was elected King of Hawaiʻi on December 19, 1863, and reigned until his death on January 20, 1891. He was succeeded by his sister Queen Liliʻuokalani, the last Hawaiian monarch.

KALANIKŪPULE   Son of Maui king Kahekili and last king of Oʻahu Island. He was captured and killed during Kamehameha's invasion of Oʻahu in 1795.

KALANIʻŌPUʻU   A descendant of ʻUmi and grandson of Kaweʻikekahialiʻiokamoku, he succeeded Alapaʻinui to become king of Hawaiʻi Island. He was king at the time of Captain Cook's visit in 1779.

KALAUNUIOHUA   Ruling chief of Hawaiʻi Island in the fifteenth century, famous for his war of attempted conquest against the other islands.

KALEIOKŪ   Priest who aided ʻUmi to usurp the kingship from Hākau, and who became ʻUmi's *kahuna nui*.

KALOLA   Daughter of Maui king Kekaulike and wife of Hawaiʻi king Kalaniʻōpuʻu, by whom she bore the royal heir Kīwalaʻō. She was also married to

Kalaniʻōpuʻu's half-brother Keōua Kupuāikalaninui, by whom she bore the high-ranking chiefess Kekuʻiapoiwa Liliha, mother of Keōpūolani.

KAMAHUALELE   A famous navigator *(kilo)* who guided Moʻikeha, Kila, and Laʻamaikahiki on their voyages to and from Kahiki.

KAMALĀLĀWALU   King of Maui in the early seventeenth century who attempted an ill-fated conquest of Hawaiʻi Island.

KAMALUOHUA   The king of Maui (or possibly only of the western part of Maui) about the early fifteenth century.

KAMEʻEIAMOKU   A chief of Kona, Hawaiʻi Island, who attacked the *Fair American* in 1790.

KAMEHAMEHA   Also known as Kamehameha I and Kamehameha the Great. Son of Keōua Kupuāikalaninui and nephew of Hawaiʻi Island king Kalaniʻōpuʻu, who gave Kamehameha control over the war god Kūkāʻilimoku. Kamehameha became ruling chief of the northwestern part of Hawaiʻi Island after defeating Kīwalaʻō, and later king over the entire island after the death of Keōua Kuahuʻula.

KAMEHAMEHA-NUI   Son of Maui king Kekaulike by his half-sister Kekuʻiapoiwanui, and nephew of Hawaiʻi Island king Alapaʻinui. Kamehameha-nui succeeded Kekaulike as king of Maui Island in the mid eighteenth century.

KANAHAOKALANI   Son of the defeated Oʻahu king Kapiʻiohokalani, only six years old when his father was killed in a war against Alapaʻinui on Molokaʻi.

KANALOA-KUAʻANA   Son of Keawenui and a ruling chief of Kona district, Hawaiʻi Island in the seventeenth century. He deposed his half-brother ʻUmiokalani to gain control over Kohala district.

KAʻOPULUPULU   High priest of Oʻahu Island under king Kahahana.

KAPIʻIOHOKALANI   King of Oʻahu Island in the mid eighteenth century. He was killed on Molokaʻi by the forces of Alapaʻinui, king of Hawaiʻi Island.

KAPOUKAHI   A famous prophet who instructed Kamehameha to build the temple of Puʻukoholā at Kawaihae, Hawaiʻi Island.

KAPUKINI-A-LĪLOA   Half-sister and royal wife of ʻUmi, king of Hawaiʻi Island. Kapukini-a-Līloa bore three sons: Keliʻiokaloa, Kapulani, and Keawenui.

KAUHIʻAIMOKU-A-KAMA   Son of Maui king Kekaulike who attempted to rebel against Kamehameha-nui. He was captured and killed by the forces of Kamehameha-nui's uncle Alapaʻinui, king of Hawaiʻi Island.

KAUHI-A-KAMA   Son of Maui king Kamalālāwalu who later succeeded his father as ruler of Maui.

KAUIKEAOULI   Younger son of Kamehameha I and brother of Liholiho, he became Kamehameha III after the death of his brother in 1825. He died December 15, 1854.

KAʻULAHEA   Ruling chief of western Maui who made a visit to Oʻahu in the time of Māʻilikūkahi, and who probably introduced the *ahupuaʻa* system to Maui.

KAUMUALIʻI   Last king of Kauaʻi Island. He was born about 1780 and died in 1824.

KAWAOKAʻŌHELE   Ruling chief of Maui Island in the sixteenth century and father of King Piʻilani.

KEALIʻIMAIKAʻI   Younger brother of Kamehameha I.

KEAWEʻIKEKAHIALIʻIOKAMOKU   A ruler of Hawaiʻi Island in the seventeenth century, noted for building the Hale-o-Keawe burial temple at Hōnaunau.

KEAWEMAUHILI   Half-brother of Hawaiʻi Island king Kalaniʻōpuʻu and an influential warrior chief.

KEAWENUI   Son of ʻUmi, king of Hawaiʻi. He became ruling chief of the Hilo region after the death of ʻUmi, and later deposed his brother Keliʻikaloa to become king of the entire island of Hawaiʻi in the early seventeenth century.

KEAWEʻOPALA   Son of Hawaiʻi Island king Alapaʻinui who briefly succeeded his father but was overthrown by Kalaniʻōpuʻu.

KEʻEAUMOKU   Powerful Hawaiʻi Island warrior chief from Kona district and supporter of Hawaiʻi Island king Kamehameha.

KEKAULIKE   King of Maui in the early eighteenth century. His royal residence was in Kaupō district, from which he launched a failed attempted to conquer Hawaiʻi Island. Among his descendants were the high-ranking chiefesses Kalola, Kekuʻiapoiwa Liliha, and Keōpūolani.

KEKŪHAUPIʻO   Famous warrior chief of Hawaiʻi Island, mentor and confidant to Kamehameha.

KEKUʻIAPOIWA   A high chiefess, wife of Keōua Kupuāikalaninui and mother of Kamehameha I.

KELIʻIOKALOA   Son of ʻUmi, king of Hawaiʻi, who became ruling chief of Kona district after the death of his father. Keliʻiokaloa was said to be an oppressive ruler, and was deposed by his brother Keawenui.

KEŌPŪOLANI   Sacred daughter of Kīwalaʻō and his half-sister Liliha Kekuʻiapoiwa, she was taken by Kamehameha to be his royal wife, bearing him his heirs Liholiho (Kamehameha II) and Kauikeaouli (Kamehameha III).

KEŌUA KUAHUʻULA   Younger son of Hawaiʻi Island king Kalaniʻōpuʻu, he became ruling chief over the southeastern part of Hawaiʻi Island after the death of his half-brother Kīwalaʻō. Keōua was killed and his body offered up in sacrifice on Kamehameha's war temple of Puʻukoholā in 1791.

KEŌUA KUPUĀIKALANINUI   Half-brother of Hawaiʻi Island king Kalaniʻōpuʻu, and father of Hawaiʻi Island king Kamehameha I.

KIHA-A-PIʻILANI   Son of Maui king Piʻilani; after the death of his father, he deposed his brother Lono-a-Piʻilani. He was the brother-in-law of Hawaiʻi Island king ʻUmi.

KILA   Third-born son of Moʻikeha and Hoʻoipoikamalanai, Kila voyaged to Kahiki to find Laʻamaikahiki and bring him to Kauaʻi to see Moʻikeha.

KĪWALAʻŌ   Oldest son of Kalaniʻōpuʻu by Kalola. Briefly succeeded Kalaniʻōpuʻu as king of Hawaiʻi Island before being killed by Keʻeaumoku at the battle of Mokuʻōhai in 1782. His daughter Keōpūolani became the sacred wife of Kamehameha I.

KOʻAʻA   Chiefly attendant of Hawaiʻi king Kalaniʻōpuʻu who greeted Captain James Cook at Kealakekua Bay in 1779.

KOʻI   Friend and companion of Hawaiʻi Island king ʻUmi, and later ruling chief of Kohala district.

KOLEAMOKU   Daughter of Maui chief Hoʻolaemakua and an early wife of Kiha-a-Piʻilani.

KŪALIʻI   King of Oʻahu Island in the late seventeenth to early eighteenth centuries.

KUMAHANA   Son of Kauaʻi and Oʻahu king Peleiʻōhōlani who briefly succeeded his father.

KUMUHONUA   First-born son of Muliʻelealiʻi, and ruling chief of Oʻahu in the later fourteenth century.

LAʻAMAIKAHIKI   A famous voyaging chief of the fourteenth century, Laʻamaikahiki was born in Kahiki and sailed with Kila to Kauaʻi to see his father, Moʻikeha. After spending some time in Hawaiʻi he returned to Kahiki.

LAʻIELOHELOHEIKAWAI   An Oʻahu high-ranking chiefess and wife of Maui king Piʻilani; mother of Kiha-a-Piʻilani and Lono-a-Piʻilani.

LANIKĀULA   Famous sorcerer priest who, among other things, warned Maui king Kamalālāwalu not to attempt to conquer Hawaiʻi Island.

LIHOLIHO   Son of Kamehameha, he succeeded to the kingship as Kamehameha II in 1819 after the death of his father. Liholiho died in London on July 14, 1824.

LILIHA KEKUʻIAPOIWA   Daughter of Keōua Kupuāikalaninui and Kalola, and hence half-sister to Kamehameha. She married Kīwalaʻō, by whom she bore the sacred chiefess Keōpūolani, who become the wife of Kamehameha I.

LILIʻUOKALANI   Born in 1838 as Lydia Kamakaʻeha, Queen Liliʻuokalani was the sister of King Kalākaua, and ruled from January 29, 1891, until being deposed on January 17, 1893. She died on November 11, 1917.

LĪLOA   Ruling chief of Hawaiʻi Island in the early sixteenth century, and descendant of Pilikaʻaiea. Līloa was the father of Hākau by his mother's sister Pinea, and the father of ʻUmi by the commoner Akahiakuleana.

LONO-A-PIʻILANI   Older son of Maui king Piʻilani, Lono inherited the kinship from his father in the early seventeenth century but was later deposed by his brother Kiha-a-Piʻilani.

LONOIKAMAKAHIKI   Son of Keawenui and one of the most famous Hawaiʻi Island kings, in the early seventeenth century. He made a lengthy voyage throughout the archipelago, and later repulsed an attempted invasion of Hawaiʻi by Maui king Kamalālāwalu.

LUʻUKIA   Granddaughter of Hikapoloa, ruling chief of Kohala, and wife of ʻOlopana. Luʻukia voyaged with ʻOlopana and Moʻikeha to Kahiki, and ended her days there.

MAHIOLOLE   A famous warrior chief of Kohala district and father of Hawaiʻi Island king Alapaʻinui.

MĀʻILIKŪKAHI   Descendant of Māweke and king of Oʻahu in the late fifteenth century. Māʻilikūkahi is famous for building up Oʻahu's irrigation works, and for innovating the *ahupuaʻa* system of land divisions.

MĀWEKE   Ruling chief of Oʻahu in the early fourteenth century and descendant of the Nānāʻulu line of chiefs. He had three sons, Muliʻelealiʻi, Keaunui, and Kalehenui.

MOʻIKEHA   A descendant of Māweke and third-born son of Muliʻelealiʻi, Moʻikeha was a famous voyaging chief of the fourteenth century who sailed to Kahiki with his wife Luʻukia, but later returned and ended his days on Kauaʻi.

MULIʻELEALIʻI   First-born son of Māweke and chief of Kona district on Oʻahu in the fourteenth century. His three sons were were Kumuhonua, ʻOlopana, and Moʻikeha.

NUNU   A priest of Waipiʻo who aided ʻUmi in overthrowing Hākau.

ʻOLOPANA   Second-born son of Muliʻelealiʻi, ʻOlopana attempted to usurp the Oʻahu chiefship from his older brother, Kumuhonua. He was forced to flee to Waipiʻo on Hawaiʻi and eventually voyaged to Kahiki, where he ended his days.

ʻOMAʻOKAMAU   Friend and companion of the Hawaiʻi Island king ʻUmi, and later ruling chief of Kaʻū district.

PĀʻAO   A voyaging priest of the fourteenth century, Pāʻao is credited with bringing the chief Pilikaʻaiea from Kahiki and installing him as the *aliʻi nui* of Hawaiʻi Island. Pāʻao is also said to have introduced the practice of human sacrifice.

PELEIʻŌHŌLANI   King of Kauaʻi in the late eighteenth century and brother of Oʻahu king Kapiʻiohokalani. After Kapiʻiohokalani was slain on Molokaʻi, Peleiʻōhōlani also became king over Oʻahu Island.

PIʻIKEA   Daughter of Maui king Piʻilani and wife of Hawaiʻi Island king ʻUmi.

PIʻILANI   King of Maui Island in the mid sixteenth century. Father of Lono-a-Piʻilani and Kiha-a-Piʻilani, and of Piʻikea, the wife of ʻUmi.

PIʻIMAIWAʻA   Friend and companion of ʻUmi, who became one of ʻUmi's most trusted warriors and ruling chief of Hāmākua district.

PILIKAʻAIEA   A chief who was brought by Pāʻao from Kahiki and installed as the *aliʻi nui* of Hawaiʻi Island, around the fourteenth century.

PINEA   The royal wife of Līloa, king of Hawaiʻi.

PUNA   Ruling chief of Kauaʻi in the fourteenth century, and father of Hoʻoipoikamalanai and Hinauʻu, who married Moʻikeha after his return from Kahiki.

ʻUMI   Son of the Hawaiʻi ruling chief Līloa by Akahiakuleana, and half-brother of Hākau. ʻUmi usurped the kingship from Hākau and unified the entire island of Hawaiʻi during the late sixteenth century.

ʻUMIOKALANI   Son of Keawenui and ruling chief of Kohala district, Hawaiʻi Island, in the seventeenth century. He was deposed by his half-brother Kanaloa-kuaʻana, the ruling chief of Kona district.

GLOSSARY OF HAWAIIAN WORDS

'aha ali'i   A council of chiefs.

ahu   A stone mound, cairn, or altar.

ahupua'a   A territorial land unit under the control of a subchief (the *ali'i 'ai ahupua'a*). Such units typically ran from the mountains to the sea.

'āina   Land.

akua   God, spirit, deity.

ali'i   Elite, member of the chiefly class.

ali'i akua   Literally, "god-king." An expression reserved for the highest-ranking *ali'i*, especially those of *pi'o* rank.

ali'i nui   Literally, "great chief." The highest-ranked chief within a polity; king.

'anā'anā   Sorcery, black magic, the practice of praying one's enemy to death.

'anu'u   A tower of wooden poles, wrapped in white barkcloth, which stood on certain kinds of temples *(heiau)*.

'aumakua   Ancestral deities; literally, "collective parents."

'awa   Kava, *Piper methysticum*, the root of which was used to prepare a psychoactive (but nonalcoholic) beverage consumed in substantial quantity by the elites.

haku'āina   A landlord, typically a *konohiki* or *ahupua'a*-level chief.

haku'ōhi'a   The main *'ōhi'a* wood image in a temple; rites for the consecration of the image.

hala The pandanus or screwpine tree *(Pandanus tectorius)* whose leaves were used to weave mats, baskets, and other artifacts.

hale House, general term for a thatched structure.

hale nauā The house in which genealogical specialists gathered to question chiefs regarding their pedigrees, and to ascertain whether they would be allowed to join the royal household.

hale noa The main dwelling house in a residential complex, in which both sexes could mingle freely (*noa* = free from taboo).

hale o Lono Temple dedicated to the god Lono (literally, "house of Lono").

hānai Adoption, an adopted child.

hānaipū A rite in which the Makahiki god is fed.

heiau The general term for a Hawaiian ritual place, where sacrifices of any kind were offered to the gods or to ancestral spirits.

heiau hoʻoulu ʻai Agricultural temple (literally, "temple to increase food").

honi The traditional Hawaiian greeting of touching of noses and sharing of breath.

hoʻokupu Tribute (literally, "to cause to grow or increase").

hula Various forms of traditional dance.

ʻieʻie An indigenous climbing shrub of the pandanus family *(Freycinetia arborea)*, the tough fibers of which were used to weave the foundations for feathered garments.

ilāmuku Executive officer, adjutant, and executioner; these individuals were responsible for maintaining the *kapu* associated with the household of a high chief or king.

ʻili A land unit, smaller in size than an *ahupuaʻa*.

ʻili kūpono A special kind of *ʻili* segment within an *ahupuaʻa*, under the direct control of the king rather than the *ahupuaʻa* chief, and whose tribute was reserved for the king.

imu The traditional Hawaiian earth oven.

kāʻai A sennit casket in the shape of a human torso and head, woven to contain the bones of certain high-ranking chiefs and kings.

kāhili Feather standard of the *aliʻi*, typically a wooden shaft (sometimes with a bone handle) ornamented with birds' feathers of red, yellow, and other colors.

kahu An attendant, guardian, or keeper.

kahuna Priest (plural *kāhuna*).

kahuna nui High priest, usually of the Kū cult, who officiated at ceremonies in the king's temple.

kāhuna pule Priests who officiated in formal temple ceremonies, especially of the Kū, Lono, and Kāne cults.

kālaimoku Counselor to the king; sometimes translated as "prime minister."

kalo Taro, *Colocasia esculenta*.

kamaʻāina Someone born to a particular locale; literally, "child of the land."

Kanaloa One of the four major gods of the Hawaiian pantheon.

Kāne A principal god of the Hawaiian pantheon; the creator god, also deity of flowing waters and irrigation, to whom taro was sacred.

kapa Barkcloth.

kapu Sacred; prohibited; forbidden.

kapu moe The prostrating taboo, obligatory before the highest-ranked chiefs.

kaukau aliʻi Lesser grades of chiefs.

kāula A prophet or seer, one who has the ability to transmit oracles.

kauwā (or *kauā*) Member of the lowest social class, sometimes referred to as "outcasts" or "slaves." *Kauwā* were suitable for offerings of human sacrifice at the *luakini* temples.

kī The ti plant, *Cordyline fruticosa*.

kilo Stargazer, reader of omens, navigator. Sometimes also called *kilo lani*.

koa Warrior; also the name of a prominent hardwood tree native to the islands (*Acacia koa*).

koʻa A fishing shrine, dedicated to the god Kūʻula.

kōʻele An area of pondfields or a garden segment worked by the common people on behalf of the *konohiki* and the chief; the production of the *kōʻele* was reserved for the use of the *konohiki* and *aliʻi*.

konohiki Land manager for an *ahupuaʻa* land unit, representing the *aliʻi ʻai ahupuaʻa* chief.

Kū One of the four principal gods of the Hawaiian pantheon; god of war.

kuaʻāina Countryside, backcountry (literally, "back [of the] land").

kuaiwi Literally, "backbone." A stone wall or alignment running up and down a slope, demarking divisions within a dryland field system.

kūʻauhau A specialist within the royal court, responsible for memorizing the genealogies and oral traditions of the *aliʻi*.

kula Dryland cultivation areas; in Mahele land claims, *kula* lands are typically distinguished from irrigated taro lands (*loʻi*).

kupuna Elder, ancestor (plural *kūpuna*).

Kūʻula   The god of fishermen, typically represented by an elongated upright stone at coastal fishing shrines.

lei niho palaoa   A special neck ornament worn by chiefs, consisting of a tongue-shaped pendant suspended by braids of human hair.

lele   The sacrificial altar on a temple.

loʻi   An irrigated pondfield for the cultivation of taro.

loko kuapa   A fishpond formed by constructing a stone wall out onto the reef flat.

Lono   One of the four principal gods of the Hawaiian pantheon; god of dryland agriculture, to whom the sweet potato was sacred.

loulu   General name applied to several species of native fan palms (*Pritchardia* spp.).

lua   A pit, such as the pit for disposing of sacrificial offerings on a temple.

luakini   A state temple dedicated to the god Kū, at which human sacrifice was performed.

luna   High or above; also applied to a foreman, overseer, or headman.

mahiole   Feathered helmet.

maika   A game in which a special bowling stone was rolled along a pitch with two upright sticks at the end.

makaʻāinana   Commoner.

Makahiki   A four-month period that commenced with the first visibility of Pleiades in November; this was the period of tribute collection, sacred to Lono, when war was forbidden.

mana   Supernatural or divine power, efficacy.

mele   Song or chant.

moku   A political district.

mokupuni   Island.

moʻo   A land unit, smaller in size than an *ʻili*.

moʻokūʻauhau   Genealogy or pedigree.

moʻolelo   Oral tradition, history.

mua   The men's eating house. The *mua* of a king was where he would hold court with other high-ranking males and advisers.

naha   One of the high ranks of chiefs.

nā kanaka   Collective term for the common people (see also *makaʻāinana*).

nā liʻi   Collective term for those of the chiefly class (see also *aliʻi*).

nī'aupi'o Offspring of a royal incestuous marriage between half-brother and sister, or full brother and sister (literally, "recurved coconut midrib").

noa Free, without *kapu*, the opposite of *kapu*.

'ōhi'a A common tree of the Hawaiian forests *(Metrosideros polymorpha)* whose trunks were used to carve temple images.

pahu A particular kind of cylindrical drum, with a tympanum of shark's skin, used during *luakini* temple ceremonies.

pahupū Cut in half; the name was given to a special cadre of Maui warriors, fiercely loyal to Kahekili, who were tattooed completely black on one half of their bodies.

pānānā Literally, "sighting wall." In the nineteenth century the term was applied to the Western mariner's compass, but apparently originally referred to walls or other constructions used to indicate cardinal directions.

pīkoi A tripping club, of wood or stone, with a cord attached.

pi'o Marriage between a brother and sister, or between half-siblings; the term also denotes the offspring of such a union. *Pi'o* unions were the exclusive privilege of the highest-ranked *ali'i*.

pōhaku Stone, rock.

pono Good, correct, just, morally upright.

pūlo'ulo'u A barkcloth-covered ball on a shaft, either carried before the king or set up in front of his residence, as a mark of taboo.

punahele A favorite, often referring to a grandchild selected by a grandparent as one to whom to pass down family knowledge.

punalua Spouses sharing a spouse, as in two brothers sharing a wife.

pu'u A hill or cinder cone.

pu'uhonua A place of refuge, where taboo breakers or defeated warriors could seek protection.

'uala Sweet potato, *Ipomoea batatas*.

'uhane Soul, spirit, or ghost.

waiwai Wealth, goods, property (literally, "water" reduplicated).

wohi A rank of chief that was exempt from the prostrating *kapu*.

SOURCES AND
FURTHER READING

The following paragraphs acknowledge, on a chapter-by-chapter basis, the primary sources that I have drawn from in this account, including references to specific quotations in the text. These notes also provide a guide to further reading for those who wish to pursue a particular topic in greater depth.

PROLOGUE: ISLANDS OUT OF TIME

Cook's arrival at Kaua'i, and his observations on the Hawaiian people, may be found in *The Journals of Captain James Cook: The Voyage of the* Resolution *and* Discovery, *1776–1780*, edited by J. C. Beaglehole (Cambridge: Hakluyt Society, Cambridge University Press, 1967); quoted passages are from pp. ccxxi, 264 n. 1, 265, 269, 279. The Native Hawaiian version of first contact is given by Samuel Kamakau in *Ruling Chiefs of Hawaii* (Honolulu: Kamehameha Schools Press, 1961); the quoted passage is from p. 92. The Hawaiian proverb of the chief as shark is from Mary Kawena Pukui, *'Ōlelo No'eau: Hawaiian Proverbs and Poetical Sayings*, Bishop Museum Special Publication, no. 71 (Honolulu: Bishop Museum Press, 1983), p. 87. The lines of the chant are from Abraham Fornander, *Fornander Collection of Hawaiian Antiquities and Folk-Lore*, Bishop Museum Memoirs, vol. 6, pt. 3 (Honolulu: Bishop Museum Press, 1920), pp. 393–94. On the Russian fort and the fascinating history of the Russian incursion on Kaua'i, see Peter R. Mills,

Hawai'i's Russian Adventure: A New Look at Old History (Honolulu: University of Hawai'i Press, 2002). A discussion of archaic states and the criteria used to define them may be found in Gary M. Feinman and Joyce Marcus, eds., *Archaic States* (Santa Fe, N.M.: School of American Research, 1998).

CHAPTER ONE: A TRAIL OF TATTOOED POTS

The story of Edward Gifford and Richard Shutler's 1952 expedition to New Caledonia is told by Christophe Sand and Patrick V. Kirch in *L'expédition archéologique d'Edward Gifford et Richard Shutler Jr. en Nouvelle-Calédonie au cours de l'année 1952*, Les Cahiers de l'Archéologie en Nouvelle-Calédonie, vol. 13 (Noumea: Service des Musées du Patrimoine de Nouvelle-Calédonie, 2002). The history of research in Lapita archaeology is recounted in Patrick V. Kirch, *The Lapita Peoples: Ancestors of the Oceanic World* (Oxford: Blackwell, 1997). This book also provides a synthesis of information on the Lapita Cultural Complex, including the ceramics, subsistence, and social organization and exchange, much of it based on the results from Mussau. The Lapita Homeland Project results are reported in Jim Allen and Chris Gosden, eds., *Report of the Lapita Homeland Project*, Occasional Papers in Prehistory, no. 20 (Canberra: Department of Prehistory, The Australian National University, 1991). The Austronesian background to Lapita is best summarized by Peter Bellwood in *The Prehistory of the Indo-Malaysian Archipelago* (Sydney: Academic Press, 1985). K. C. Chang's excavations at Tap'enk'eng were reported in *Fengpitou, Tapenkeng, and the Prehistory of Taiwan*, Yale University Publications in Anthropology, no. 73 (New Haven, Conn.: Department of Anthropology, 1969). The early dugout canoes at the site of Kuahuqiao, as well as other early Austronesian remains in South China, are described in Tianlong Jiao, ed., *Lost Maritime Cultures: China and the Pacific* (Honolulu: Bishop Museum Press, 2007). My 1985 and subsequent discoveries in Mussau, at Talepakemalai and other sites, have been summarized in my 1997 book *The Lapita Peoples*, and described in more detail in Patrick V. Kirch, ed., *Lapita and Its Transformations in Near Oceania*, Archaeological Research Facility Contribution no. 59 (Berkeley: University of California, 2001). The archaeological investigation of Niuatoputapu, including its Lapita site of Lolokoka, is reported in Patrick V. Kirch, *Niuatoputapu: The Prehistory of a Polynesian Chiefdom* (Seattle: Burke Museum, 1988). An account of the August 1, 2002, ceremony at Koné, along with photos of the event, is provided by Christophe Sand in *Pacific Archaeology: Assessments and Prospects*, Les Cahiers de l'Archéologie en Nouvelle-Calédonie, vol. 15 (Noumea: Service des Musées du Patrimoine de Nouvelle-Calédonie, 2003).

CHAPTER TWO: EAST FROM HAWAIKI

The Samoan myth of Tangaroa is recounted in various sources, including Margaret Mead, *Social Organization of Manua*, Bishop Museum Bulletin 76 (Honolulu: Bishop Museum Press, 1930). Our archaeological investigations at the To'aga site on 'Ofu Island are documented in P. V. Kirch and T. L. Hunt, eds., *The To'aga Site: Three Millennia of Polynesian Occupation in the Manu'a Islands, American Samoa*, Contributions of the Archaeological Research Facility, no. 51 (Berkeley: University of California, 1993). The world of Hawaiki and ancestral Polynesia is extensively reconstructed in Patrick V. Kirch and Roger C. Green, *Hawaiki, Ancestral Polynesia: An Essay in Historical Anthropology* (Cambridge: Cambridge University Press, 2001). For details of Proto-Polynesian language, one can also consult Jeff Marck, *Topics in Polynesian Language and Culture History*, Pacific Linguistics, no. 504 (Canberra: Australian National University, 2000). Robert Suggs's discoveries in the Marquesas are vividly recounted in *The Hidden Worlds of Polynesia* (New York: Harcourt, Brace, and World, 1962). A scientific report on his research is Robert Carl Suggs, *The Archaeology of Nuku Hiva, Marquesas Islands, French Polynesia*, American Museum of Natural History Anthropological Papers, vol. 49, pt. 1 (New York: American Museum of Natural History, 1961). For a review of the most recent research in Eastern Polynesia, see Patrick V. Kirch and Jennifer Kahn, "Advances in Polynesian Prehistory: A Review and Assessment of the Past Decade (1993–2004)," *Journal of Archaeological Research*, 15 (2007): 191–238. The early sweet potato finds from Mangaia Island are described in Jon Hather and Patrick V. Kirch, "Prehistoric Sweet Potato *(Ipomoea batatas)* from Mangaia Island, Central Polynesia," *Antiquity* 65 (1991): 887–93. For a recent synthesis of archaeological, linguistic, and human biological evidence for the broad patterns of human dispersal across the Pacific, see Patrick V. Kirch, "The Peopling of the Pacific: A Holistic Anthropological Perspective," *Annual Review of Anthropology* 39 (2010): 131–48.

CHAPTER THREE: FOLLOW THE GOLDEN PLOVER

My account of the preparations for and undertaking of the first voyage to Hawai'i involve a certain degree of narrative license. However, they are thoroughly grounded in the archaeological and ethnographic evidence indicating what such a voyage must have entailed. The Polynesian words and names I have used in this chapter are based on reconstructions of Proto–Eastern Polynesian language. An excellent treatise on Polynesian and Pacific voyaging and voyaging canoes is *Vaka Moana: Voyages of the Ancestors*, ed. K. R. Howe (Auckland: David Bateman Ltd., 2006). An older but nonetheless indispensable work on Oceanic canoes is by A. C. Haddon and James Hornell, *Canoes of Oceania*, Bernice P. Bishop Museum Special Publications, nos. 27, 28, and 29 (Honolulu: Bishop Museum Press, 1936, repr. 1975). The Pacific golden plover *(Pluvialis fulva)* and its migratory habits are described by Anne Gouni and Thierry Zysman in *Oiseaux du Fenua* (Papeʻete: Tethys Editions, 2007). The canoe-hauling chant from Hao atoll is from Kenneth P. Emory, *Material Culture of the Tuamotu Atoll* (Honolulu: Bishop Museum Press, 1975), p. 138; I have slightly modified the translation to better render the Polynesian syntax. The quote from Captain James Cook is from J. C. Beaglehole, ed., *The Journals of Captain James Cook on His Voyages of Discovery*, vol. 1: *The Voyage of the* Endeavor *1768–1771*, Hakluyt Society Extra Series no. 34 (Cambridge: Cambridge University Press, 1955), p. 154. The story of Cook and Tupaia has been expertly recounted by Anne Salmond, *Aphrodite's Island: The European Discovery of Tahiti* (Berkeley: University of California Press, 2009). Abraham Fornander's classic work on Polynesian traditions is his three-volume *An Account of the Polynesian Race: Its Origins and Migrations* (London: Trübner, 1878–85). S. Percy Smith's theory of Polynesian migrations is put forward in *Hawaiki: The Original Home of the Maori; With a Sketch of Polynesian History*, 4th ed. (Auckland: Whitcombe & Tombs, Ltd., 1921). Andrew Sharp presented his thesis of drift voyaging in *Ancient Voyagers in the Pacific*, Polynesian Society Memoir, no. 32 (Wellington: The Polynesian Society, 1956). Te Rangi Hiroa's synthesis of Polynesian history was published under his European name, Peter H. Buck, *Vikings of the Sunrise* (Philadelphia: J. B. Lippincott, 1938). The quote from Thor Heyerdahl is from his book *Fatu-Hiva: Back to Nature* (New York: Doubleday, 1975), p. 217. An excellent account of how Thor Heyerdahl developed his theory of South American origins of the Polynesians, and of the *Kon-Tiki* voyage, is provided by Axel Andersson, *A Hero for the Atomic Age: Thor Heyerdahl and the* Kon-Tiki *Expedition* (Oxford: Peter Lang, 2010). I have based my narrative account of the voyage to Hawaiʻi to some degree on the actual experiences of the *Hōkūleʻa* on her experimental voyages. These are recounted in Ben Finney's *Voyage of Rediscovery: A Cultural Odyssey through Polynesia* (Berkeley: University of California Press, 1994). Other important sources on Polynesian voyaging and wayfinding include David Lewis, *We, the Navigators* (Honolulu: University of Hawai'i Press, 1972); and Geoff Irwin, *The Prehistoric Exploration and Colonization of the Pacific* (Cambridge: Cambridge University Press, 1992).

CHAPTER FOUR: VOYAGES INTO THE PAST

The story of Abraham Fornander is told by Eleanor Harmon Davis in *Abraham Fornander: A Biography* (Honolulu: University Hawai'i Press, 1979). Roger Rose deals with the founding and early years of the Bishop Museum in *A Museum to Instruct and Delight* (Honolulu: Bishop Museum Press, 1980). John Stokes's career is discussed by Tom Dye in "Tales of Two Cultures: Traditional Historical and Archaeological Interpretations of Hawaiian Prehistory," *Bishop Museum Occasional Papers*, 29 (1989): 3–22, as well as in Dye's introduction to his edition of John F. J. Stokes, *Heiau of the Island of Hawai'i*, Bishop Museum Bulletin in Anthropology, no. 2 (Honolulu: Bishop Museum Press, 1991). Edward S. C. Handy laid out his theory of Polynesian origins in "The Problem of Polynesian Origins," *Bishop Museum Occasional Papers*, 9 (1930): 1–27. Te Rangi Hiroa's correspondence with Apirana Ngata was edited by M. P. K. Sorrenson, *Na To Hoa Aroha, From Your Dear Friend: The Correspondence between Sir Apirana Ngata and Sir Peter Buck, 1925–50*, 3 vols. (Auckland: Auckland University Press, 1986–88). A biography of Kenneth Emory, by his longtime friend Honolulu newspaper columnist Bob Krauss, is *Keneti: South Seas Adventures of Kenneth Emory* (Honolulu: University of Hawai'i Press, 1988); quoted passage is from p. 338. A historical overview of Pacific archaeology, including the theories of Te Rangi Hiroa and the emergence of modern archaeology after World War II, can be found in chapter 1 of Patrick V. Kirch, *On the Road of the Winds: An Archaeological History of the Pacific Islands before European Contact* (Berkeley: University of California Press, 2000). The Hawaiian fishhook analysis was published by Kenneth P. Emory, William J. Bonk, and Yosihiko H. Sinoto, *Hawaiian Archaeology: Fishhooks*, Bernice P. Bishop Museum Special Publication, no. 47 (Honolulu: Bishop Museum Press, 1959). The Hālawa Valley Project from 1969 to 1970 produced important finds regarding the prehistory of Moloka'i, as described in Patrick V. Kirch and Marion Kelly, eds., *Prehistory and Ecology in a Windward Hawaiian Valley: Halawa Valley, Molokai*, Pacific Anthropological Records, no. 24 (Honolulu: Bishop Museum Press, 1975).

CHAPTER FIVE: THE SANDS OF WAIMĀNALO

For a thorough account of Hawaiian geology and landscape formation, see Gordon A. Macdonald and Agatin T. Abbott, *Volcanoes in the Sea: The Geology of Hawaii* (Honolulu: University of Hawai'i Press, 1970). A comprehensive review of Hawaiian natural history is also provided by Alan C. Ziegler in his *Hawaiian Natural History, Ecology, and Evolution* (Honolulu: University of Hawai'i Press, 2002). The 1967 excavations at the Bellows dune site, O-18, were described by Richard J. Pearson, Patrick V. Kirch, and Michael Pietrusewsky, "An Early Prehistoric Site at Bellows Beach, Waimanalo, Oahu, Hawaiian Islands," *Archaeology and Physical Anthropology in Oceania*, 6 (1971): 204–34. The site is also discussed and illustrated in Patrick V. Kirch, *Feathered Gods and Fishhooks: An Introduction to Hawaiian Archaeology and Prehistory* (Honolulu: University of Hawai'i Press, 1985), pp. 69–74. The redating and interpretation of the chronology of the Bellows dune site is presented by Tom Dye and Jeffrey Pantaleo, "Age of the O18 Site, Hawai'i," *Archaeology in Oceania*, 45 (2010): 113–19. The question of when Hawai'i was first settled by Polynesians is discussed by Patrick V. Kirch in "When Did the Polynesians Settle Hawai'i: A Review of 150 Years of Scholarly Inquiry and a Tentative Answer," *Hawaiian Archaeology*, 12 (2011): 3–26.

CHAPTER SIX: FLIGHTLESS DUCKS AND PALM FORESTS

The coring of 'Ukoa and Kawainui marshes and results of the pollen analysis of their sediments are summarized by J. Stephen Athens, "Hawaiian Native Lowland Vegetation in Prehistory," in Patrick V. Kirch and T. L. Hunt, eds., *Historical Ecology in the Pacific Islands* (Honolulu: University of Hawai'i Press, 1997), pp. 248–70. Additional results of the pollen analysis may be found in J. Stephen Athens and Jerome V. Ward, "Environmental Change and Prehistoric Polynesian Settlement in Hawai'i," *Asian Perspectives*, 32 (1993): 205–23. The initial finds of fossilized birds in Hawai'i were described by Storrs L. Olson and Helen F. James in *Prodromus of the Fossil Avifauna of the Hawaiian Islands*, Smithsonian Contributions to Zoology, no. 365 (Washington, D.C.: Smithsonian Institution, 1984). More detailed accounts of the *moa nalo* and other extinct birds are given in Helen James and Storrs Olson, *Descriptions of Thirty-two New Species of Birds from the Hawaiian Islands*, Ornithological Monographs, no. 45 (Washington, D.C.: American Ornithologists Union, 1991). The study of fossilized land snails from the Barber's Point sites is described by Carl C. Christensen and Patrick V. Kirch,

"Non-Marine Mollusks and Ecological Change at Barbers Point, Oʻahu, Hawaiʻi," *Bishop Museum Occasional Papers*, 26 (1986): 52–80. Investigations of the ʻEwa Plain sinkholes, including dating of Polynesian-introduced rats, are presented by J. Stephen Athens, H. D. Tuggle, J. V. Ward, and D. J. Welch, "Avifaunal Extinctions, Vegetation Change, and Polynesian Impacts in Prehistoric Hawaiʻi," *Archaeology in Oceania*, 37 (2002): 57–78. Alan C. Ziegler provides a good summary of Hawaiian flightless birds and fossil sites in chapter 22 of his *Hawaiian Natural History, Ecology, and Evolution* (Honolulu: University of Hawaiʻi Press, 2002). The nature of Hawaiian ecosystems prior to the advent of humans, based on recent archaeological and paleontological evidence, is also summarized by John L. Culliney, *Islands in a Far Sea: The Fate of Nature in Hawaiʻi* (Honolulu: University of Hawaiʻi Press, 2006).

CHAPTER SEVEN: VOYAGING CHIEFS FROM KAHIKI

Kepelino's comments on astronomy and *pānānā* are to be found in Martha Beckwith's *Kepelino's Traditions of Hawaii*, Bishop Museum Bulletin 95 (Honolulu: Bishop Museum Press, 1932). The dictionary definitions of *pānānā* are from Mary Kawena Pukui and Samuel H. Elbert, *Hawaiian Dictionary* (Honolulu: University of Hawaiʻi Press, 1986), p. 313, and from Lorrin Andrews, *A Dictionary of the Hawaiian Language* (Honolulu: Island Heritage Publishing, 2003, facsimile of 1865 edition), p. 456. The legends of Moʻikeha, ʻOlopana, Kila, and Laʻamaikahiki are recounted in various sources; I have used a combination of them in the version given here. Abraham Fornander's *An Account of the Polynesian Race*, 3 vols. (London: Trübner, 1878–85), vol. 2, gives the main story line, but a more detailed version of parts of the tradition can be found in Abraham Fornander, *Fornander Collection of Hawaiian Antiquities and Folk-Lore*, ed. T. G. Thrum, Bernice P. Bishop Museum Memoirs, vol. 4, pt. 1 (Honolulu: Bishop Museum Press, 1916); the quoted passage is from p. 128. Other useful sources for and commentary on these traditions include Samuel Kamakau's *Tales and Traditions of the People of Old* (Honolulu: Bishop Museum Press, 1991); Martha Beckwith's *Hawaiian Mythology* (New Haven, Conn.: Yale University Press, 1940); and Te Rangi Hiroa's *Vikings of the Sunrise* (Philadelphia: J. B. Lippincott, 1938). *The Legends and Myths of Hawaiʻi*, attributed to the last king, David Kalākaua (but probably ghost-written by R. M. Daggett [repr. Honolulu: Mutual Publishing, 1990]), also gives a version of the Laʻamaikahiki tradition, including the famous triple marriage. The probable location of Moaulanuiākea as being in the Punaʻauia district of Tahiti is discussed by Edward S. C. Handy in *History and Culture in the Society Islands*, Bishop Museum Bulletin 79 (Honolulu: Bishop Museum Press, 1930). The chant "Eia Hawai'i" and Kila's speech to Lu'ukia are quoted from Abraham Fornander, *Fornander Collection of Hawaiian Antiquities and Folk-Lore*, Bernice P. Bishop Museum Memoirs, vol. 4, pt. 1, pp. 20–21 and 124. The results of XRF analysis tracing adz chips from Mangaia Island to the Tatanga-matau quarry in Samoa were published by M. I. Weisler and P. V. Kirch, "Interisland and Interarchipelago Transport of Stone Tools in Prehistoric Polynesia," *Proceedings of the National Academy of Sciences, USA*, 93 (1996): 1381–85. Adz C7727 from Napuka Island is described by Kenneth Emory in *Material Culture of the Tuamotu Archipelago* (Honolulu: Bishop Museum Press, 1975), and its chemical composition and sourcing are reported by K. D. Collerson and M. I. Weisler, "Stone Adz Compositions and the Extent of Ancient Polynesian Voyaging and Trade," *Science*, 317 (2007): 1907–11.

CHAPTER EIGHT: MĀ'ILIKŪKAHI, O'AHU'S SACRED KING

The *moʻolelo* of Mā'ilikūkahi is given in its most detailed version by Samuel Kamakau in *Tales and Traditions of the People of Old* (Honolulu: Bishop Museum Press, 1991). Abraham Fornander's *Ancient History of the Hawaiian People* (repr. Honolulu: Mutual Publishing, 1996) is an additional important source on Mā'ilikūkahi and O'ahu in the fifteenth century. Elspeth Sterling and Catherine Summers compiled information on the famous birthing place of Kūkaniloko in *Sites of O'ahu* (Honolulu: Bishop Museum Press, 1978); the site is also discussed in my book *Legacy of the Landscape*, with photographs by Thérèse Babineau (Honolulu: University of Hawai'i Press, 1996). The story of Kalaunuiohua of Hawai'i, and his attempted conquest of the islands, can be found in David Malo's *Hawaiian Antiquities*, Bishop Museum Special Publication, no. 2 (Honolulu: Bishop Museum Press, 1951), pp. 251–54. A discussion of the Hawaiian *ahupua'a* system with citations to key sources may be found in Patrick V. Kirch, *How Chiefs Became Kings: Divine Kingship and the Rise of Archaic States in Ancient Hawai'i* (Berkeley: University of California Press, 2010), pp. 47–51 and 66–69.

CHAPTER NINE: THE WATERS OF KĀNE

The importance of taro in Hawaiian culture, along with the traditional methods of farming taro and other crops in Hawai'i, is thoroughly treated in Edward S. C.

Handy's *Native Planters in Old Hawaii: Their Life, Lore, and Environment*, Bishop Museum Bulletin 233 (Honolulu: Bishop Museum Press, 1972). The "Water of Kane" *mele* is quoted from Nathaniel Emerson's *Unwritten Literature of Hawaii: The Sacred Songs of the Hula*, Bureau of American Ethnology Bulletin, no. 38 (Washington, D.C.: Smithsonian Institution, 1909), pp. 257–59; note that I have slightly modified the English translation. Our pioneering 1970 study of irrigated agricultural terraces in the upper valley of Mākaha was reported in detail in D. E. Yen, P. V. Kirch, P. Rosendahl, and T. Riley, "Prehistoric Agriculture in the Upper Valley of Makaha, Oahu," *Makaha Valley Historical Project, Interim Report no. 3*, E. J. Ladd and D. E. Yen, eds. (Honolulu: Bishop Museum Press, 1972). Jane Allen's study of the development of major irrigation works in Kaneʻohe is reported in *Five Upland ʻIli: Archaeological and Historical Investigations in the Kaneʻohe Interchange, Interstate Highway H-3, Island of Oʻahu*, Bishop Museum Anthropology Department Report, no. 87–1 (Honolulu: Bishop Museum Press, 1987). Allen also discusses key aspects of her research in "The Role of Agriculture in the Evolution of the Pre-Contact Hawaiian State," *Asian Perspectives*, 30 (1991): 117–32. Another important study of valley irrigation on Oʻahu is by Matthew Spriggs and Patrick V. Kirch, "*ʻAuwai, Kanawai,* and *Waiwai*: Irrigation in Kawailoa-Uka," in Patrick V. Kirch and Marshall Sahlins, *Anahulu: The Anthropology of History in the Kingdom of Hawaiʻi* (Chicago: University of Chicago Press, 1992), vol. 2, pp. 118–64. The Karl Wittfogel quote is from *Oriental Despotism: A Comparative Study of Total Power* (New Haven, Conn.: Yale University Press, 1957), pp. 22 and 246. Timothy Earle's book is *How Chiefs Come to Power: The Political Economy in Prehistory* (Stanford, Calif.: Stanford University Press, 1997); quoted passage from p. 82. The GIS modeling of irrigation and dryland agriculture across the Hawaiian archipelago is presented in Thegn Ladefoged, Patrick Kirch, Sam Gon III, Oliver Chadwick, Anthony Hartshorn, and Peter Vitousek, "Opportunities and Constraints for Prehistoric Intensive Agriculture in the Hawaiian Archipelago," *Journal of Archaeological Science*, 36 (2009): 2374–83, and by Natalie Kurashima and Patrick V. Kirch in "Geospatial Modeling of Precontact Hawaiian Production Systems on Molokaʻi Island, Hawaiian Islands," *Journal of Archaeological Science* 38 (2011): 3662–74.

CHAPTER TEN, "LIKE SHOALS OF FISH"

The quote from Cook, "I have no where . . . ," is from J. C. Beaglehole, ed., *The Journals of Captain James Cook: The Voyage of the* Resolution *and* Discovery,

1776–1780 (Cambridge: Hakluyt Society, Cambridge University Press, 1967), pp. 490–91. Lieutenant King's initial population estimate is from the same source, pp. 619–20. His later, revised estimate was published in the official Admiralty account of the voyage. On the effect of introduced diseases on the Hawaiian population, see O. A. Bushnell, *The Gifts of Civilization: Germs and Genocide in Hawai'i* (Honolulu: University of Hawai'i Press, 1993). Robert Schmitt's lower estimate for initial Hawaiian population, along with George Youngson's estimate and the missionary census of 1831–32, is presented in Schmitt's book, *Demographic Statistics of Hawaii, 1778–1965* (Honolulu: University of Hawai'i Press, 1968). The challenge to Schmitt and the historical demographers came from David E. Stannard, *Before the Horror: The Population of Hawai'i on the Eve of Western Contact* (Honolulu: Social Science Research Institute, University of Hawai'i, 1989); the quoted passage is from p. 143. The Malthus quote is from *An Essay on the Principle of Population* (London: J. Johnson, 1798), p. 61. Theories of Malthus, and of Verhulst, and the logistic equation and its variants are well described in G. Evelyn Hutchinson's *An Introduction to Population Ecology* (New Haven, Conn.: Yale University Press, 1978). The theory of island colonization is presented in Robert H. MacArthur and Edward O. Wilson, *The Theory of Island Biogeography*, Monographs in Population Biology, 1 (Princeton, N.J.: Princeton University Press, 1967). Raymond Firth's observations on population control on Tikopia are presented in chapter 12 of his book *We, the Tikopia* (New York: American Book Company, 1936). Ester Boserup's book on innovation and agricultural change is *The Conditions of Agricultural Growth: The Economics of Agrarian Change under Population Pressure* (Chicago: Aldine, 1965). Archaeological efforts to track population growth and change in Hawai'i are discussed in Patrick V. Kirch, "'Like Shoals of Fish': Archaeology and Population in Pre-Contact Hawai'i," in P. V. Kirch and J.-L. Rallu, eds., *The Growth and Collapse of Pacific Island Societies* (Honolulu: University of Hawai'i Press, 2007), pp. 52–69. My ideas on demographic change in Polynesian societies, including a detailed discussion of the evidence from Hawai'i, were presented in chapter 5 of Patrick V. Kirch, *The Evolution of the Polynesian Chiefdoms* (Cambridge: Cambridge University Press, 1984). Tom Dye and Eric Komori published their radiocarbon-date-based model of Hawaiian population in "A Pre-Censal Population History of Hawai'i," *New Zealand Journal of Archaeology*, 14 (1992): 113–28. My analysis of the population of Kahikinui on Maui is presented in Patrick V. Kirch, "Paleodemography in Kahikinui, Maui: An Archaeological Approach," in P. V. Kirch and J.-L. Rallu, eds., *The Growth and Collapse of Pacific Island Societies* (Honolulu: University of Hawai'i Press, 2007), pp. 90–107. The most recent archaeological data on ancient Hawaiian population is synthesized in Patrick V. Kirch, *How Chiefs Became Kings: Divine Kingship and the Rise of Archaic States in Ancient Hawai'i* (Berkeley: University of California Press, 2010).

CHAPTER ELEVEN: 'UMI THE UNIFIER

Charles Wilkes's description of the Ahu a 'Umi may be found in his *Narrative of the United States Exploring Expedition during the years 1838 . . . 1842* (London: Ingram, Cooke, and Co., 1842); quoted passages are from p. 142. The proposal that the Ahu a 'Umi functioned as an astronomical observatory was put forward by A. M. DaSilva and R. K. Johnson in "Ahu a 'Umi Heiau: A Native Hawaiian Astronomical and Directional Register," *Annals of the New York Academy of Sciences*, 385 (1982): 313–31. The story of 'Umi is perhaps the most famous *mo'olelo* in the Hawaiian traditional corpus; three primary sources are Abraham Fornander, *Fornander Collection of Hawaiian Antiquities and Folk-Lore*, Bernice P. Bishop Museum Memoirs, vol. 4, pt. 2 (Honolulu: Bishop Museum Press, 1917), pp. 178–235; quoted passages are from pp. 182, 184, 188, 198, 200, and 202; David Malo, *Hawaiian Antiquities*, Bernice P. Bishop Museum Special Publication, no. 2 (Honolulu: Bishop Museum Press, 1951); quoted passage is from p. 262; and Samuel Kamakau, *Ruling Chiefs of Hawai'i* (Honolulu: Kamehameha Schools Press, 1961), pp. 1–21; quoted passage is from p. 14. The history is also recounted by Abraham Fornander in *An Account of the Polynesian Race*, 3 vols. (London: Trübner, 1878–85), vol. 2, pp. 95 et seq.; by Padraic Colum in *The Bright Islands* (New Haven, Conn.: Yale University Press, 1925), pp. 27–44; and by Martha Beckwith in *Hawaiian Mythology* (Honolulu: University of Hawai'i Press, 1940), pp. 389–92. I have drawn from all of these sources for my own rendering of the 'Umi story. My description of the royal compound and temples of Waipi'o is based on Ross Cordy, *Exalted Sits the Chief* (Honolulu: Mutual Publishing, 2000).

CHAPTER TWELVE, 'UMI'S DRYLAND GARDENS

The quote from Menzies's journal is from *Hawaii Nei 128 Years Ago: Journal of Archibald Menzies* (Honolulu: W. F. Wilson, 1920), p. 52. Father Bond's remarks are from Ethel M. Damon, *Father Bond of Kohala: A Chronicle of Pioneer Life in Hawaii* (Honolulu: The Friend, 1927), pp. 159 and 207. T. Stell Newman's work on the Lapakahi field system, which includes a detailed analysis of the early ethnohistorical records, is contained in his *Hawaiian Fishing and Farming on the Island of Hawaii in A.D. 1778* (Honolulu: Division of State Parks, Department of Land and Natural Resources, State of Hawai'i, n.d. [1970]). Results of the University of Hawai'i archaeology field school at Lapakahi were summarized by H. David Tuggle and P. Bion Griffin, eds., *Lapakahi, Hawaii: Archaeological Studies*, Asian and Pacific Archaeology Series, no. 5 (Honolulu: Social Science Research Institute, University of Hawai'i, 1973). Paul Rosendahl's detailed mapping of the Lapakahi uplands, and his excavations in the small field shelters, are presented in his article "Aboriginal Hawaiian Structural Remains and Settlement Patterns in the Upland Agricultural Zone at Lapakahi, Island of Hawai'i," *Hawaiian Archaeology*, 3 (1994): 14–70. The *pīkoi* type of tripping club is described by Te Rangi Hiroa (Peter H. Buck) in *Arts and Crafts of Hawai'i*, Bishop Museum Special Publication, no. 45 (Honolulu: Bishop Museum Press, 1957), pp. 455–60. Kamalālāwalu's war against Hawai'i is described in several sources, including S. M. Kamakau, *Ruling Chiefs of Hawai'i* (Honolulu: Kamehameha Schools Press, 1961), pp. 55–61, and Abraham Fornander, *Fornander Collection of Hawaiian Antiquities and Folk-Lore*, Bishop Museum Memoirs, vol. 4, pt. 2 (Honolulu: Bishop Museum Press, 1917), pp. 338–50. The early sweet potato remains from Kohala are reported by T. N. Ladefoged, Michael W. Graves, and James H. Coil in "The Introduction of Sweet Potato in Polynesia: Early Remains in Hawai'i," *Journal of the Polynesian Society*, 114 (2005): 359–73. The radiocarbon dating of the Kohala field system is described in Thegn N. Ladefoged and Michael W. Graves, "Variable Development of Dryland Agriculture in Hawaii," *Current Anthropology*, 49 (2008): 771–802. The results of the Hawai'i Biocomplexity Project are summarized in Patrick V. Kirch, ed., *Roots of Conflict: Soils, Agriculture, and Sociopolitical Evolution in Ancient Hawai'i* (Santa Fe, N.M.: School of Advanced Research, 2010). Additional insights on Hawaiian soils and how these affected the distribution of indigenous cultivation may be found in P. M. Vitousek, T. N. Ladefoged, P. V. Kirch, A. S. Hartshorn, M. W. Graves, S. C. Hotchkiss, S. Tuljapurkar, and O. A. Chadwick, "Soils, Agriculture, and Society in Precontact Hawai'i," *Science*, 304 (2004): 1665–69. I also develop the arguments about intensification of Hawaiian dryland agriculture, and present more detailed evidence to support these, in my book *How Chiefs Became Kings: Divine Kingship and the Rise of Archaic States in Ancient Hawai'i* (Berkeley: University of California Press, 2010). The quote from Marshall Sahlins is from *Stone Age Economics* (Chicago: Aldine, 1972), p. 82. The progressive subdivision of *ahupua'a* territories in Kohala is demonstrated by Thegn Ladefoged and Michael Graves in "The Formation of Hawaiian

Territories," in Ian Lilley, ed., *Archaeology of Oceania* (Oxford: Blackwell, 2006), pp. 259–83.

CHAPTER THIRTEEN: THE HOUSE OF PIʻILANI

An excellent summary of Hawaiian theology, including the pantheon of gods, is given by Valerio Valeri, *Kingship and Sacrifice: Ritual and Society in Ancient Hawaiʻi* (Chicago: University of Chicago Press, 1985). Samuel Kamakau gives an account of the Hawaiian priesthood in *Ka Poʻe Kahiko: The People of Old*, Bernice P. Bishop Museum Special Publication, no. 51 (Honolulu: Bishop Museum Press, 1964). The tradition of Pāʻao and the introduction of the cult of Kū and of human sacrifice is recounted by various sources, including Samuel Kamakau's *Tales and Traditions of the People of Old: Nā Moʻolelo a ka Poʻe Kahiko* (Honolulu: Bishop Museum Press, 1991). The quote from Abraham Fornander is from *An Account of the Polynesian Race*, 3 vols. (London: Trübner, 1878–85), vol. 2, p. 101. Piʻilanihale *heiau* is described and illustrated in Patrick V. Kirch and Thérèse I. Babineau, *Legacy of the Landscape: An Illustrated Guide to Hawaiian Archaeological Sites* (Honolulu: University of Hawaiʻi Press, 1996). Michael Kolb reported his excavations at Piʻilanihale in "Monumental Grandeur and Political Florescence in Pre-Contact Hawaiʻi: Excavations at Piʻilanihale Heiau, Maui," *Archaeology in Oceania*, 34 (1999): 73–82. For the story of the Piʻilani line of chiefs, I follow the version given by Samuel Kamakau in *Ruling Chiefs of Hawaiʻi* (Honolulu: Kamehameha Schools Press, 1961); quoted passages are from pp. 25, 27, and 28. The story of the encounter with the lizard goddess Kihawahine was published by an anonymous author in the Hawaiian language newspaper *Ka Nupepa Kuokoa* ("Ka Moolelo o Kihapiilani," February 9, 1884). A copy of this article, and a manuscript translation by Hal Wright, were among the papers of the late Charles Pili Keʻau of Maui. I am grateful to Keʻau's daughter Bernice for providing me with copies of these documents, which add a fascinating coda to the *moʻolelo* of Kiha-a-Piʻilani. The quote from Samuel Kamakau is from *The Works of the People of Old, Na Hana o ka Poʻe Kahiko*, Bishop Museum Special Publication, no. 61 (Honolulu: Bishop Museum Press, 1976), p. 129. My study of the Kahikinui *heiau* and their orientations is presented in Patrick V. Kirch, "Temple Sites in Kahikinui, Maui, Hawaiian Islands: Their Orientations Decoded," *Antiquity*, 78 (2004): 102–14. The results of the coral dating study are published in Patrick V. Kirch and Warren D. Sharp, "Coral ^{230}Th Dating of the Imposition of a Ritual Control Hierarchy in Precontact Hawaii," *Science*, 307 (2004): 102–04.

CHAPTER FOURTEEN: "LIKE A SHARK THAT TRAVELS ON THE LAND"

Bruce G. Trigger, *Understanding Early Civilizations* (Cambridge: Cambridge University Press, 2003). The David Malo quotes are from *Hawaiian Antiquities*, Bishop Museum Special Publication, no. 2 (Honolulu: Bishop Museum Press, 1951), pp. 57 and 60. Samwell's description of the Hawaiian feather cloaks is taken from J. C. Beaglehole, ed., *The Journals of Captain James Cook: The Voyage of the* Resolution *and* Discovery, *1776–1780*, 2 vols. (Cambridge: Hakluyt Society, Cambridge University Press, 1967), p. 1179. Adrienne Kaeppler, "Polynesia and Micronesia," In A. L. Kaeppler, C. Kaufmann, and D. Newton, eds., *Oceanic Art* (New York: Harry N. Abrams, 1997), pp. 21–155. Lilikalā Kameʻeleihiwa, *Native Land and Foreign Desires* (Honolulu: Bishop Museum Press, 1992). William H. Davenport, *Piʻo: An Enquiry into the Marriage of Brothers and Sisters and Other Close Relatives in Old Hawaiʻi* (New York: University Press of America, 1994). Samuel Kamakau, *Ka Poʻe Kahiko: The People of Old*, trans. M. K. Pukui, Bernice P. Bishop Museum Special Publication, no. 51 (Honolulu: Bishop Museum Press, 1964), p. 4. The lines of the chant are from Abraham Fornander, *Fornander Collection of Hawaiian Antiquities and Folk-Lore*, Bernice P. Bishop Museum Memoirs, vol. 6, pt. 3 (Honolulu: Bishop Museum Press, 1920), pp. 393–94. Irving Goldman, *Ancient Polynesian Society* (Chicago: University of Chicago Press, 1970), quoted passage is from pp. 544–45. Patrick Vinton Kirch, *How Chiefs Became Kings: Divine Kingship and the Rise of Archaic States in Ancient Hawaiʻi* (Berkeley: University of California Press, 2010). Ernst Mayr, *This Is Biology: The Science of the Living World* (Cambridge, Mass.: Harvard University Press, 1997), quoted passage is from p. 67. E. DeMarrais, L. J. Castillo, and T. Earle, "Ideology, Materialization and Power Strategies," *Current Anthropology*, 37 (1996): 15–31. Colin Renfrew, "Introduction: Peer Polity Interaction and Socio-political Change," in C. Renfrew and J. Cherry, eds., *Peer Polity Interaction and Socio-Political Change* (Cambridge: Cambridge University Press, 1986), pp. 1–18.

CHAPTER FIFTEEN: THE ALTAR OF KŪ

David Malo's account of the Makahiki is given in *Hawaiian Antiquities*, Bernice P. Bishop Museum Special Publication, no. 2 (Honolulu: Bishop Museum Press, 1951); the quote is from p. 142. Samuel Kamakau describes the Makahiki and procession of the Lono god in *Ka Poʻe Kahiko: The People of Old* (Honolulu: Bishop Museum Press, 1964), p. 21. The political histories of kingly succession and wars of territorial conquest given in this chapter draw from Samuel Kamakau's *Ruling Chiefs of Hawai'i* (Honolulu: Kamehameha Schools Press, 1961); quoted passages are from pp. 46, 70–71 and 72, supplemented by Abraham Fornander, *An Account of the Polynesian Race*, 3 vols. (London: Trübner, 1878–85), vol. 2. The agricultural field system of Kaupō, along with its great temples, is described in P. V. Kirch, J. Holson, and A. Baer, "Intensive Dryland Agriculture in Kaupō, Maui, Hawaiian Islands," *Asian Perspectives*, 48 (2010): 265–90. The Kawela archaeological survey on Moloka'i is reported in Marshall I. Weisler and Patrick V. Kirch, "The Structure of Settlement Space in a Polynesian Chiefdom: Kawela, Moloka'i, Hawaiian Islands," *New Zealand Journal of Archaeology*, 7 (1985): 129–58.

CHAPTER SIXTEEN: THE RETURN OF LONO

The main sources for Kalani'ōpu'u's wars against Kahekili are Abraham Fornander's *An Account of the Polynesian Race*, 3 vols. (London: Trübner, 1878–85), vol. 2, and Samuel Kamakau's *Ruling Chiefs of Hawai'i* (Honolulu: Kamehameha Schools Press, 1961); quoted passages from pp. 85, 88, 98 and 102–03. Additional details, especially of the warrior Kekūhaupi'o, are given by Stephen L. Desha, *Kamehameha and His Warrior Kekūhaupi'o* (Honolulu: Kamehameha Schools Press, 2000). All quotes from the journals of Captain James Cook and Lieutenant James King are from J. C. Beaglehole, ed., *The Journals of Captain James Cook: The Voyage of the* Resolution *and* Discovery, *1776–1780* (Cambridge: Hakluyt Society, Cambridge University Press, 1967); quoted passages are from pp. 473–74, 475, 476, 478 n. 1, 499, 503–07, 512, 517, 521, 528, 551, 560–61. The most complete summary of the Makahiki rituals is provided by Valerio Valeri in *Kingship and Sacrifice: Ritual and Society in Ancient Hawaii* (Chicago: University of Chicago Press, 1985). Accounts of the Makahiki rituals are also given by David Malo, *Hawaiian Antiquities* (Honolulu: Bishop Museum Press, 1951), by John Papa 'Ī'ī, *Fragments of Hawaiian History* (Honolulu: Bishop Museum Press, 1959), pp. 70–77, and Samuel Kamakau, *Ka Po'e Kahiko* (Honolulu: Bishop Museum Press, 1964), quoted passage from pp. 20–21. The quoted passage from Kelou Kamakau is from Abraham Fornander, *Fornander Collection of Hawaiian Antiquities and Folk-Lore*, Bernice P. Bishop Museum Memoirs, vol. 6, pt. 1 (Honolulu: Bishop Museum Press, 1919), p. 42. The interpretation of Captain Cook as the returning god Lono, put forward convincingly by Sahlins but also supported by the Native Hawaiian traditions, spurred a major debate in late twentieth-century anthropology between Marshall Sahlins and Gananath Obeyesekere. The quote from Marshall Sahlins is from *How "Natives" Think: About Captain Cook for Example* (Chicago: University of Chicago Press, 1995), p. 55. Sahlins also analyzes Cook's encounter with the Hawaiians in *Historical Metaphors and Mythical Realities: Structure in the Early History of the Sandwich Islands Kingdom* (Ann Arbor: University of Michigan Press, 1981). Gananath Obeyesekere's critique of Sahlins's interpretation of Cook as Lono is presented in *The Apotheosis of Captain Cook* (Princeton, N.J.: Princeton University Press, 1992); his arguments are extensively refuted by Sahlins in *How "Natives" Think*. The William Ellis quote is from the *Journal of William Ellis: Narrative of a Tour of Hawaii, or Owhyhee* (repr. Honolulu: Advertiser Publishing Company, 1963 [orig. 1827]), p. 87.

CHAPTER SEVENTEEN: PROPHECY AND SACRIFICE

The history of Kamehameha and his rise to power has been recounted in numerous books. Kamakau's *Ruling Chiefs of Hawai'i* (Honolulu: Kamehameha Schools Press, 1961) remains an authoritative source from the Hawaiian point of view; quoted passages are from pp. 121, 128, 132, and 150. Stephen L. Desha's *Kamehameha and His Warrior Kekūhaupi'o* (Honolulu: Kamehameha Schools Press, 2000) adds many details not in Kamakau, but some of these may be later embellishments; quoted passages are from pp. 85 and 109. Volume 2 of Abraham Fornander's *Account of the Polynesian Race*, 3 vols. (London: Trübner, 1878–85)—the quotation regarding Keawemauhili is from p. 312—also summarizes the main events. I have drawn on all of these in my rendition of the story. The accounts vary about the exact manner of Kīwala'ō's death. Desha (p. 136) says that Ke'eaumoku disemboweled him, whereas Kamakau (p. 121) and Fornander (p. 310) write that Ke'eaumoku slit his throat with the *lei o mano*. The quote from Lieutenant King describing Kamehameha is from J. C. Beaglehole, ed., *The Journals of Captain James Cook: The Voyage of the* Resolution *and* Discovery, *1776–1780* (Cambridge: Hakluyt Society, Cambridge University Press, 1967), pp. 512–13. For specific dates and actions of the earliest European ships in Hawai'i, see Ralph S. Kuykendall, *The Hawaiian Kingdom*, vol. 1: *1778–1854, Foundation and Transformation* (Honolulu: University Press of Hawai'i, 1938). Peter H. Buck, *Explorers of the Pacific* (Honolulu: Bishop Museum Press, 1953) also provides a useful summary of early European voyages to Hawai'i. The quote "a magnanimous monarch . . . " is from Marshall Sahlins, *Historical Ethnography*, vol. 1 of Patrick V. Kirch and Marshall Sahlins, *Anahulu: The Anthropology of History in the Kingdom of Hawai'i* (Chicago: University of Chicago Press, 1992), p. 42.

EPILOGUE: HAWAI'I IN WORLD HISTORY

Robert Carneiro's circumscription theory was outlined in his classic paper, "A Theory of the Origin of the State," *Science*, 169 (1970): 733–38. Charles Spencer and Else Redmond's ideas about social evolution are summarized in "Primary State Formation in Mesoamerica," *Annual Review of Anthropology*, 33 (2004): 173–99. The theory of peer polity interaction is discussed from various perspectives in Colin Renfrew and John Cherry, eds., *Peer Polity Interaction and Socio-Political Change* (Cambridge: Cambridge University Press, 1986). William Rathje outlined his trade model for the development of Maya civilization in "Praise the Gods and Pass the Metates: A Hypothesis of the Development of Lowland Rainforest Civilizations in Mesoamerica," in M. P. Leone, ed., *Contemporary Archaeology* (Carbondale: Southern Illinois University Press, 1972), pp. 365–92. David Webster, "Warfare and the Evolution of the State: A Reconsideration," *American Antiquity*, 40 (1975): 464–70. Timothy Earle's comparison of Hawai'i, Peru, and Denmark can be found in his book *How Chiefs Come to Power: The Political Economy in Prehistory* (Stanford, Calif.: Stanford University Press, 1997). Joyce Marcus and Kent V. Flannery trace the long-term evolution of Mesoamerican society in *Zapotec Civilization: How Urban Society Evolved in Mexico's Oaxaca Valley* (London: Thames and Hudson, 1996).

INDEX

adoption, 178
adzes, 25, 26, 35, 42, 50, 76, 93, 94, 126–130, 173
agency, 228–229
agriculture: development of, 141, 150–153, 201, 231, 294–295; intensification of, 159, 191, 198–201, 231, 298; population and, 162, 163, 198–199, 294; rainfall and, 196; ritual and, 216; social evolution and, 227–228, 294–295; soils and, 196–197; 'Umi and, 182–183. *See also* field systems, irrigation, sweet potatoes, taro
'aha ali'i, 134, 137
Ahia, 274
Ahu a 'Umi, 173–175, 185, 233
'Ahu'ena, Hawai'i Is., *285fig*, 286
ahupua'a, 139–141, 142, 152, 186, 190, 199–200, 203, 215, 223, 232, 295. *See also* land tenure
'Aihakoko, 211
'āina, 141, 222–223, 225
Akahiakuleana, 177–178, 180
'Ālapa, 249

Alapa'inui, 236, 238–245, 248, 267
'Aleunuihāhā Channel, 117, 193, 201, 209, 238, 247, 283
ali'i: birthing places of, 132–133; bodily practices of, 298; class of, 191, 200, 201, 295; featherwork and, 203, 247; female, 234; and religion, 202; training of, 180; tribute collection by, 253. *See also* chiefs, god-kings, kingship, nā li'i
ali'i 'ai ahupua'a, 140
ali'i 'ai moku, 139, 185
ali'i akua, xi, 5, 201, 204, 217–218. *See also* divine kingship, god-kings
ali'i kapu, 131, 137
ali'i nui, 4, 123, 136, 138, 217
Allen, Jane, 151–152
Allen, Jim, 23, 30
aloha, 224, 228
'anā'anā, 204, 245. *See also* kahuna
Anahulu Valley, O'ahu Is., 151
ancestral Polynesian culture, 40, 43–46. *See also* Hawaiki, Proto-Polynesian

335

anthropology, 58, 68. *See also* archaeology
'anu'u tower, 10, 11*fig*, 179, 257. *See also* temples
archaeology, 17–18, 36, 43, 68, 72, 73, 76, 78–81, 164, 190, 212
archaic states, 6–7, 15–16, 170, 196, 289–290
Athens, Steve, 101–103, 109
Austronesian: canoes, 26, 30, 33; expansion, 30, 35; genetics, 31; houses, 29; language, 25, 26; peoples, 23, 25; social organization, 34; tattooing, 31–33
'awa. *See* kava

bananas, 25, 44, 150, 260
Banks, Sir Joseph, 56–57
Barber's Point, O'ahu Island, 106–109
barkcloth, 43, 51, 55, 176, 203, 224, 231, 253, 267
Bellows Dune Site, O'ahu Is., 85–89, 88*fig*, 92–96
Bingham, Hiram, 287
birds, 105–110, 173, 218–219, 247, 259. *See also* feathers
birthing places, 132. *See also* Kūkaniloko
Bishop, Charles Reed, 69
Bishop Museum, xii, 22–23, 29, 34, 47, 59, 62, 68, 69, 71, 73, 75, 79–80, 85, 87, 93, 101, 105, 106, 147, 151, 189, 219, 233
Bismarck Archipelago, 21, 26, 30, 33, 36
Bligh, William, 263, 275
Bonk, William J., 78
Boserup, Ester, 162, 163, 166, 228
breadfruit, 5, 44, 51, 53, 66, 124–125
Brigham, William T., 69–71
Brookfield, Harold, 198
Buck, Sir Peter. *See* Hiroa, Te Rangi
Burley, David, 35

calendar: of ancient Hawaiki, 46; lunar, 184, 252; rising of Pleiades and, 215; ritual cycle, 202–203, 252–254. *See also* Makahiki
canoes: ancient Austronesian, 26, 31; chant for launching, 54–55; double-hulled, 5, 48, 50, 235, 259; early Polynesian, 48; Lapita, 33; manufacture of, 50–52; names, 54; outrigger, 25–26, 31, 34; sails, 51; war, 210, 283. *See also Hōkūle'a*, voyages, way-finding
Carneiro, Robert, 291, 293
causation, historical, 155, 226–227, 292
Chadwick, Oliver, 195–196
Chang, Kwang-Chih, 24, 87
chickens, 25, 33, 44, 48, 54
chiefdom, 16, 227, 293. *See also* archaic states
chiefs, 7, 45, 176, 219, 223, 226. See also *ali'i*, divine kingship, god-kings
China, 4, 6, 15, 24–25, 153, 220. *See also* Taiwan
Chiu, Scarlett, 30, 32
Christensen, Carl, 107–108, 109
Christmas Island, 2
circumscription, 291
class distinctions, 295. *See also* social change
climate, 228
cloaks, feathered, 8, 175, 203, 218, 229, 232, 243, 247, 249, 251, 259, 261, 262, 298. *See also* featherwork
coconut grater, 94
coconuts, 44, 48, 51, 53, 137, 143–144, 232, 260
commoners, 7, 142. See also *maka'āinana*
Cook, Captain James: arrival at Kealakekua Bay, 255; arrival off Maui Island, 250–251; attempts to take Kalani'ōpu'u hostage, 262; explores Waimea, 10–11; on Hawaiian featherwork, 218; death of, 262–264; discovery of Kaua'i, 3–5, 9; disposition of remains of, 263–264; Hawaiian ideas

about, 250; at Kealakekua Bay, 156; as Lono, 46, 250, 252, 254–264; meets Kalani'ōpu'u, 251, 259; views on Polynesian origins, 11–13, 56; voyage to Tahiti, 1–2
coral offerings, 213, 215–216
coral reefs, 83
cultural resource management, 80. *See also* archaeology

Davenport, William, 220
Davis, Issac, 277, 280
demography. *See* population
disease, 157–158, 168
divine kingship, 4, 5, 7, 170, 201, 217–218, 289–290, 296
dogs, 5, 25, 33, 44, 48, 54, 182, 203, 253
drought, 169, 228, 299
dryland cultivation, 154, 155, 166, 188–194, 195–201, 202, 203, 209, 228, 251, 296–297, 299. *See also* field systems
Dye, Thomas, 95, 166–167

Earle, Timothy, 153, 229, 291, 293
earth oven, 43, 44, 91, 144
Easter Island, 26, 35, 48, 49, 63
Eastern Polynesia, 47, 48, 50
economic cycles, 297, 299
Egypt, xi, 4, 6, 7, 15, 177, 218, 220
Ehukaimalino, 185
Ellis, William, 264
Eloaua Island, 27. *See also* Mussau Islands
Emerson, Nathaniel, 145
Emory, Kenneth, 47, 54, 75–78, 80–81, 94, 128–129, 148, 162
'Ewa, O'ahu Is., 106, 107, 118

Fair American, 276–277
famine, 201, 208, 299
feathers, 175, 181, 203–204, 218–219, 224, 231–232, 235, 247, 253, 259, 298, 299
featherwork, 177, 203–204, 218–219, 249, 251, 259, 298, 299

Feinman, Gary, 6
Field, Julie, 191–193
field systems, 188–194, 202, 260, 296–297. *See also* agriculture, dryland cultivation, sweet potato, taro
Fiji, 22, 34, 39, 47
Finney, Ben, 61, 63
fishhooks, 25, 35, 42, 47, 76, 77–78, 86, 93, 94
fishing: for *aku*, 182, 254; ancestral Polynesian, 44; Lapita, 31; at South Point, 77
fishponds, 99, 138, 201
Flannery, Kent, 292
flightless birds, 105–110
Fornander, Abraham, 58, 68, 73, 126, 175, 206, 234, 272
fortifications, 136, 210, 211*fig*, 239–242, 284
fur trade, 275–276, 284

geckos, 54, 108
Geertz, Clifford, 296
genealogies, 7–8, 17, 73, 219, 222, 229, 301
Gifford, Edward, 22, 36–37, 47
god-kings, xi, 8, 204, 217–218, 220–221, 229, 233, 243, 258, 269. *See also* divine kingship
golden plover, 5, 39, 53
Golson, Jack, 22
Graves, Michael, 194, 197, 199
Green, Roger, 22–23, 31, 34, 43, 78–79, 147, 163
Gregory, Herbert E., 71, 75
Goldman, Irving, 225, 229

H-3 freeway, 151
Ha'atuatua, 47
Haka, 136, 142
Hākau, 177, 180–184, 191, 221, 269
haku'ōhi'a, 183
Hālawa, Hawai'i Is., 268, 270, 272

Hālawa Valley, Molokaʻi Is., 79–80, 90, 93, 94, 163, 224, 235
Haleakalā, Maui Is., 5, 66, 75, 112, 115, 167, 197, 212, 215, 237, 250
Halekiʻi, Maui Is., 240, *279fig*
Halemaʻumaʻu, Hawaiʻi Is., 265, 281
Hale-o-Keawe, 236, 270–272, *271fig*
Hale o Līloa, Hawaiʻi Is., 181
Halley's Comet, 267
Hāmākua district, Hawaiʻi Is., 122, 154, 178, 180, 183, 185
Hāna district, Maui Is., 204, 206, 209–210, *211fig*, 237, 247, 248, 254, 277, 279
Hanalei Valley, Kauaʻi Is., 146
Hanamauloa, Maui Is., 114–115
Handy, Edward S. C., 71–72, 80, 120, 126, 143, 189
Hawaiʻi Biocomplexity Project, 195, 200
Hawaiʻi Island, 83, 119–120, *176map*, 204, 231, 233
Hawaiʻiloa, 9
Hawaiki, 5, 39–40, 43–46, 48, 50, 51, 91, 138, 219, 222–223, 293
heiau: of Hawaiʻi, 70, 174; *hoʻoʻuluʻai*, 140, 213; of Kahikinui, Maui, 213–216; at Kaupō, Maui, 237–238; for Kū, 204; of Molokaʻi, 70; rituals, 229; at Waimea, Kauaʻi, 10. See also *luakini*, temples
helmets, 8, 177, 203, 249. See also featherwork
Heyerdahl, Thor, 59–60, 61, 63, 126
Hikapoloa, 119–120
Hikiau, Hawaiʻi Is., 254, *256fig*, 257–258, 259, 275
Hikinaakalā, Kauaʻi Is., *125fig*
Hiroa, Te Rangi, 5, 58, 59, 72–73, 76, 80, 121, 126, 192
Hōkūleʻa, 61–63, 126, 128
Holoʻae, 246, 250, 270, 274
Hommon, Robert, 148, 163, 164–165
Hōnaunau, Hawaiʻi Is., 236, 245, 270, *271fig*, 272

Honolulu, 284–285
Honuaʻula, Maui Is., 179, 184, 207, 249, 276
hoʻokupu, 139, 142, 191, 202–203, 207, 216, 219, 251, 254, 255, 286. See also tribute
Hoʻolaemakua, 209–210
horticulture, 25, 30. See also agriculture
houses, 44, 179, 190
Hualālai, Hawaiʻi Is., 173, 186, 233
Huapouleilei, 134–135
hula, 18
human sacrifice, xi, 7, 8, 9, 15, 16, 69, 142, 183, 184, 204, 210–211, 213, 221, 229, 269–270, 278, 284, 289, 296
hunting-and-gathering, 30, 33
hydraulic economy, 153. See also irrigation

ʻĪao Valley, Maui Is., 280
ideology, 229
ʻĪʻī, John Papa, 68
ʻili, 140, 142, 186, 190, 223. See also *ahupuaʻa*, land tenure
ʻĪmakakoloa, 269–270
Inca, xi, 6, 59, 220
incest, royal, xi, 7, 176–177, 185, 220–221, 289. See also *piʻo*
inland expansion hypothesis, 164. See also population
intensification. See agriculture, field systems, irrigation
irrigation: archaic states and, 289; areas suited to, 154; dating of, 152; and economy, 153, 201; Kāne and, 203; in Koʻolaupoko district, Oʻahu, 150–151; at Luluku, Oʻahu, 151–153; management of, 141; on Maui, 206; pondfield, 144–146; and population, 169, 198; and social evolution, 294–295; terraces, 152; at Waimea, Kauaʻi, 10, in Waipiʻo Valley, Hawaiʻi, 119–120, 179; in western Maui, 155. See also agriculture, taro

James, Helen, 106, 107
Japan, 7
Ka'ahumanu, 287
kā'ai, 181, 263–264
Ka'alakea, Rev. Kawika, 113–116
Ka'awaloa, Hawai'i Is., 262, *264fig*
Kahahana, 278
Kaha'i-a-Ho'okamali'i, 66, 124–125, 126, 131
Kahakahakea, Hawai'i Is., 87
Kahaunokama'ahala, 176
Kahekili, 248–250, 261, 276, 277–281, 283
Kahiki, 9, 18, 115, 118, 120, 127, 137, 204, 221, 226, 254. *See also* Tahiti
Kahikinui, Maui Is., 5, 112–118, 167–168, 174, 194–199, 212–216
kāhili, 235, 259, 298. *See also* featherwork
Kaho'olawe Island, 71, 103, 117, 124, 129–130
kahuna, 135, 202, 207–208. *See also kāula*, priests
kahuna nui, 185, 207. *See also* priests
Kaiana, 275–276
Kaiholena, Hawai'i Is., 195
Kaikilani, 234
kāinanga, 222–223. *See also 'āina*
Kakohe, 183–184
Kākuhihewa, 235
Kalākaua, King David, 124
Kalaku'i'aha, 273
Kalanikūpule, 279–280, 283–284
Kalani'ōpu'u, 243, 244–246, 247–250, 250, 251, 254, 258–264, 267, 268–269, 290
Kalaunuiohua, 134–135, 141, 296
Kaleiokū, 182–185
kalo. *See* taro
Kalola, 246, 247, 249–250, 268, 280
Kamahualele, 121, 123
Kamakahonu, Hawai'i Is., 210, 223, *285fig*, 286

Kamakau, Samuel, 17, 68, 73, 133, 184, 204, 206, 214, 217, 221, 232, 239, 243, 253, 255, 266, 267–268, 278
Kamalālāwalu, 193–194, 235
Kame'eleihiwa, Lilikalā, 220, 224
Kamehameha I, King: altercation with Kīwala'ō, 272–273; assumes control of war god, 269; attacks Maui, 279–280; attempted abduction of, 245; at battle of Moku'ōhai, 273–275; birth of, 267–268; conquers O'ahu, 193, 283–284; constructs Pu'ukoholā *heiau*, 281–282; death of, 286–287; feathered cloak of, 219; kills Ahia, 274; meets Captain Cook, 268; at Pākini *heiau*, 270; sacrifices Keōua, 283; saves Kekūhaupi'o, 248; secures Keōpūolani, 280; sends message to Kahekili, 280; struggle against King Kaumuali'i, 15; struggles against Keōua, 265; trading policy of, 276–277, 284–285; youth of, 268
Kamehameha II, King, 280
Kamehameha III, King, 139, 287
Kamehameha-nui, 240, 243–244, 246, 247–248
Kanahaokalani, 242
Kanaloa, 39, 203, 213
Kanaloa kahoolaweensis, 103–104, 110
Kanaloa-kua'ana, 234, 235
Kāne: association with Kaua'i and O'ahu, 155; chant for, 145–146; as creator god, 144–145; cult of, 142, 203; irrigation and, 203; temples and, 69, 140, 214–215; and water, 145–146
Kānemalohemo, Maui Is., 238
Kāne'ohe, O'ahu Is., 66, 150–151, 169, 243
Ka 'Ohana o Kahikinui, 112
Ka'opulupulu, 278
Kapi'iohokalani, 240, 242
Kapoukahi, 281
kapu, 8, 179, 217, 223, 298
kapu moe, 5, 217, 249, 258, 298

Kapukapuākea, Oʻahu Is., 120, 137
kauā, 221
Kaʻū district, Hawaiʻi Is., 193, 199, 234, 236, 245, 254, 265–267, 269, 275, 282
Kauaʻi Island, 6, 83, 120–125, 131, 135, 198, 201, 203, 284, 294
Kauhiʻaimoku-a-Kama, 243–244
Kauhi-a-Kama, 193–194
Kaʻuiki, Maui Is., 210, *211fig*, 247
kāula, 134, 135
Kaʻulahea, 142
Kaumualiʻi, King, 15, 284
Kaupō district, Maui Is., 112, 113, 199, 212, 214, 236–238, 248
kava, 17, 46, 91, 150, 183, 208, 251, 258, 273
Kawaihae, Hawaiʻi Is., 277, 282–283
Kawainui Marsh, Oʻahu Is., 101
Kawaokaʻōhele, 206
Kawela, Molokaʻi Is., 239–242, 283
Ke Ala i Kahiki, Kahoʻolawe Is., 124, 127, 129
Kealakekua Bay, Hawaiʻi Is., 156–157, 168, 189, 236, 254, 255, 258–264, *264fig*, 268
Keawemauhili, 267, 272, 274, 275, 277, 279, 281
Keawenui, 185, 233–234
Keaweʻopala, 245–246
Keʻeaumoku, 245, 274–275
Kekaulike, 213, 214, 236–238, 239–240, 246, 248, 267
Kekūhaupiʻo, 248, 249, 262, 268, 270, 272, 273
Kekuʻiapoiwanui, 240, 248, 267
Keliʻikea, 257, 259
Keliʻiokaloa, 224, 232, 233
Keoneʻōʻio, Maui Is., 249
Keōpūolani, 280, 287
Keōua Kuahuʻula, 265–266, 267, 275, 277, 281–283
Keōua Kupuaikalaninui, 243, 244–245
Kepelino, 214

Kiha-a-Piʻilani, 207–211, 217, 220
Kīhei, Maui Is., 240, 249–250
Kila, 9, 123, 126
Kīlauea, Hawaiʻi Is., 265, *266fig*
King, Lieutenant, 3, 156–157, 168, 251, 256–257, 259, 260, 261, 263, 268, 286
kingship, 7, 8, 217–218, 221. See also *aliʻi akua*, *aliʻi nui*, divine kingship, god-kings
kinship, 7, 222–223, 224
Kīpahulu district, Maui Is., 247
Kīpapa, Oʻahu Is., 142
Kīwalaʻō, 249–250, 268–275
koʻa, 213
Koʻaʻa, 255–258
kōʻele, 224, 225, 231
Kohala district, Hawaiʻi Is., 82, 154, 164, 169, 186, 187–201, 204, 234, 236, 267–268
Kohala field system, 188–201
Koʻi, 178–179
Kolb, Michael, 206
Koleamoku, 209
Kona district, Hawaiʻi Is., 185, 188, 193, 199, 234
Kondo, Yoshio, 107
konohiki, 141, 201, 223–224, 225, 232, 286, 295
Kon-Tiki, 59–60, 126
Koʻolaupoko district, Oʻahu Is., 150–151
Kū: cult of, 142, 203–204, 233, 297–298; and human sacrifice, 9, 69, 183, 221, 270, 297; images of, 219, 233; priests of, 204; symbols of, 203; temples of, 204, 215, 216, 252; as war god, 8, 155, 179, 203, 233
Kualiʻi, 151, 242
Kualoa, Oʻahu Is., 124, 151, 278
Kūkāʻilimoku, 179, 181, 204, 221, 232–233, 253, 259, 269, 280, 281, 298. See also Kū
Kūkaniloko, Oʻahu Is., 132–134, *133fig*, 137

Kula, Maui Is., 207, 208–209
Kuliʻouʻou rockshelter, Oʻahu Is., 76, 89
kūlolo, 137, 144, 253
Kumahana, 278
Kumuhonua, 118–119
Kūnuiākea, 258, 260
Kūʻula, 213

Laʻamaikahiki, 118, 123–124, 126, 127, 129–130, 134
Ladefoged, Thegn, 191–192, 194, 197, 199
Lahaina, Maui Is., 207, 244, 283
Lahainaluna, Maui Is., 17
Lānaʻi Island, 129
land snails, 105, 107–108, 110
land tenure: early Hawaiian, 139–142; Māʻilikūkahi and, 153; transformations in, 222–225; ʻUmi and, 186. See also *ahupuaʻa*, *kōʻele*, Mahele
language: Austronesian, 25, 26; Papuan, 30–31; Polynesian, 35; Proto-Polynesian, 43
Lapakahi, Hawaiʻi Is., 164, 166, 190, 191
Lapita: cultural complex, 23; discovery of, 22–23; dispersal, 34, 52; fishing, 31; pottery, 22, 27–30, 31–33, *32fig*; in Reef Islands, 33–34; in Samoa, 35, 42; site of, 36–37; stilt houses, 28–29, *29fig*; tattooing, 31–33; in Tonga, 34–35; triple-I model of, 31; use of obsidian, 28, 34
Lapita Homeland Project, 21, 23, 30
Laupāhoehoe, Hawaiʻi Is., 182
Lee, Charlotte, 195
lei niho palaoa, 89, 175, 177–178, 253
Libby, Willard, 73–75, 80, 92
Liholiho, 280, 287
Līhuʻe, Oʻahu Is., 131–132, 136, 212
Liliuʻokalani, Queen, 287
Līloa, 175–183, 206, 269
linguistics, 43
Linton, Ralph, 71–72

Little Ice Age, 228
lizard goddess, 208
Loʻaloʻa, Maui Is., 214, 237–238
Lokoʻea Pond, Oʻahu Is., 99
Lolokoka site, 35
Lono: as agriculture god, 8, 140, 155; association with Hawaiʻi and Maui, 155; Captain Cook and, 46, 250; image of, 235, 253–254; Makahiki and, 202–203, 232, 252–254; and seasons, 9, 202–203; and sweet potato, 155, 203, 252, 297; temples of, 215, 216, 254, 298
Lono-a-Piʻilani, 207–211
Lonoikamakahiki, 194, 218, 234–236
loulu palms, 101–102, 110, 133
luakini, 15, 179, 183, 204, 210, 213, 221, 254, 269. See also *heiau*
Luluku, Oʻahu Is., 151–152
Luʻukia, 120–121, 123

MacArthur, Robert, 161–162
Mahele, Great, 139, 224, 287
Mailekini, Hawaiʻi Is., 281
Māʻilikūkahi, 131–142, 146, 148, 152–153, 155, 204, 220, 221, 223, 278, 295
makaʻāinana, 7, 141–142, 163, 175, 186, 200, 201, 209, 219, 222–223, 225, 230, 231, 258, 269, 286, 300. See also social organization
Mākaha Valley, Oʻahu Is., 147–150, 163
Makahiki, 139, 183, 202–203, 215, 219, 223, 232–233, 238, 251–254, 259, 264. See also *hoʻokupu*
Makaliʻi (Pleiades), 9, 46, 139, 215, 238, 251–252
Makapuʻu, Oʻahu Is., 84
Makiloa, Hawaiʻi Is., 191, 193–194, 195
malaria, 31
Malo, David, 17, 68, 73, 135, 214, 218, 226, 232
Malthus, Thomas, 6, 159–160, 162, 166, 198, 228
Mangaia Island, 127–128

Mangareva, 48, 96, 128
Manu'a Islands, 38–39, 57
Marcus, Joyce, 6, 292
Marin, Francisco de Paula, 284
Marquesas Islands, 23, 47, 48, 50, 59, 66, 71, 86, 93, 94, 293
marriage, 176–177, 220–221, 236, 296, 297. *See also* incest, royal; *pi'o*
Marx, Karl, 227
materialization, 229, 291–292
mats, 43, 44, 51
Mau Piailug, 62
Mauna Kea, Hawai'i Is., 6, 66, 82, 117, 173, 187
Mauna Loa, Hawai'i Is., 6, 82, 117, 173, 187, 266
Maui Island, 83, 112, 135, 153–154, 193–194, 231, 236, *237map*, 250–251
Māweke, 118, 124, 126, 136, 278
Mayr, Ernst, 61, 226–227, 292, 297
Meares, John, 275
mele, 18
Menzies, Archibald, 188–189
Mesoamerica, 4, 6, 15, 218
Mesopotamia, 153
Metcalfe, Simon, 276–277
Mills, Peter, 14
missionaries, 287
moa nalo, 109, 110, 111
Moa-ula-nui-ākea, 120–121, 123, 131, 134, 137
Mo'ikeha, 9, 118–124, 126
moku, 139, 142. *See also* land tenure
Mokulau, Maui Is., 238, 239, 240
Moku'ōhai, Hawai'i Is., 273–275
Moloka'i Island, 66, 70, 82, 105, 106, 135, 201, 203, 207, 216, 224, 239–242, 277, 294
Mo'okini, Hawai'i Is., 70, 122, 204
mo'olelo: xiv, 125; collections of, 68; as history, xii; insider perspective of, 18; of long-distance voyages, 126–127; of 'Umi, 175;

Mo'omomi, Moloka'i Is., 105
Mo'orea Island, 48, 75, 78
Muli'eleali'i, 118
museums, 69
Mussau Islands, 21, 23, 24, 26–30, 32*fig*

Nae'ole, 268
Nāholokū, Maui Is., 237
nā kanaka, 191, 219, 224, 230, 300. *See also maka'āinana*
nā li'i, 142, 191, 202, 219, 224, 230, 300. *See also ali'i*
Nānā'ulu, 118, 120, 134, 137, 219, 278
National Science Foundation (NSF), 194–195
navigation. *See* way-finding
Nā Wai Ehā, Maui Is., 206
Near Oceania, 24, 35
Necker Island, 75, 83
New Albion, 2, 8, 11, 250, 275
New Archaeology, 79, 162
New Caledonia, 22, 34, 36, 47
New Guinea, 26, 29, 33
Newman, T. Stell, 189–190
New Zealand, 12, 35, 49
nī'aupi'o, 220, 280. *See also pi'o*
Nihoa Island, 75, 83
Ni'ihau Island, 11, 14, 235
Niuatoputapu Island, 34–35
Nunu, 183–184
Nu'uanu, O'ahu Is., 283

O'ahu Island, 6, 83, 118, 120, 124, 131–142, 132*map*, 150–155, 198, 201, 203, 242–244, 277–278, 294
'Ofu Island, 40–43
'Olopana, 118–123, 126
Olowalu massacre, 276
Olson, Storrs, 106–107
'Oma'okamau, 178–179, 182–185
Oneawa, O'ahu Is., 243
oral traditions, 17, 58, 68, 73. *See also mo'olelo*

outrigger canoe. *See* canoes

Pāʻao, 9, 69, 142, 175, 204, 221, 229
pahu drum, 123–124, 134, 137, 179
pahupū, 248
Pakaʻalana, Hawaiʻi Is., 179, 180
Pākini, Hawaiʻi Is., 269–270
palm forests, 104–105. See also *loulu*
palynology. *See* pollen analysis
Panakiwuk rockshelter, 24
pānānā, *114fig*, 115–118, 130
pandanus, 51–52, 91–92
Papa, 220
Papaʻenaʻena, Oʻahu Is., 284
Papua New Guinea, 21, 36
Pauahi, 69
Pearson, Richard, 86–89, 92–96, 189–190
peer polity interaction, 230, 291, 297
Pele, 6, 265, 281
Peleiʻōhōlani, 242–244, 278
Pietrusewsky, Michael, 93, 94
Piʻikea, 185, 206, 209–210, 220, 235
Piʻilani, 185, 206–207, 216, 217, 220
Piʻilanihale, Maui Is., 204–206, *205fig*, 211, 223, 237
Piʻimaiwaʻa, 178–179, 182–185, 210
pigs, 4, 5, 25, 33, 44, 48, 54, 139, 182, 191, 203, 297
pīkoi, 192, 194, 195
Pilikaʻaiea, 69, 175, 204
Pinea, 176–177, 180
piʻo, 175, 177, 179, 181, 185, 220–221, 233, 240, 248, 280, 297. *See also* incest, royal
Pleiades, 46, 141, 202, 215, 251–252, 254; see also Makaliʻi
poi, 143–144, 179, 223
poi pounders, 144
Polaris, 65
political economy, 170, 191, 196, 198, 199, 202. *See also* staple economy, wealth economy
pollen analysis, 100–104

Polynesian: ancestral culture, 40; canoe technology, 48; gods, 39–40; homeland, 36, 47, 48; navigation, 58; origins, 59–60, 72; Triangle, 35–36, 40
pondfields. *See* irrigation
pono, 224, 228
population, Hawaiian: and agriculture, 163; archaeological estimates of, 163–166; and carrying capacity, 159; as causal factor, 155; colonizing, 161; disease and, 157–158, 168, 286; Dye-Komori estimate of, 166–167; dynamics, 195; growth, 160–162, 165–166, 168–170, 191, 198–199, 200, 227, 294; of Hawaiʻi at contact, 6, 156, 168, 286, 300; inland expansion of, 164–165; of Kahikinui, 167–168; of Kohala, 189; of leeward Hawaiʻi Island, 165–166; Lt. King's estimate of, 156–157; logistic equation and, 161; missionary census, 157, 168; pressure, 160, 163, 226; of Oʻahu, 152–153, 155; rates of growth, 165; and social change, 169–170, 227, 294
Portlock, Nathaniel, 151, 275
pottery: cord-marked, 24; decline of, 47; decoration of, 31–33; early Austronesian, 26; Lapita, 22, 27, 31–33, *32fig*, 35; in Marquesas, 47; Polynesian, 41, 42–43
prestige goods, 7
priests, 9, 135, 182–183, 185, 202, 204, 207–208, 213, 229, 246, 254, 257, 258. See also *kahuna*, *kāula*
pristine states, 6. *See also* archaic states
Pritchardia. See *loulu* palms
Proto-Austronesian, 25. *See also* Austronesian
Proto-Polynesian: calendar, 46; cosmogony, 45–46; culture, 23; language, 40, 43; rituals, 46; spiritual concepts in, 45–46; words for social groups, 44–45

pūloʻuloʻu, 176, 298
Puna district, Hawaiʻi Is., 234
Puʻu Aliʻi, Hawaiʻi Is., 77–78, 92
puʻuhonua, 241
Puʻu Kehena, Hawaiʻi Is., 187–188
Puʻukoholā, Hawaiʻi Is., 281–281, *282fig*
Puʻu Waʻawaʻa, Hawaiʻi Is., 234

radiocarbon dating: of Barber's Point, 108; of Bellows Dune Site, 92, 95–96; discovery of, 71–73; and Hawaiian archaeology, 80; of irrigation features, 147, 149–150; of Kahikinui, 197; of Kawainui Marsh, 102; of Kohala field system, 197; of Kuliʻouʻou rockshelter, 76; of Lapakahi, 191, 197; of Lapita, 22, 28; method, 74–75, 95; of Niuatoputapu, 34; and population estimation, 166–167; of Puʻu Aliʻi, 78; of rat bone, 109; of sweet potatoes, 49, 197; of Talepakemalai, 28; of Toʻaga, 42; of ʻUkoʻa Pond, 102
Raʻiatea Island, 2, 52, 56, 120
rainfall, 196, 299
Rapa Nui. *See* Easter Island
rats, 5, 54, 108, 109
Redmond, Else, 291
Reef Islands, 33–34
religion, 202–204, 216. *See also* Kāne, Kū, Lono, priests, temples
Renfrew, Colin, 230, 291, 293
ritual homicide. *See* human sacrifice
Rongo, 39. *See also* Lono
Rosendahl, Paul, 147, 164, 166, 190–191, 197
royal centers, 7, 235
royal incest. *See* incest, royal
Russian fort, 14–15

sacred stones, 55
Samwell, David, 218
Sahlins, Marshall, 199, 225, 263
Samoa, 34, 36, 47, 50, 128
sandalwood trade, 286
Sandwich Islands, 11
Schmitt, Robert, 158, 168
sea level, changes in, 27, 41, 99
secondary state formation, 16. *See also* archaic states
sennit, 51–52
settlement patterns, 78–79, 163
sharks, 8, 221
Sharp, Andrew, 60, 61, 63
Sharp, Warren, 215
shell valuables, 28
Shutler, Richard, Jr., 36
Sinoto, Yosihiko, 47, 77–78, 86, 93, 94, 148
social change, 138–142, 169–170, 225–230, 291–301
social evolution, 289, 291–292, 299
social organization: ancestral Polynesian, 44–45, 138; Austronesian, 34; early Hawaiian, 138–139, 141–142; hierarchy in, 169; land and, 223–224; population and, 169–170
Society Islands. *See* Moʻorea, Raʻiatea, Tahiti
Soehren, Lloyd, 78, 85–87, 94
soil nutrients, 167, 187, 195–197, 200, 299
South America, 49, 59
Southern Cross, 62, 65, 117
Spanish in the Pacific, 8–9
Spencer, Charles, 291
Stannard, David, 158–159, 168
staple economy, 232, 298. *See also* wealth economy
star paths, 62. *See also* way-finding
states, 226–230. *See also* archaic states
status rivalry, 228–229
Stokes, John F. G., 70–73, 214
sugarcane, 5, 44, 193, 199, 200
Suggs, Robert, 47
surplus, 25, 170, 191, 199, 201, 224, 231, 289, 295, 297. See also *hoʻokupu*

sweet potatoes, 10, 49, 53, 79, 184, 188, 191, 193, 195, 203, 208–209, 215, 231, 237, 248, 252, 260, 296, 297. See also dryland culivation, field systems, Kohala field system

taboo. See *kapu*
Tahiti Island, 1–2, 5, 16, 23, 48, 52, 56, 57, 61, 63, 75, 78, 94, 115, 120, 127, 156
Taiwan, 24, 25, 26, 35. See also China
Talepakemalai, 27–30, 29*fig*, 31, 32*fig*
Tane, 39. See also Kāne
Tangaroa, 39–40, 45
Tap'enk'eng site, Taiwan, 24, 25
Taputapuātea, 57, 120
taro, 5, 10, 25, 30, 34, 44, 99, 137, 143–146, 188, 191, 193, 200, 203, 231, 237, 260, 296. See also irrigation
tattooing: of Kahekili, 248; of Lapita pottery, 31–33, 37; as punishment, 235
Ta'u Island, 38
taxation, 7, 8, 231–232, 295. See also *ho'okupu*, tribute
temples: changes in, 69–70; constructed by 'Umi, 185; coral dating of, 215–216; at Hōnaunau, Hawai'i, 270–271; of Kahikinui, Maui, 213–216; at Kapukapuākea, 137; in Kaupō, Maui, 237–238; for Kū, 204; on Maui, 204–206; orientations of, 213–215, 238; *pahu* drums at, 123–124, 134; rituals on, 257–258; sepulchral, 271; at Taputapuātea, Tahiti, 120; at Waimea, Kaua'i, 10, 11*fig*; of Waipi'o, Hawai'i, 179; for war, 8, 204. See also 'Ahu'ena, Ahu a 'Umi, Haleki'i, Hale o Keawe, *heiau*, Hikiau, *ko'a*, Lo'alo'a, *luakini*, Kānehemomalo, Mo'okini, Pākini, Pi'ilanihale, Pu'ukoholā, Waha'ula
territorial expansion, 232. See also war
Thambetochen, 106, 108–109
Tikopia, 89, 161–162

To'aga site, 41–43
Tonga, 12, 22, 23, 34, 36, 39, 47, 50
trade, 291
trails, 190
tribute, 7, 8, 139, 216, 253, 291. See also *ho'okupu*, *kō'ele*, Makahiki, taxation
Trigger, Bruce, 218
Tu, 39. See also Kū
Tuamotu Islands, 23, 52, 54, 60, 63, 75, 126, 128–129
Tuljapurkar, Shripad, 195
Tupaia, 56–58, 63
turmeric, 55

'Uko'a Pond, O'ahu Is., 99–101, 102, 103, 106
'Umi ('Umi-a-Līloa), 171–186, 188, 191, 193, 196, 199, 206, 209–211, 216, 217, 220, 233, 296–297
'Umiokalani, 234
'Upolu Point, Hawai'i Is., 117
U. S. Exploring Expedition, 174
U/Th dating, 215–216

vaka moana. See canoes
vegetation change, 102–105, 108, 109–110
Verhulst, Pierre-François, 160–161, 165
Vitousek, Peter, 194–196
volcanic eruptions, 265–266
voyages: ancient Austronesian, 26; of colonization, 53–54, 84; drift, 60, 126; experimental, 61, 126; between Hawai'i and Kahiki, 18, 117, 120–126; of *Hōkūle'a*, 61–63, 126; Lapita, 33–34; Micronesian, 62; Polynesian, 40, 56–58; reasons for, 52. See also canoes, way-finding

Wa'ahia, 134–135
Waha'ula, Hawai'i Is., 70, 204
Wahiawā, O'ahu Is., 131, 137
Wai-'Ahukini, Hawai'i, Is., 270–271

Wai'ale'ale, Mt., Kaua'i Is., 3, 13
Waialua, O'ahu Is., 99, 118, 120, 131–132
Wai'anae district, O'ahu Is., 118, 136
Waihe'e, Maui Is., 66, 206, 207
Waikīkī, O'ahu Is., 84, 118, 135, 137, 141, 142, 242, 278, 279, 283, 284
Wailua, Kaua'i Is., 122, 124, *125fig*
Wailuku, Maui Is., 247, 249, *279fig*
Waimānalo, O'ahu Is., 66, 84–85. *See also* Bellows Dune Site
Waimea, Hawai'i Is., 194, 235
Waimea, Kaua'i Is., 3, *11fig*, 16
Waipi'o Valley, Hawai'i Is., 82, 119–120, 119*fig*, 155, 176–186, 188, 191, 223, 269, 281
Waipunalei, Hawai'i Is., 181–183
Wākea, 144–145, 220
wangka. *See* canoes
war: against Mā'ilikūkahi, 142; of Alapa'inui against Maui and O'ahu, 239–244; forbidden during Makahiki, 251; of Kalani'ōpu'u against Keawe'opala, 245–246; of Kalani'ōpu'u against Maui, 247–250; of Kalaunuiohua, 134–135; of Kamalālāwalu against Hawai'i, 235; of Kamehameha against Kalanikūpule, 283–284; of Kamehameha against Kīwala'ō, 273–275; of Kanaloa-kua'ana against 'Umiokalani, 234; of Keawenui against Keli'iokaloa, 233–234; of Kekaulike against Hawai'i, 238–239; of Kiha-a-Pi'ilani against Lono-a-Pi'ilani, 210–211; population and, 160; social evolution and, 299–300; of 'Umi against Hākau, 183–184; use of western firearms in, 277
Ward, Jerome, 101–103
warriors, 193–194
way-finding, 56–58, 62, 116
wealth economy, 7, 232, 247, 289, 298–299. *See also* staple economy
Webber, John, 10, *11fig*, 259, 260
Webster, Daniel, 291
Weisler, Marshall, 48, 127–129, 239
Western Polynesia, 48
Wilson, Edward O., 161–162
Wittfogel, Karl, 153
writing, 301

X-ray fluorescence, 127

yams, 11, 25, 30, 34, 44. *See also* sweet potatoes
Yen, Douglas, 79, 147–149, 163
Young, John, 277, 280, 284

Ziegler, Alan, 105, 107

Text:	10.25/14 Fournier
Display:	Fournier
Compositor:	BookMatters, Berkeley
Printer and Binder:	Maple-Vail Book Manufacturing Group